World Water Resources

Volume 10

Series Editor

V. P. Singh, Department of Biological and Agricultural Engineering,
Texas A&M University, College Station, USA

This series aims to publish books, monographs and contributed volumes on water resources of the world, with particular focus per volume on water resources of a particular country or region. With the freshwater supplies becoming an increasingly important and scarce commodity, it is important to have under one cover up to date literature published on water resources and their management, e.g. lessons learnt or details from one river basin may be quite useful for other basins. Also, it is important that national and international river basins are managed, keeping each country's interest and environment in mind. The need for dialog is being heightened by climate change and global warming. It is hoped that the Series will make a contribution to this dialog. The volumes in the series ideally would follow a "Three Part" approach as outlined below:

In the chapters in the first Part *Sources of Freshwater* would be covered, like water resources of river basins; water resources of lake basins, including surface water and under river flow; groundwater; desalination and snow cover/ice caps. In the second Part the chapters would include topics like: *Water Use and Consumption*, e.g. irrigation, industrial, domestic, recreational etc. In the third Part in different chapters more miscellaneous items can be covered like impacts of anthropogenic effects on water resources; impact of global warning and climate change on water resources; river basin management; river compacts and treaties; lake basin management; national development and water resources management; peace and water resources; economics of water resources development; water resources and civilization; politics and water resources; water-energy-food nexus; water security and sustainability; large water resources projects; ancient water works; and challenges for the future. Authored and edited volumes are welcomed to the series. Editor or co-editors would solicit colleagues to write chapters that make up the edited book. For an edited book, it is anticipated that there would be about 12–15 chapters in a book of about 300 pages. Books in the Series could also be authored by one person or several co-authors without inviting others to prepare separate chapters. The volumes in the Series would tend to follow the "Three Part" approach as outlined above. Topics that are of current interest can be added as well.

Readership

Readers would be university researchers, governmental agencies, NGOs, research institutes, and industry. It is also envisaged that conservation groups and those interested in water resources management would find some of the books of great interest. Comments or suggestions for future volumes are welcomed.

Series Editor:

Vijay P. Singh, Department of Biological and Agricultural Engineering & Zachry Department of Civil Engineering, Texas A&M University, TX, USA.

Email: vsingh@tamu.edu

Mariam Iyabo Adeoba · Sesan Abiodun Aransiola ·
Olusola Ololade · Olabisi Peter Abioye
Editors

Water Resources Management in Nigeria

Springer

Editors
Mariam Iyabo Adeoba
Department of Mechanical Bioresources and Biomedical Engineering
University of South Africa
Johannesburg, Gauteng, South Africa

Olusola Ololade
Centre for Environmental Management
University of the Free State
Bloemfontein, South Africa

Sesan Abiodun Aransiola
Department of Microbiology
University of Abuja
Abuja, Nigeria

Olabisi Peter Abioye
Department of Microbiology
Federal University of Technology
Minna, Nigeria

ISSN 2509-7385 ISSN 2509-7393 (electronic)
World Water Resources
ISBN 978-3-032-22812-3 ISBN 978-3-032-22813-0 (eBook)
https://doi.org/10.1007/978-3-032-22813-0

This Springer imprint is published by the registered company Springer Nature Switzerland AG
The registered company address is: Gewerbestrasse 11, 6330 Cham, Switzerland

Foreword

Water is a fundamental resource underpinning human survival, economic development, and ecological integrity. Yet, its management in the context of climate variability, rapid urbanization, and socio-economic disparities remains one of the most complex challenges of the 21st century. This book provides a rigorous and multidisciplinary assessment of water resource governance, sustainability, and resilience, with a focus on Nigeria, a country whose hydrological diversity and developmental trajectory embody the worldwide water paradox: abundance among shortage.

The volume is designed to provide a comprehensive study of scientific, technical, legal, and policy issues. It begins by establishing the hydrological and geographical foundations of Nigeria's water systems and then proceeds to explore climate-induced perturbations in hydrological cycles, the intricate interdependencies within the water-energy nexus, and the socio-legal frameworks shaping resource allocation and equity. Subsequent chapters address critical themes such as water quality, sanitation, and public health, while highlighting systemic governance gaps and infrastructure challenges that impede progress toward Sustainable Development Goals (SDGs) 6, 7, and 13.

What distinguishes this work is its integrative approach. By synthesizing empirical evidence, policy analysis, and case studies, the authors illuminate the interconnectedness of water security with energy systems, climate adaptation, and socio-economic development. The book does more than just diagnose problems; it also proposes concrete answers, ranging from technological advances like solar-powered water systems and smart irrigation to nature-based solutions and participatory governance models that can inform both national and global policy debates.

This book serves as a critical resource for scholars, practitioners, and policymakers in advancing knowledge and practice in integrated water resources management. The results reverberate beyond Nigeria, suggesting lessons applicable to other regions dealing with similar difficulties of resource scarcity, institutional fragmentation, and climatic vulnerability. By foregrounding equity, resilience, and sustainability, the authors contribute meaningfully to the evolving literature on water governance and environmental law, while reinforcing the imperative for interdisciplinary collaboration in addressing global water crises.

I commend the authors for their scholarly rigour and commitment to bridging research and practice. This volume will undoubtedly enrich academic discourse and guide transformative action toward sustainable water futures.

Dr. Opeyemi A. Oyewo
Department of Chemical and Materials Engineering, College of Science, Engineering and Technology
University of South Africa (UNISA)
Pretoria, South Africa

Preface

Water resource management is a cornerstone of sustainable development, yet it remains one of the most pressing challenges in regions experiencing rapid population growth, urbanization, and climate variability. Nigeria exemplifies this complexity: a country endowed with significant water resources but confronted by persistent issues of scarcity, pollution, and governance inefficiencies. These challenges have profound implications for public health, food security, energy systems, and socio-economic development.

This book was developed to provide a comprehensive and interdisciplinary perspective on water resource sustainability in Nigeria, while situating the discussion within global frameworks such as the United Nations Sustainable Development Goals (SDGs) and the Paris Climate Agreement. It integrates insights from hydrology, environmental science, engineering, law, and policy analysis to offer a holistic understanding of water-related issues and their interconnections with energy, climate resilience, and socio-economic equity.

The intended audience includes researchers, graduate students, policymakers, and practitioners in water resources management, environmental law, and sustainable development. By combining empirical research with policy-oriented recommendations, this volume aims to bridge the gap between theory and practice, offering actionable strategies for decision-makers and development partners. While the primary focus is Nigeria, the insights presented here have broader relevance for other regions facing similar challenges.

We extend our appreciation to the reviewers and editorial team at Springer for their guidance in ensuring that this book meets the highest standards of academic rigor and clarity.

It is our hope that this book will serve as a valuable resource for advancing scholarship and informing policy in the quest for sustainable water governance. By fostering interdisciplinary dialogue and collaborative action, we aspire to contribute meaningfully to global efforts toward water security and climate resilience.

Johannesburg, South Africa — Mariam Iyabo Adeoba
Abuja, Nigeria — Sesan Abiodun Aransiola
Bloemfontein, South Africa — Olusola Ololade
Minna, Nigeria — Olabisi Peter Abioye

Acknowledgments

The completion of this book was made possible through the collective efforts and support of numerous individuals and institutions. We express our deepest gratitude to our contributing authors for their scholarly rigor and commitment to advancing knowledge in the field of water resource management. Their expertise and insights have enriched the quality and depth of this volume.

We are particularly thankful to the reviewers and editorial team at Springer for their invaluable guidance and constructive feedback throughout the publication process. Their dedication ensured that this work meets the highest standards of academic excellence.

Special appreciation goes to our colleagues and students who assisted with data collection, analysis, and manuscript preparation. We also recognize the role of international partners and development organizations whose resources and technical expertise informed several chapters of this book.

Finally, we extend heartfelt thanks to our families and friends for their patience and encouragement during the demanding process of writing and editing this work.

Johannesburg, South Africa — Mariam Iyabo Adeoba, Ph.D.
Abuja, Nigeria — Sesan Abiodun Aransiola, Ph.D.
Bloemfontein, South Africa — Olusola Ololade, Ph.D., Professor
Minna, Nigeria — Olabisi Peter Abioye, Ph.D., Professor
Ogbomoso, Nigeria — Olumuyiwa Idowu Ojo, D.Tech., Professor

Contents

Editors and Contributors

About the Editors

Dr. Mariam Iyabo Adeoba is a post-doctoral fellow researcher at the Department of Mechanical, Bioresources and Biomedical Engineering at the University of South Africa (UNISA). She is also a visiting lecturer at the Pan African University Institute for Life and Earth Sciences (PAULESI), University of Ibadan, where she teaches Water Resources Management and Environmental Sustainability. She holds a Ph.D. in Zoology and integrates her multidisciplinary expertise in bioresources utilization, renewable energy, and ecological resilience to promote sustainable development in Africa. Dr. Mariam Iyabo Adeoba has extensive expertise in environmental sustainability, water resources management, renewable energy, and bioresources systems. She is affiliated with the University of South Africa (UNISA), where her work spans teaching, postgraduate supervision, and interdisciplinary research at the interface of ecology, engineering, and sustainable development. Her research interests include water–energy–environment nexus studies, nature-based solutions, wastewater treatment and reuse, bioenergy, and the application of emerging technologies such as IoT and data-driven systems for sustainable resource management in Africa. Dr. Adeoba has published widely, contributed to policy-relevant discussions, and actively participates in academic conferences, editorial and review activities. She is also engaged in mentoring early-career researchers and advancing solutions aligned with the UN Sustainable Development Goals, particularly in water security, clean energy, and environmental resilience. At UNISA, Dr. Adeoba contributes to the Waste-to-Energy Research Initiative, focusing on the conversion of agricultural and industrial waste into sustainable bioenergy systems. She is currently working on characterization and potential of some aquaculture waste to biogas and biomethane potential. She is also examining the water footprint of bioenergy generation and its implications for industrial and policy frameworks in South Africa.

Through teaching, research, and capacity building, Dr. Adeoba advances the Sustainable Development Goals 6 (Clean Water) and 7 (Clean Energy), supporting Africa's transition toward a circular and low-carbon economy.

Dr. Sesan Abiodun Aransiola is a lecturer at the Department of Microbiology, University of Abuja, and Philomath University, Abuja, Nigeria. He obtained his B.Tech., M.Tech., and Ph.D. in Environmental Microbiology from the Federal University of Technology, Minna, Nigeria. He has demonstrated his research expert in the production of vermicasts from vermicomposting of organic wastes to assist plants in the remediation of polluted soil with heavy metals. His expertise spans environmental health, phytoremediation, vermicomposting, marine resources and bioremediation of pollutants in the soil. Aransiola's research is deeply rooted in innovative solutions for soil and water remediation, with significant contributions to microbial biotechnology and waste management. He is an award-winning researcher of different grant including TETFUND/DR&D/CE/UNU/ABUJA/IBR/2025/VOL.1. He has over 130 publications including book chapters, research and review articles of good international repute with high impact factor. He has co-edited scientific books of global interest which include *microbial biotechnology for bioenergy, marine bioresources: prospects and obstacles, marine bioresources: prospects and obstacles, marine microbial products: applications and opportunities* (Elsevier). *Ecological interplays in microbial enzymology, prospects for soil regeneration and its impact on environmental protection, Soil microbiome in green technology sustainability, marine bioprospecting for sustainable*

blue-bioeconomy, vermitechnology: economic, environmental and agricultural sustainability, white pollution: biodiversity and hazards in marine plastisphere, wetland ecosystems: conservation strategies, policy management and applications (Springer). *Microbial biofilms applications and control, marine greens: environmental, agricultural, industrial and biomedical applications, environmental footprint of bioplastic additives, emerging contaminants in food and food products, phytoremediation in food safety: risks and prospects, biotoxins in food: threats and benefits, fuel cells in environmental sustainability, biotechnology of climate-smart agriculture in food security and sustainable farming* (CRC Press) with many more in the process. Dr. Aransiola has been a review editor and as guest editor, and he hosted two special issues with Elsevier (Research in Microbiology) and Frontiers *on biofilms—community, structure, applications and mitigation and effects of microplastics on soil ecosystems*, respectively. Previously before joining academic, he rose to the rank of an assistant chief scientific officer at National Biotechnology Research and Development Agency, Nigeria, where he was involved in the uses of vinasse (a by-product of ethanol) as bio-fertilizer to reclaim polluted soil for agricultural purposes. Dr. Aransiola's academic journey is complemented by robust research, teaching career (with proper mentoring and supervising both undergraduate and post graduate students), and active involvement in community services, where he leverages his knowledge to foster sustainability. His contributions continue to impact both academia and the broader environmental sciences community. He is a member of Nigerian Society for Microbiology, Association of Environmental Impact Assessment of Nigeria (Affiliation of International Association for Impact Assessment) and American Society for Microbiology.

Olusola Ololade Ph.D. is a research fellow at the Centre for Environmental Management, University of the Free State (Bloemfontein, South Africa). She previously served as an acting director and associate professor at the Centre. Her expertise spans land-use and land-cover analysis, integrated water resources management, the water–energy–food nexus, mining and sustainability, and sustainable development. Professor Ololade has authored and co-authored over 50 peer-reviewed journal articles, 17 book chapters, and three peer-reviewed conference proceedings. She has served as a principal investigator or co-principal investigator on more than six completed and ongoing research projects. She is a member of the editorial boards of four international journals and regularly reviews manuscripts for over 40 international journals. She has delivered more than 25 conference presentations, including 17 at international conferences, with invited roles as both a keynote and plenary speaker. Professor Ololade has successfully supervised and graduated 31 master's students and six doctoral candidates.

Olabisi Peter Abioye is a professor of environmental microbiology at the Department of Microbiology, Federal University of Technology, Minna, Nigeria. He graduated in this same department in the year 2002 with first-class honors and was retained as a lecturer. He obtained his Master's Degree (M.Tech.) in 2006 from this school before proceeding for his Ph.D. program at the University of Malaya, Kuala Lumpur, Malaysia, and completed this degree in 2011. His research interest includes environmental microbiology/biotechnology and bioremediation of environment contaminated with both organic and inorganic pollutants. He has been a lecturer since 2002 at the Federal University of Technology, Minna, Nigeria, and has since taught at undergraduate and postgraduate levels. He has supervised numerous students at different levels of degree and has vast administrative experience. He has more than 70 publications, including chapters in books, conference proceedings, peer reviews, and research papers. He has served as an external examiner in number of universities both nationally and internationally and belongs to many professional societies.

Contributors

Olabisi Peter Abioye Department of Microbiology, School of Life Sciences, Federal University of Technology, Minna, Nigeria;
Department of Public Health, School of Life Sciences, Federal University of Technology, Minna, Nigeria

Francis Olawale Abulude Environmental and Sustainable Research Group, Science and Education Development Institute, Akure, Nigeria

Adewuyi Felix Adedokun Department of Mechanical Engineering, University of Ibadan, Ibadan, Nigeria

Aisha Eniola Adegbite Faculty of Science, Department of Microbiology, University of Abuja, Abuja, Nigeria

Mariam Iyabo Adeoba Department of Mechanical, Bioresources and Biomedical Engineering, University of South Africa, Pretoria, South Africa;
UNISA Biomedical Research Group, Department of Mechanical, Bioresources and Biomedical Engineering, College of Science, Engineering and Technology (CSET), University of South Africa, Johannesburg, Florida, South Africa

Samuel Adeyanju Department of Biological Science, Faculty of Science, University of Abuja, Abuja, Nigeria

Ja'afar Malam Ahmed Department of Zoology, Usmanu Danfodiyo University, Sokoto, Nigeria

Stephen Olubusoye Ajagbe Department of Wildlife and Ecotourism, Forestry Research Institute of Nigeria, Ibadan, Nigeria

Ajulor National Biotechnology Research and Development Agency, Abuja, Nigeria

Bolaji Oluwatosin Akanbi Department of Microbiology, Faculty of Life Science, University of Abuja, Abuja, Nigeria

Joshua Ogboghena Akhigbe Faculty of Science, Department of Microbiology, University of Abuja, Abuja, Nigeria

Akinyinka Akinnusotu Department of Science Laboratory Technology, University of Medical Sciences (UNIMED), Ondo, Nigeria

Lawrence Enejeta Akpoterai Institute for Climate Change and Adaptation (ICCA), University of Nairobi, Nairobi, Kenya

Adejoke Blessing Aransiola Department of Surveying and Geoinformatics, Faculty of Environmental Science, University of Abuja, Abuja, Nigeria

Sesan Abiodun Aransiola Department of Microbiology, Faculty of Science, University of Abuja, Abuja, Nigeria;

Centre for Artificial Intelligence & Multidisciplinary Innovation Studies, Department of Auditing, College of Accounting Sciences, University of South Africa, Pretoria, South Africa;
Department of Microbiology, Faculty of Life Science, University of Abuja, Abuja, Nigeria

Enoch Bessah Department of Agricultural and Biosystems Engineering, Kwame Nkrumah University of Science and Technology, Kumasi, Ghana

Cynthia Ihotu Danboyi Department of Veterinary Medicine, Faculty of Veterinary Medicine, University of Abuja, Abuja, Nigeria

Victor Adetunji Ebo Department of Civil Engineering, Ladoke Akintola University of Technology, Ogbomoso, Nigeria

Oyebola Adebola Elemide Department of Agricultural and Bio-Environmental Engineering, Federal College of Agriculture, Akure, Nigeria

Oluwaseun Fadeyi Centre for Environmental Management, Faculty of Natural and Agricultural Sciences, University of the Free State, Bloemfontein, South Africa

Opeyemi Caleb Fatayo Pan African University Life and Earth Sciences Institute, Ibadan, Nigeria

Iyanuoluwa Busola Feyijin Department of Environmental Health Sciences, University of Ibadan, Ibadan, Nigeria

Victor Olakunle Ibidunmoye Department of Mechanical Engineering, University of Ibadan, Ibadan, Nigeria

Nosa Idusuyi Department of Mechanical Engineering, University of Ibadan, Ibadan, Nigeria

Ayokunle Isaac Ilesanmi Department of Fisheries and Aquaculture, University of Agriculture and Environmental Sciences, Umuagwo, Imo State, Nigeria

Oluwafemi Oladipupo Kolajo Department of Agricultural and Environmental Engineering, University of Ibadan, Ibadan, Nigeria

Naga Raju Maddela Departamento de Ciencias Biológicas, Facultad la Ciencias de la Salud, Universidad Técnica de Manabí, Portoviejo, Manabí, Ecuador

Ibrahim Muhammad Magami Department of Biology, Usmanu Danfodiyo University, Sokoto, Nigeria

Olumuyiwa Adewale Obijole Department of Chemistry, Adeyemi Federal University of Education, Ondo, Nigeria

Olawunmi Opeyemi Obisesan Department of Clinical Legal Education, School of Law, Kampala International University, Kampala, Uganda

Enovwo E. Odjegba Department of Water Resources Management and Agrometeorology, Federal University of Agriculture, Abeokuta, Nigeria

Olalekan Joseph Ogunniyi Department of Industrial and Engineering Management, College of Science, Engineering and Technology, University of South Africa, Johannesburg, South Africa

Jeremiah Oludele Ojediran VC Office, Bells University of Technology, Ota, Ogun State, Nigeria

Olumuyiwa Idowu Ojo Department of Agricultural Engineering, Ladoke Akintola University of Technology, Ogbomoso, Nigeria

Ojonoma Loretta Okwute Department of Microbiology, Faculty of Science, University of Abuja, Abuja, Nigeria

Damilola Rebecca Oladejo Department of Biochemistry, Federal University, Oye-Ekiti, Nigeria

Gabriel O. Oladipo Department of Science Laboratory Technology, D.S Adegbenro ICT Polytechnic, Itori-Ewekoro, Ogun State, Nigeria

Olushola Ahmed Olaleye Department of Zoology, Faculty of Science, University of Johannesburg, Johannesburg, South Africa

Rebecca Adepate Olaoye Department of Civil Engineering, Ladoke Akintola University of Technology, Ogbomoso, Nigeria

Alimot Olawumi Olatunji Department of Civil Engineering, Ladoke Akintola University of Technology, Ogbomoso, Nigeria

Olusola Ololade Centre for Environmental Management, Faculty of Natural and Agricultural Sciences, University of the Free State, Bloemfontein, South Africa

Sunday Emmanuel Olusola Depatment of Fisheries and Aquaculture Technology, School of Agriculture, Food and Natural Resources, Olusegun Agagu University of Science and Technology, Okitipupa, Nigeria

Samuel Dare Oluwagbayide Department of Agricultural and Bio-Environmental Engineering, The Federal Polytechnic, Ilaro, Ogun State, Nigeria

Ayodetimi Omoniyi Centre of Excellence in Agricultural Development and Sustainable Environment (CEADESE), Federal University of Agriculture, Abeokuta, Nigeria

Thomas Aghogho Orhadahwe Department of Mechanical Engineering, University of Ibadan, Ibadan, Nigeria

Rasaq Rahman Owolabi Department of Geological Sciences, Achievers University, Owo, Ondo State, Nigeria

Juliana Okwena Pondei Department of Microbiology, Faculty of Life Science, University of Abuja, Abuja, Nigeria

Nosakhare Praise National Biotechnology Research and Development Agency, Abuja, Nigeria

Aishat Olorunkemi Salisu Department of Microbiology, Faculty of Life Science, University of Abuja, Abuja, Nigeria

Stella Suanu National Biotechnology Research and Development Agency, Abuja, Nigeria

Amos Musa Sumakada Department of Microbiology, Faculty of Life Science, University of Abuja, Abuja, Nigeria

Bridget Tomilayo Tola-Winjobi Department of Agricultural Engineering, Abiola Ajimobi Technical University, Ibadan, Nigeria

Amina Yahaya Faculty of Science, Department of Microbiology, University of Abuja, Abuja, Nigeria

Leh-Togi Zobeashia National Biotechnology Research and Development Agency, Abuja, Nigeria

Part I
Foundations and Context

Chapter 1
Water Resources in Nigeria: General Overview

Cynthia Ihotu Danboyi, Samuel Adeyanju, Adejoke Blessing Aransiola, Ojonoma Loretta Okwute, and Sesan Abiodun Aransiola

Abstract Nigeria which is a part of Africa has man-made and natural water sources. These sources comprise of lakes, streams, rivers, and underground water sources across various regions. Bakolori, Kainji and Jebba are part of the preeminent lakes created by man, while Benue, Cross river and Niger are substantial river systems. These regime of waters are vital for providing water for industrial and domestic use, generation of electricity and agronomy. Even so, numerous difficulties exist, such as bad water control, weather changes, pollution of water, continuous dry season, and exponential growth in population. The review implies that, to totally tackle these challenges, there must be a combined effort involving various collaborators, including local communities, and adjustable elucidation should follow through to ensure a dependable and immaculate water supply for posterity.

Keywords Sustainable management · Irrigation · Water scarcity · Climate change · Water resources · Hydroelectric power

C. I. Danboyi (✉)
Department of Veterinary Medicine, Faculty of Veterinary Medicine, University of Abuja, Abuja, Nigeria
e-mail: cynthiadanboyi05@gmail.com

S. Adeyanju
Department of Biological Science, Faculty of Science, University of Abuja, Abuja, Nigeria

A. B. Aransiola
Department of Surveying and Geoinformatics, Faculty of Environmental Science, University of Abuja, Abuja, Nigeria

O. L. Okwute · S. A. Aransiola
Department of Microbiology, Faculty of Science, University of Abuja, Abuja, Nigeria

M. I. Adeoba et al. (eds.), *Water Resources Management in Nigeria*, World Water Resources 10, https://doi.org/10.1007/978-3-032-22813-0_1

1.1 Geographical Content

1.1.1 Nigeria's Location in West Africa

Nigeria spans a total space of 923,768 km^2 (356,669 sq mi) and is situated in Western Africa moving forward towards the Gulf of Guinea, making it the world's 32nd largest country (Omokaro et al., 2024). It lies between latitudes 4°0′ and 14°0′ North, and longitudes 2°2′ and 14°30′ East (Fig. 1.1). Nigeria is bordered by the Republic of Benin to the west, Chad and Cameroon to the east, the Niger Republic to the north, the Gulf of Guinea to the south, and Lake Chad to the northeast (Fig. 1.1). The country is a multiethnic and culturally rich society divided into 36 self-governing states (World Bank Group, 2017). Nigeria experiences two main seasons: dry and rainy. The country also has six different ecological regions: the Mangrove zone, Forest zone, Montane zone, Guinea savannah zone, Sudan savannah zone, and Sahel savannah zone (Popoola et al., 2021).

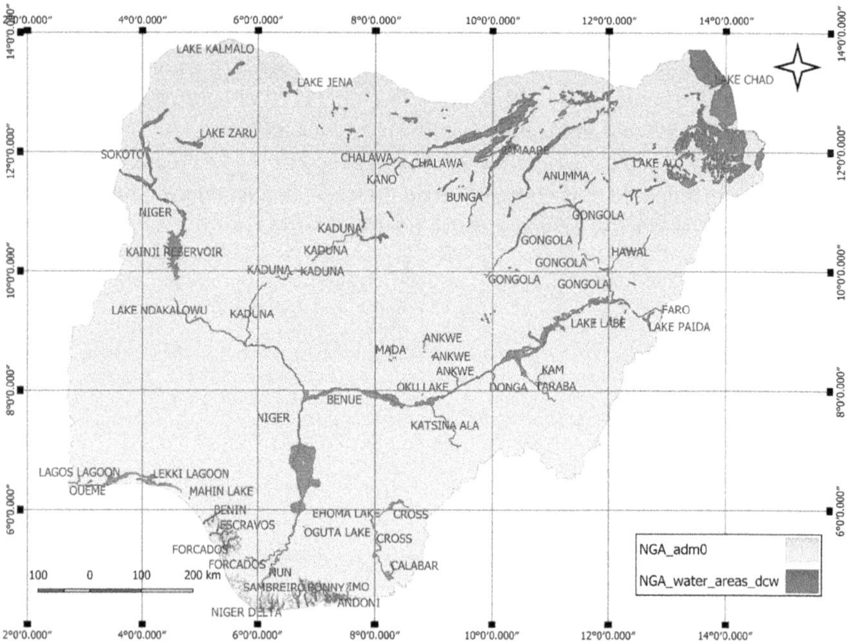

Fig. 1.1 Nigeria map with some water areas

1.2 Water Resource Availability in Nigeria

According to the National Water Policy of the Federal Republic of Nigeria (2014), Nigeria has a lot of water resources. There is surface water almost everywhere in the country, and there is also a lot of underground water. The total amount of surface water is about 267.3 billion cubic meters, and the potential for underground water is around 51.9 billion cubic meters.

1.2.1 Surface Water Resources

1.2.1.1 Niger River Basin

In Nigeria, the largest stream network is the Niger River Basin taking up to about 120,197 km^2, sweeping from the North-Central to the South-South demesne. The basin lies between latitudes 90 30′ 0″ N and 40 28′ 0″ N, and longitudes 50 0′ 0″ E and 90 0′ 0″ E (Adeyemi & Komolafe, 2025). The drainage area consisting of different climate region comprises of a complex network of smaller river systems. According to Adeyemi and Komolafe (2025), the rainfall of the area which has a sweltering climate fluctuates from north to south. Less than 400 mm of rainfall is received in most areas while few other areas get more than 1200 mm of rain. 21 °C to 32 °C is the annual mean temperature in the basin. Rivers frequently brim over and result in massive flooding which usually happen between April and October which is the raining season in Nigeria (Adeyemi & Komolafe, 2025).

1.2.1.2 Lake Chad Basin

At the southern part of the Sahara Desert, Lake Chad is found at the midpoint of the Sahel region. Lake Chad is a source of support close to 50 million people offering them food and water. According to Pharm-Duc et al. (2020), Water do not spurt out in the section around Lake Chad. With drought in the northern parts, the average weather condition changes as you move from south to north. The Lake Chad has gotten lesser resulting in the drop in the water level in that area. The reason of the reduction in the water level is since the water is used excessively and the extensive drought period. Lake Chad in the 1960s covered about 25,000 km^2, approximately covering one-fourth of that in Nigeria. The demand for water around the lake increased due to the farming communities surrounding the lake, as stated in (Joint Environmental Audit, 2015).

1.2.1.3 Cross River Basin

Southeast Nigeria and western Cameroon are home to the Cross River Basin. It begins in Cameroon and travels across Nigeria before emptying into the Atlantic Ocean. Folorunshona and Awosika (2014) estimate the basin's area to be approximately 53,590 km^2. The bulk of the basin is in Nigeria, between latitudes 4°28′ and 6°55′ North and longitudes 7°50′ and 9°28′ East, according to the Water Supply and Sanitation Reform Programme II (2016). Farms, factories and homes, all use the basin's water. But it also has problems including denuding, washing of the topsoil, and pollution from businesses and towns.

1.2.1.4 Other Significant Rivers

The Benue River Basin (BRB) is another substantial surface upwelling. Cameroon being the second biggest river encompasses a large amount of the Niger River Basin (NRB). As stated by (Nonki et al., 2019), inland navigation, sprinkling, and hydroelectric power generation are the activities which the basin supports. Moreover, the Sokoto–Rima River Basin (SRRB) includes Nigeria, Benin, and Niger (Raji et al., 2021). The SRRB encloses the states of Katsina, Zamfara, Kebbi, Sokoto and Katsina and prolongs southeast toward Niger State in Nigeria. In this basin, the dams such as Bakolori and Goronyo are necessary for affusion, network of water, and agronomic support (Achugbu et al., 2022).

1.2.2 Lakes and Reservoirs

1.2.2.1 Kainji Lake

As reported by (Raji et al., 2011), The concoction of the Kainji Lake which was the first lake created by man in Nigeria was predominantly made to generate hydroelectric power. The construction of the Kainji lake began in March 1964 and was finalized in December 1968 (Ogunkoya et al., 2023; Saminu & Zayyanu, 2023). In the lake, energy production, cultivation, angling, and flood prevention are carried out in the lake (Ogunkoya et al., 2023). Some human activities such as dumping of refuse into the river and atmospheric changes has impacted the functioning of the Kainji lake which in turn affects Niger river (Oyerinde, 2016).

1.2.2.2 Jebba Lake

Jebba Lake is in the North Central part of Nigeria. It was made in 1983 when a dam was built across the upper Jebba basin. The main reason for creating the lake was to produce electricity through hydroelectric power. However, the lake also offers chances for fishing, farming using water from the lake, and moving boats and ships (Adelakun et al., 2017).

1.2.2.3 Bakolori Reservoir

Originally situated in Sokoto State in Northwestern Nigeria, the Bakolori dam had the coordinates 12°30′43″N and 6°11′0″E. It was there before Zamfara State was established in 1995. In 1978, the dam was finished, and by 1981, its reservoir was full. On the Sokoto River, a tributary of the Rima River that eventually empties into the Niger River, is the Bakolori reservoir, a notable reservoir. According to Shuaibu (2024), the Bakolori irrigation project is supported by the water from this dam.

1.2.3 Groundwater

Surface water is often polluted, making groundwater (water found beneath the earth's surface, within the soil and rock layers) the preferred option for human consumption, as it offers reliable and cleaner source. This is because surface water is often polluted, causing uncertainty and dissatisfaction for users. In the urban and rural region, groundwater is extremely important which is also the main source of water for about third-fifth (60%) of the population. Even with its importance, the facts surrounding how it is used and nurtured is sparse, and to completely understand its status, methodic monitoring is key (Omokaro et al., 2024).

The first type of geological area which covers 50% of the country in which groundwater mostly flow is the basement complex. In this first part, rocky formations which have been eroded such as deeper cracks, fractured rocks and regolith, reserve water in them. Water is not held or produced much in this type of groundwater sources. Sedimentary area which covers the other 50% of the country is the second type of geological area. Basins such as Dahomey, Niger Delta, Chad, Sokoto, Anambra and Benue Trough which are in this area has more than one layer of groundwater. In this part, they are quite the opposite of the basement complex as they store more water and are more productive (Egbeuri et al., 2025).

1.2.4 Role in Water Supply, Irrigation and Hydroelectric Power

1.2.4.1 Irrigation

Irrigation is a significant approach which is necessary and key to that is essential to conserving and securing one of the basic needs of man which is food Annually.

Dams such as the Goronyo and Bakolori in Sokoto and the Ede-Erinle reserve in Osun state comprises a fraction of the numerous dams in Nigeria which is more than 200 in number (Omokaro et al., 2024). The dams and reserve are primarily used to distribute and share water for municipal and irrigation. The processing of food industrially and the productivity in agronomy are dependent on the water that these dams produce. If there is constant availability and supply of water, production levels will increase exponentially (Youdeowei et al., 2019).

1.2.4.2 Hydroelectricity

For hydropower generation and production, three biggest dams in Nigeria which are Jebba, Kainji on the River Niger, and Shiroro on the River Kaduna are mainly used (Adelakun et al., 2017; Ogunkoya et al., 2023). The dams which are in the states of Kebbi, Niger, Kwarra, and Plateau, supply at least half of the electricity used in Nigeria (Youdeowei et al., 2019).

1.2.4.3 Domestic Use

In the southwest part of Nigeria, Dams like the Asejire and Eleyele dams were constructed to give uncontaminated water for use by the local communities. Even though many towns and villages rely on both surface water sources such as rivers, lakes, and dams, as well as groundwater from wells and boreholes, for their daily needs, according to (Youdeowei et al., 2019).

1.3 Challenges in Water Resource Management

1.3.1 Water Scarcity

Water is one of the world's most precious resources, yet it is becoming increasingly scarce in many areas (Aransiola et al., 2016; Babaniyi et al., 2025). For example, Nigeria is experiencing significant water shortages, especially in the Nigerian part of the African Sahel, where local communities face difficulties in accessing enough and clean drinking water. According to a report by the Brookings Institution, Lake Chad has decreased in size by nearly 90% since its peak in the 1960s (Sow, 2017). The Falkenmark Water Stress Index defines water scarcity based on the amount of renewable freshwater available per person per year. A country is in a state of water stress when the availability drops below 1700 cubic meters per person per year, and severe water scarcity is identified when it falls below 500 cubic meters per person per year (Mekonnen & Hoekstra, 2016).

1.3.1.1 Causes of Water Scarcity

Climate Change

Climate change means the Earth's climate is changing over a long time, mostly because humans are putting more greenhouse gases into the air (Anagnostopoulou et al., 2021; Gimba et al., 2024). Due to higher temperatures and constant occurrence of overflooding and drought, the atmospheric condition in Nigeria is hard to predict because of these changes (Nnam et al., 2025). Climate changes highly impact the obtainability of fresh water in Nigeria which is noticeably lowered due to these changes (Adebisi-Adelani et al., 2014). Droughts tends to worsen and endure in some places because of the changes in the patterns of precipitation. Water available for farming, fishing, irrigation, drinking and other necessary use is reduced (Alo et al., 2017).

Population Growth

The growth in the population is one of the major issues surrounding the constant supply of water. As the natality rate exponentially supersedes the mortality rate, the requisition for uncontaminated water increases and there is deficiency in the amount of water already (Gu et al., 2021). Nigeria which is the largest country in Africa is the best case to see how the rate of water supply directly affects population growth. Nigeria's water system and supply have strived to level up to the country's increasing demand having over 213 million people (Ben, 2023). It continues to get difficult to get pure and dependable water especially in those places where the population is as high as the demand for water because the population keeps increasing and the system cannot handle the demand (Okeke et al., 2021).

Drought

Several environmental issues and dryness is allied with drought. The effect of climate change which affect water supply around the globe is showed in the 2021 report from the Intergovernmental Panel on Climate Change (IPCC) (Masson-Delmotte et al., 2023; Nnam et al., 2025). When water becomes infrequent in dry places and there is an increase in temperature, it results in drought. Drought becomes worse when there is an increase in evaporation and transpiration which is due to the increase in temperature. At intervals, underground water sources and channels can get unsafe in a lengthy period of dryness, risking the health of the people because the water is polluted (Ibrahim & Emeka, 2024).

1.3.1.2 Impact of Water Scarcity

Agriculture

Farmers on a normal condition (safe and constant water) should be more productive in their business earning a lot of money but due to the low amount of water for the farming process, Farmers lose a lot of money. Rice and maize production in Nigeria were cut by 30% due to the presence of drought as seen in a study from the International Institute of Tropical Agriculture (IITA) in 2023. Majority of the population will starve because of shortage of food resulting in price increment and all this will occur if nothing is done immediately (Aransiola et al., 2024; Woomer et al., 2025).

Industries

Many industries do rely on groundwater to ensure smooth flow of business. The absence of water is a critical issue that affects many economies and productivity. Industries, like pure water or bottle water production industry, the agricultural industry, healthcare industry, energy production, and other industries need water for their continuous productivity. If water scarcity persists or the water is unsafe for use, it can lower the yield, increase the cost of living because of increase in price, can also bring production to a stop and in some cases, some of the industries can shut down completely (Adeniyi et al., 2024).

Domestic Use

Nigeria has a good considerable amount of water resource, yet it has not been effectively used to fulfill the needs of the people. However, the main factor determining water security is not the availability of water resources, but rather the consistent access to a safe and clean water supply. It is estimated that over 65 million people in Nigeria lack access to a safe or clean water supply (Majuru et al., 2011). Most people, especially those in rural areas, rely on unsafe water sources, which makes them vulnerable to water-borne diseases. These diseases cause the deaths of over 194,000 people each year, with children under five being the most affected (Arokoyu & Ukpere, 2014).

1.3.2 Water Pollution

In Nigeria, Water pollution is an extreme case as the country cannot access safe and uncontaminated water to use. A good amount of unclean water is released into the water bodies which is polluted from different sources such as plastic bottles, domestic waste, nylons, etc. A certain amount of this water is not properly allayed.

This unchecked actions now results in the release of high levels of toxins, harmful chemicals and other things into the water, according to a report by the United Nations (Banerjee, 2017).

1.3.2.1 Sources of Water Pollution

Industrial Source

According to Boussaha and Laifa (2017), the major threat to the ecosystem and a key contributor to water pollution is industrial waste. Different types of chemicals and dyes used in many industries, such as Medical, food, textile, makeup, leather and paper, pollutes the water bodies by releasing the materials that are left after the main product has been used. For instance, the paper industry uses a significant number of dyes and energy (Svensson & Berntsson, 2014). This industry also discharges waste into water, which harms aquatic life by creating toxic colors, harmful substances, and slime (Arif et al., 2020).

Agricultural Source

In farming areas, the runoff has a negative result on the quality of the water because of the chemical substances released into water bodies. Chemicals such as herbicides, pesticides, rodenticide, fungicide, and insecticide are used a lot in this area. Improper disposing of chemical waste can cause chemical poisoning of the water. Poor soil management which is a farming method is another issue that can cause water pollution (Wato et al., 2020). When pesticide leftovers move into ponds, rivers, and lakes through surface water or spray, they can cause sudden poisoning, like fish dying, or long-term harm, where animals are exposed to low levels that aren't immediately deadly (Mishra et al., 2023).

Domestic Waste

Human activities have been shown to affect the quality of groundwater (Elemile et al., 2019). Most of the harmful microbes in water come from animal and human waste (Takal & Quaye-ballard, 2018). Many of these sources are not controlled because they are managed by private people, and steps to make sure water is safe and lasts long are often not taken. Water from these sources can easily become dirty because of waste from human activities, leaking sewage, and runoff from areas with polluted soil (Saria & Thomas, 2012).

1.3.2.2 Impact of Water Pollution on Water Quality and Health

Our health is something to be cherished and if the water is polluted, it can cause adverse effect on our health and give room for many other diseases. Harmful chemicals, bacteria and germs can be found in polluted water and when ingested, can cause great harm to one's health. This makes the water unfit for use. These dangerous substances have the potential to cause significant harm once they enter our bodies. Swimming in contaminated water or drinking contaminated water can make people sick. The fish and other sea food that we consume, which originate from contaminated areas, may also contain these dangerous substances (Pona et al., 2021).

1.3.3 Climate Change

1.3.3.1 Effects on Precipitation Patterns and Water Availability

In Nigeria, the region in the south tends to have lasting and more tempestuous season while the region in the north has concise and a moderate season which points to a drier condition establishing the difference in rainy season from the south to the north (Salami et al., 2025). About one-quarter precipitation in the Sahel region in Nigeria has decreased over the last 30 years (Amanchukwu et al., 2015). Between 1960 and 2007, Lake Chad which was dried up abated in size from roughly 4000 km^2 to around 3000 km^2, is imputed to the effects of global warming in that area (Elisha et al., 2017). Climate change has also affected the availability of water in Nigeria by the decrease in precipitation in the northern region, resulting to dryness and reduced subsurface water level, while inundation and waterflood in the southern parts increases. The results of these changes is seen in water shortage and pollution of water, impacting asepsis, drinking water, and agribusiness (Heider, 2019).

1.3.3.2 Adaptation and Mitigation Strategies

Adaptation

The leading to the creation of the National Adaptation Strategy and Plan of Action on Climate Change for Nigeria was initiated by the Building Nigeria's Response to Climate Change (BNRCC) and was funded by the Canadian International Development Agency (CIDA). According to Woodley et al. (2011), the following reasons was why the primary projects were created which include to restore the biosphere through afforestation, to ameliorate access to better sources of water, to relinquish communities from taking forest products, to suggest farming in water as an alternate source of income, to provide wood stoves which are fuel efficient and to increase food security by using improved varieties.

Mitigation Strategies

1. **Agriculture and forestry:** In farming, the use of eco-climate agriculture is used to reduce the effect of global warming. A powerful greenhouse gas which is Methane (CH4) is produced in large quantities by livestock (Dioha & Kumar, 2020). Feeding animals more concentrated feeds, afforestation and storing of manure and fertilizers in cooler locations to decrease the presence of nitrous oxide emissions and methane (Nwobodo et al., 2022).
2. **Waste management**: Human activities around the world are a major contributor to the emission of greenhouse gases from the waste industry. This issue of waste from industries is a severe issue in a country like Nigeria where the system that manages waste is futile and inefficient (Malley et al., 2021). One of the ways to reduce greenhouse gas emissions is the use of bioenergy since it is carbon neutral can effectively be a good source of energy and properly collect waste (Elum & Momodu, 2017). Converting waste into useful material and energy can be helpful in conserving natural resources (Okafor et al., 2022).

1.4 Water Resource Management Framework

1.4.1 Policies and Regulations

A full-bodied legislative framework and an all-inclusive strategy has been developed by Nigeria but the effectiveness is confined by execution problems (Ngene et al., 2021).

1.4.1.1 National Water Policy

In 2016, the background for the water resource management provided by the national water policy in Nigeria was last revised. Some of the ideas that was explained in the policy include equity, the economic advantage of water, eco-friendly, Stakeholder participation and integrated water resources management (IWRM), which is condensed as "People, Environment, Water and Sustainable Development." Hydroelectric power generation, cleanliness, hydration, climate change, quality control of water and water supply are the key topics it addresses. The target and goal of The National Water Policy is to provide the public with clean and uncontaminated water, protect life in water and to manage water resources to create a balance to the future demands and current ones. Furthermore, it also bespeaks to concoct portable water provision to increase economic participation and to reduce indigence (FMWR, 2016).

1.4.1.2 Institutional Framework

The Federal Ministry of Water Resources (FMWR) oversees significant water resource development projects and the allocation of water among states (Idu, 2015). FMWR consist of sixteen (16) parastatals and agencies, including twelve (12) River Basin Development Authorities (RBDAs), the Nigeria Hydrological Services Agency (NIHSA), the Nigeria Integrated Water Resources Management Commission (NIWRMC), the Gurara Water Management Authority (GWMA), and the National Water Resources Institute (NWRI). The Federal Ministry of Water Resources (FMWR) is responsible for developing national policies and strategy recommendations for the public water supply.

However, each state oversees passing and implementing national rules and regulations, as well as creating the institutional structure for service delivery. Although not every state has a distinct ministry of water resources, state ministries of water resources oversee the laws, regulations, and supervision that are passed down to manage the local governments. For instance, there is no distinct ministry of water resources and rural development (RSWRRD) in Rivers State. As a result, although certain institutions and regulations related to the water supply are in place nationwide, there is a great deal of variation in the administration and provision of water among jurisdictions (WHO, 2013).

1.5 Water Governance

1.5.1 Stakeholder Involvement

Water is a problem that affects people in a specific area and involves many different groups of people at different levels, such as within river basins, local government areas, states, regions, the country, and even internationally. When there is not a good system in place to manage how different policies and government levels work together, it becomes hard for policymakers to create and carry out effective changes to water management (Ogwezzy, 2020). Nigeria handles water resource management through a system where different government departments or organizations look after various aspects of water use and management. The country is divided into eight main areas based on water systems. The Federal Ministry of Water Resources is the main body in charge of water-related issues such as developing water sources, providing clean water and sanitation, managing irrigation and drainage, and dealing with floods and erosion. This ministry works with several organizations like the River Basin Development Authorities, the Nigerian Integrated Water Resources Management Commission, the Nigerian Hydrological Services Agency, and the National Water Research Institute. The Federal Ministry of Environment, on the other hand, takes care of environmental concerns related to water resources (Adedeji & Ako, 2009).

1.5.2 Community Participation in Water Management

In rural Nigeria, communities often face severe water shortages, underscoring the need to address water resource management at the local level. Globally, water stress and food insecurity are pressing issues, as food production (predominantly occurring in rural areas) is the largest consumer of freshwater (Brauman et al., 2013). One important model for putting in place efficient rural water supply systems is community-based management (Miller et al., 2019). Water has cultural, spiritual, and socioeconomic value in African contexts. Therefore, it is essential to create plans that allow rural communities to independently manage their water resources. To prevent waste and contamination, the government of Nigeria must assist local communities in managing water supply systems. To achieve SDG 6's objective of guaranteeing sustainable water management for all, it is essential to implement strategies that encourage community involvement in water management. At the local level, this strategy ought to support integrated water resource management (Arimoro & Musa, 2020). Communities should be encouraged to set up borehole systems through group contributions, with government support in the form of cost subsidies or technical assistance, given the government's incapacity to meet rural water supply needs. Community members should oversee these initiatives with little political intervention (Arimoro & Musa, 2020).

1.6 Conclusion

Nigeria has many supplies of water, which is necessary for day-to-day domestic use, commercial enterprise, production of energy and agronomy. But the management of this water is inadequate. Rivers and subsurface water are deteriorating due to pollution, excessive water use, and weather changes. To effectively manage water in Nigeria, a new strategy that unites various groups and involves residents is required. It's critical to upgrade water management systems, construct better infrastructure, and ensure that environmental regulations are adhered to. These actions are essential to ensuring the nation's continued access to clean water and fostering its development.

References

Achugbu, I. C., Olufayo, A. A., Balogun, I. A., Dudhia, J., McAllister, M., Adefisan, E. A., & Naabil, E. (2022). Potential effects of land use land cover change on streamflow over the Sokoto Rima River Basin. *Heliyon, 8*(7), e09779, ISSN 2405-8440. https://doi.org/10.1016/j.heliyon.2022.e09779

Adebisi-Adelani, O., & Oyesola, O. B. (2014). Adaptation strategies of citrus and tomato farmers towards the effect of climate change in Nigeria. *Vulnerability of agriculture, water and fisheries to climate change: Toward sustainable adaptation strategies* (pp. 157–165). Springer.

Adedeji, A. A., & Ako, R. T. (2009). Towards achieving the United Nations' millennium development goals: The imperative of reforming water pollution control and waste management laws in Nigeria. *Desalination, 248*(1–3), 642–649.
Adelakun, K. M., Mustapha, M. K., Amali, R. P., & Mohammed, N. (2017). Seasonal variation in nutritional quality of catfish (Clarias gariepinus) from Upper Jebba Basin, Nigeria. *Journal Nutrition Food Science, 7*, 622. https://doi.org/10.4172/2155-9600.1000622
Adeniyi, D., Dekeyser, K., & Ekundayo, M. (2024). The horticulture sector in southwest Nigeria: A political economy analysis.
Adeyemi, A. B. & Komolafe, A. A. (2025) Flood hazard zones prediction using machine-learning-based geospatial approach in lower Niger River Basin, Nigeria. *Natural Hazards Research*, 399–412, ISSN 2666-5921. https://doi.org/10.1016/j.nhres.2025.01.002
Alo, A. O., Baines, R., Conway, J., & Cannon, N. (2017). The impacts of climate change on agriculture in developing countries: A case study of Oyo state, Nigeria. *The International Journal of Climate Change: Impacts and Responses, 9*(2), 1–21.
Amanchukwu, R. N., Amadi-Ali, T. G., & Ololube, N. P. (2015). Climate change education in Nigeria: The role of curriculum review. *Education, 5*(3), 71–79.
Anagnostopoulou, C., & Skoulikaris, C. (Eds.). (2021). *Impact of climate-change on water resources*. MDPI.
Aransiola, S. A., Evinemi, O. T., Falade, N. O., & Falese, T. T. (2016). Assessment of water contamination in Nigeria-review. *Journal of Basic and Applied Research International, 19*(1), 62–76. http://www.ikpress.org/abstract/5846
Aransiola, S. A., Babaniyi, B. R., Aransiola, A. B., Maddela, N. R. (Eds.). (2024). In Prospects for soil regeneration and its impact on environmental protection. (1st ed.). Springer. https://link.springer.com/book/9783031532696
Arif, A., Malik, M. F., Liaqat, S., Aslam, A., Mumtaz, K., Afzal, A.,... & Javed, R. (2020). 3. Water pollution and industries. *Pure and Applied Biology (PAB), 9*(4), 2214–2224.
Arimoro, A. E., & Musa, H. (2020). Towards sustainable water resource management in rural Nigeria: The role of communities. *Journal of Sustainable Development Law and Policy (The), 11*(1), 1–17.
Arokoyu, S. B., & Ukpere, D. R. T. (2014). Access to safe water supply and sanitation in lower Orashi River Basin, Rivers State, Nigeria. *ARPN Journal of Science and Technology, 4*(11), 639–646.
Babaniyi, B. R., Babaniyi, E. E., Aransiola, S. A., & Maddela, N. R. (Eds.). (2025). Wetland ecosystems: Conservation strategies, policy management and applications. Springer Nature. https://link.springer.com/book/9783031919817
Banerjee, D. K. (2017). Water pollution and human health.
Ben, V. E. (2023). Population growth and the Nigerian environment: Challenges and opportunities. In Academic practitioners research for sustainable development goals in Africa. IkotEkpene, ICIDR Publishing House
Boussaha, S., & Laifa, A. (2017). Wadi Bounamoussa's waters quality in the north-east of Algeria: Statistical treatment of some physical and chemical parameters. *Journal of Water and Land Development, 34*(1), 77.
Brauman, K. A., Siebert, S., & Foley, J. A. (2013). Improvements in crop water productivity increase water sustainability and food security—A global analysis. *Environmental Research Letters, 8*(2), Article 024030.
Dioha, M. O., & Kumar, A. (2020). Exploring greenhouse gas mitigation strategies for agriculture in Africa: The case of Nigeria. *Ambio, 49*(9), 1549–1566.
Egbueri, J. C., Agbasi, J. C., Onuba, L. N., Nweke, N. D., Uwajingba, H. C., & Abba, S. I. (2025). Groundwater development within the Nigerian crystalline and sedimentary aquifers: Challenges and opportunities. In S. Ali, & A. Negm (Eds.), Groundwater in developing countries. Springer Water. Springer. https://doi.org/10.1007/978-3-031-79122-2_13

Elemile, O. O., Raphael, D. O., Omole, D. O., Oloruntoba, E. O., Ajayi, E. O., & Ohwavborua, N. A. (2019). Assessment of the impact of abattoir effluent on the quality of groundwater in a residential area of Omu-Aran, Nigeria. *Environmental Sciences Europe, 31*(1), 16.

Elisha, I., Sawa, B. A., & Ejeh, U. L. (2017). Evidence of climate change and adaptation strategies among grain farmers in Sokoto State, Nigeria. *IOSR Journal of Environmental Science, Toxicology and Food Technology (IOSR-JESTFT), 11*(3), 1–7.

Elum, Z. A., & Momodu, A. S. (2017). Climate change mitigation and renewable energy for sustainable development in Nigeria: A discourse approach. *Renewable and Sustainable Energy Reviews, 76*, 72–80.

Federal Ministry of Water Resources (FMWR). (2016). National Water Resources Policy

Folorunshona, R., & Awosika, L. (2014). Morphological characteristics of the Bonny and Cross River (Calabar) estuaries in Nigeria: Implications for navigation and environmental hazards. In S. Diop, J. Barusseau, C. Descamps (Eds.), The land/ocean interactions in the coastal zone of West and Central Africa. (pp. 87–96). Springer.

Gimba, U. A., Aransiola, S. A., Oyewole, O. A., & Maddela, N. R. (2024). Marine greens: Roles in climate change and global warming mitigations. In Aransiola, S. A., Oyewole, O. A., Maddela, N. R., Ijah, U. J. J., & Manjunatha, B. (Eds.), *Marine greens: Environmental, agricultural, industrial and biomedical applications*. CRC Press, ISBN: 9781032438856. https://www.routledge.com/Marine-Greens-Environmental-Agricultural-Industrial-and-Biomedical-Applications/Aransiola-Oyewole-Maddela-Josiah-Ijah-Manjunatha/p/book/9781032438856

Gu, D., Andreev, K., & Dupre, M. E. (2021). Major trends in population growth around the world. *China CDC Weekly, 3*(28), 604.

Haider, H. (2019). Climate change in Nigeria: Impacts and responses.

Ibrahim, U. M., & Emeka, A. L. (2024). Impacts of climate change on groundwater resources in a semi-arid Region: A case study of Damaturu. *Yobe State. Communication in Physical Sciences, 11*(3), 450–462.

Idu, A. J. (2015). Threats to water resources development in Nigeria. *Journal of Geology and Geophysics, 4*(3). https://doi.org/10.4172/2329-6755.1000205.

Joint Environmental Audit. (2015). Joint environmental audit on the drying up of Lake Chad.

Majuru, B., Mokoena, M. M., Jagals, P., & Hunter, P. R. (2011). Health impact of small-community water supply reliability. *International Journal of Hygiene and Environmental Health, 214*(2), 162–166.

Malley, C. S., Omotosho, D., Bappa, B., Jibril, A., Tarfa, P., Roman, M., & Lefèvre, E. N. (2021). Integration of climate change mitigation and sustainable development planning: Lessons from a national planning process in Nigeria. *Environmental Science and Policy, 125*, 66–75.

Masson-Delmotte, V., Zhai, P., Pirani, A., Connors, S. L., Péan, C., Berger, S.,... & Zhu, B. (2023). Summary for policymakers. *Climate change 2021: The physical science basis. Contribution of working group I to the sixth assessment report of the intergovernmental panel on climate change* (pp. 3–32).

Mekonnen, M. M., & Hoekstra, A. Y. (2016). Four billion people facing severe water scarcity. *Science Advances, 2*(2), e1500323. https://doi.org/10.1126/sciadv.1500323

Miller, M., Cronk, R., Klug, T., Kelly, E. R., Behnke, N., & Bartram, J. (2019). External support programs to improve rural drinking water service sustainability: A systematic review. *Science of the Total Environment, 670*, 717–731.

Mishra, S., Sahu, A., Dungdung, M., Ahmed, S., & Baitharu, I. (2023). Pesticide pollution in freshwater and its impact on community health. In *Current developments in biotechnology and bioengineering* (pp. 33–52). Elsevier.

Ngene, B. U., Nwafor, C. O., Bamigboye, G. O., Ogbiye, A. S., Ogundare, J. O., & Akpan, V. E. (2021). Assessment of water resources development and exploitation in Nigeria: A review of integrated water resources management approach. *Heliyon, 7*(1).

Nnam, V., Odumosu, J., Lamidi, S., Aransiola, A. B. (2025). Spatio-temporal monitoring of desertification in Sokoto state using remote sensing and vegetation indices. *Journal of Spatial Information Sciences, 2*(2), 30–49. ISSN: 2354-3361. https://doi.org/10.5281/zenodo.15667752

Nonki, R. M., Lenouo, A., Lennard, C. J., et al. (2019). Assessing climate change impacts on water resources in the Benue River Basin, Northern Camroon. *Environ Earth Science, 78*, 606. https://doi.org/10.1007/s12665-019-8614-4

Nwobodo, C. E., Nwokolo, B., Iwuchukwu, J. C., Ohagwu, V. A., & Ozioko, R. I. (2022). Determinants of ruminant farmers' use of sustainable production practices for climate change adaptation and mitigation in Enugu State, Nigeria. *Frontiers in Veterinary Science, 9*, Article 735139.

Ogunkoya, O. O. (2023). Kainji Dam and Lake. In A. Faniran, L. K. Jeje, O. A. Fashae, & A. O. Olusola (Eds.), Landscapes and landforms of Nigeria. *World geomorphological landscapes*. Springer. https://doi.org/10.1007/978-3-031-17972-3_11

Ogwezzy, M. D. (2020). Water governance and service delivery in the water sector in Niger-delta.

Okafor, C. C., Nzekwe, C. A., Ajaero, C. C., Ibekwe, J. C., & Otunomo, F. A. (2022). Biomass utilization for energy production in Nigeria: A review. *Cleaner Energy Systems, 3*, Article 100043.

Okeke, F. O., Uzuegbunam, F. O., Nnaemeka-Okeke, R. C., & Ezema, E. C. (2021, October). People, people, everywhere; the architectural design response for Enugu city, Nigeria. In Proceedings of the DII-2021 international conference: "Building smart, resilient and sustainable infrastructure in developing countries (pp. 104–129).

Omokaro, G. O., Airueghian, E., & Ikiouenigha, M. I. (2024). Water resources, pollution, integrated management and practices in Nigeria—An overview. *American Journal of Environmental Economics, 3*(1), 5–18. (link unavailable)

Oyerinde, G. T. (2016). Climate change in the Niger Basin on hydrological properties and functions of Kainji Lake, West Africa (Doctoral dissertation, University of Abomey-Calavi).

Pham-Duc, B., Sylvestre, F., Papa, F., et al. (2020). The Lake Chad hydrology under current climate change. *Science and Reports, 10*, 5498. https://doi.org/10.1038/s41598-020-62417-w

Pona, H. T., Xiaoli, D., Ayantobo, O. O., & Tetteh, N. D. (2021). Environmental health situation in Nigeria: current status and future needs. *Heliyon, 7*(3).

Popoola, K. O., Jerneck, A., & Ajayi, S. A. (2021). Climate variability and rural livelihood security: Impacts and implications. In N. Oguge, D. Ayal, L. Adeleke, & I. da Silva (Eds.), African handbook of climate change adaptation. Springer. https://doi.org/10.1007/978-3-030-45106-6_200

Raji, A., Okaeme, A. N. & Ibeun, M. O. (2011). Forty years on Lake Kainji fisheries research. New Bussa, Espatek Nigeria Limited. ISBN 978-177-074-0.

Raji, S. A., Odunuga, S., & Fasona, M. (2021). Spatial appraisal of seasonal water yield of the Sokoto-Rima Basin. *Sakarya University Journal of Science, 25*(4), 950–968. https://doi.org/10.16984/saufenbilder.800302

Salami, A. A., Olanrewaju, R. M., Bakare, K. O., & Babatunde, O. R. (2025). Spatial distribution of rainfall in Nigeria. *Arabian Journal of Geosciences, 18*(1), 29.

Saminu, A., & Zayyanu, A. S. (2023). Analysis of stream inflow and peak flow of Kainji Lake using stochastic models. *European Journal of Engineering and Technology Research, 8*(3). https://doi.org/10.24018/ejeng.2023.8.3.3023

Saria, J. A., & Thomas, I. M. (2012). Water quality in selected shallow wells in Dar es Salaam. *Huria: Journal of the Open University of Tanzania, 11*, 13–24.

Shuaibu, A. M (2024). Bathymetric survey and volumetric analysis of Bakolori dam reservoir Northwest Nigeria. *African Scientific Reports, 3*(2), 154. https://doi.org/10.46481/asr.2024.3.2.154

Sow, M. (2017). Figure of the week: The shrinking Lake Chad. *Brookings Institution, 9*.

Svensson, E., & Berntsson, T. (2014). The effect of long lead times for planning of energy efficiency and bio refinery technologies at a pulp mill. *Renewable Energy, 61*, 12–16.

Takal, J. K., & Quaye-Ballard, J. A. (2018). Bacteriological contamination of groundwater in relation to septic tanks location in Ashanti Region, Ghana. *Cogent Environmental Science, 4*(1), 1556197 (2018)

Water Supply and Sanitation Reform Programme Phase II: Cross River. (2016). Retrieved from https://wsssrp.org/crossriver

Wato, T., Amare, M., Bonga, E., Demand, B. B. O., & Coalition, B. B. R. (2020). The agricultural water pollution and its minimization strategies—A review. *Journal Resourcing Deviation Management, 64*, 10–22.

Woodley, E. (2011). Building Nigeria's response to climate change: Pilot projects for community-based adaptation in Nigeria. In *Experiences of climate change adaptation in Africa* (pp. 297–315). Springer.

Woomer, P. L., Lewis, S., Mulei, W. E. L. I. S. S. A., Adenmosum, A., Dashiell, K., & Sanginga, N. (2025). African agricultural transformation: Perspectives from the international institute of tropical agriculture.

World Bank Group. (2017). The world bank annual report. http://odlsystem2.utm.my/1681/1/AnnualReport2017_BNM.pdf

World Health Organization (WHO). (2013). Nigeria sanitation, drinking water and hygiene status overview. Retrieved from https://cdn.who.int/media/docs/default-source/wash-documents/glaas/glaas-2013-14/glaas-2013-14-country-

Youdeowei, P. O., Nwankwoala, H. O., Desai, D. D. (2019). Dam structures and types in Nigeria: Sustainability and effectiveness. *Water Conservation and Management, 3*(1), 20–26. https://doi.org/10.26480/wcm.01.2019.20.26.

Chapter 2
Climate Change and Its Impact on Water Resources

Leh-Togi Zobeashia, Stella Suanu, Ajulor, and Nosakhare Praise

Abstract Climate change is a major global challenge, dramatically affecting water resource, hydrological systems and undermining water security worldwide. Water, an essential natural resource, is highly susceptible to changes in certain factors such as temperature, precipitation, and atmospheric circulation, making it an important indicator of climate impacts. The chapter explores the complex interactions between climate variability and hydrological cycles, showing how climate-induced warming (greenhouse gases) alters precipitation patterns, evaporation rates, and snow and glacier dynamics. Applying global climate models coupled with hydrological forecasts, the analysis addresses changes to ground and surface water resources. Vital topics include shifts in water availability, such as evolving precipitation intensity and seasonality, glacier retreat in regions such as Asia and South America, and variability in rainfall across sub-Saharan Africa. Water quality is also significantly impacted, with a rising temperature fueling algal blooms, salinization of coaster aquifers, increased sediment loads from extreme rainfall, and emerging pollutants posing escalating threats. The chapter further evaluates hydrological extremes including droughts floods, cyclones, and river flow variability, and their cascading impacts on water urban supplies agriculture, public health, energy generation, and natural ecosystems. Discussions focus on vulnerabilities in regions including Africa, Asia, Europe, the Americas, and Small Island Developing State. The chapter also reviews mitigation and adaptation strategies aimed at reinforcing resilience and facilitating sustainable water resource management.

Keywords Water resource · Hydrological system · Climate change · Water quality · Sustainable water management

L.-T. Zobeashia (✉) · S. Suanu · Ajulor · N. Praise
National Biotechnology Research and Development Agency, Abuja, Nigeria
e-mail: lehtogistella@gmail.com

M. I. Adeoba et al. (eds.), *Water Resources Management in Nigeria*, World Water Resources 10, https://doi.org/10.1007/978-3-032-22813-0_2

2.1 Introduction

Climate change is a global challenge of the twenty-first century, which is driven predominantly by anthropogenic emissions of greenhouse gases such as methane and carbon dioxide (IPCC, 2023a, 2023b, 2023c, 2023d). The following rise in global temperatures, approximately 1.1 °C since pre-industrial times, is unprecedented in historical records and is associated with profound and multifaceted impacts on natural and human systems worldwide (UN, 2024; IPCC, 2023a, 2023b, 2023c, 2023d). Climate change scale and complexity transcend beyond regional and sectoral boundaries, demanding global cooperative action. The phenomenon is not merely focusing on environmental problems but a comprehensive threat to sustainable development, affecting energy, food security, biodiversity, health, and water resources. These consequences are intertwined with economic, social, and political dimensions, increasing vulnerabilities and inequalities globally (Feulner, 2015). Given the accelerating pace of climate change phenomena and the uncertainty embedded in future projections, adapting to its effects while concurrently mitigating its causes has become an urgent imperative for humanity.

Water is important natural resources, essential for life, socioeconomic development, and ecosystem sustainability. Although water covers more than 70% of Earth's surface, only about 2.5% is freshwater, and a fraction of this is accessible for human use (World Bank, 2025). Freshwater resources are instrumental for drinking water, sanitation, agriculture, industrial processes, hydropower, and sustaining aquatic and terrestrial ecosystems. However, these resources are becoming increasingly constrained due to urbanization, growth in population, pollution, and unsustainable water management practices. Climate change builds an additional layer of complexity by altering the quality, availability, and temporal distribution of water worldwide (Marinova et al., 2025; WHO, 2019). Changes in precipitation patterns, increasing temperatures, and more frequent and severe extreme weather events exacerbate water scarcity and variability, disproportionately impacting vulnerable populations and ecosystems. The sustainability of water resources is thus intricately linked to climate resilience, requiring integrated governance frameworks and adaptive management strategies tailored to regional and local contexts (Ringler et al., 2022).

The hydrological cycle, the natural circulation of water through precipitation, runoff, infiltration, evaporation, and transpiration, is directly influenced by climatic variables, making it highly sensitive to climate variability and change. Rising temperatures accelerate atmospheric moisture capacity, intensify evaporation rates, and alter precipitation regimes, leading to more extreme hydrological events such as floods and droughts (Yang et al., 2021). Shifts in snowmelt timing and glacial retreat further influence streamflow patterns, with profound implications for water storage and availability in many regions. Groundwater recharge and quality are also affected, with potential long-term consequences for water security. These modifications to the hydrological cycle challenge traditional water resource management paradigms, which historically relied on the assumption of stationarity that is, that past hydroclimatic patterns provide reliable guidance for future conditions (Brekke et al., 2009). Under climate change, systems require flexible design and operation models that account for a wider range of variability and uncertainty. Incorporating climate projections, paleoclimate data, and stochastic simulation methods are increasingly recognized as essential for improved water resources planning and risk management (Brekke et al., 2009; Yang et al., 2021).

Water resources management faces a dual challenge posed by climate change and increased water demand due to population growth and economic development. The spatial and temporal mismatch between water supply and demand is intensifying, necessitating innovative governance approaches and adaptive strategies to maintain water security. Integrated Water Resources Management (IWRM) has emerged as a widely advocated framework aiming to harmonize social, economic, and environmental objectives through participatory governance and coordinated management across sectors and scales (Chivambo, 2025). Nevertheless, the integration of climate considerations into water management remains uneven and often insufficient, hampered by institutional fragmentation, resource limitations, and challenges in data availability and capacity building. Strengthening water governance systems, fostering multi-stakeholder collaboration, and promoting community engagement are imperative to enhance resilience against climate-induced water risks (Chivambo, 2025). Moreover, bridging knowledge gaps and institutional learning across local, national, and global platforms is critical for effective policy implementation and adaptive capacity building. The consequences of climate change on water resources are geographically and socioeconomically heterogeneous. Regions dependent on snowpack and glacier melt, such as the Himalayas and Andes, face altered flow regimes impacting agriculture, hydropower, and livelihoods. Arid and semi-arid regions confront heightened water scarcity, exacerbated by rising evapotranspiration and diminished rainfall. Conversely, some regions may experience increased precipitation, raising flood risks and water quality concerns (Marinova et al., 2025; Noh et al., 2024). Vulnerable communities, often with limited adaptive capacities, bear the brunt of these impacts, highlighting issues of equity and justice in water resource management under climate change. Ensuring equitable access and sustainable use of water resources constitutes a critical pillar in global efforts toward climate adaptation and sustainable development goals. This chapter aims to provide an integrative and comprehensive assessment of the multifaceted impacts of climate change on water resources globally. It synthesizes advances in scientific understanding, evaluates observed trends and projected scenarios, and discusses implications for water availability, quality, and extreme hydrological events. The analysis spans diverse geographic contexts and scales, emphasizing region-specific vulnerabilities and adaptation needs.

2.2 Climate Change and Hydrological Systems of Water Resources

Climate change is primarily driven by the buildup of greenhouse gases (GHGs) in the atmosphere, which disrupts Earth's energy equilibrium and accelerates global warming. The core process behind this is the greenhouse effect, in which gases like nitrous oxide (N_2O), carbon dioxide (CO_2), methane (CH_4), and water vapor absorb and re-emit longwave radiation released by Earth's surface, trapping heat in the atmosphere (Royal Society, 2015). This mechanism naturally sustains habitable conditions on Earth by maintaining temperatures suitable for life. Yet, since the Industrial Revolution, human activities have greatly raised the concentrations of these gases, most notably CO_2, which has increased from pre-industrial levels of about 280 ppm to over 420 ppm today (Friedlingstein et al., 2024). The strengthened greenhouse effect produces positive radiative forcing, with global average surface temperatures now about 1.1 °C above pre-industrial values and expected to climb further depending on emissions pathways (IPCC, 2023a, 2023b, 2023c, 2023d). This warming influences physical, chemical, and biological systems globally, forming the scientific foundation of climate change impacts, including those on water cycles. Major climatic factors regulate precipitation and evaporation trends, thereby shaping hydrological systems regionally and globally. Solar radiation, atmospheric circulation, and temperature interact dynamically to control the water cycle. For instance, solar energy fuels evaporation and water vapor formation, while temperature gradients drive pressure differences that generate winds and storm patterns (NASA GPM, 2020). At the equator, intense solar heating causes warm air to rise, triggering strong evaporation and convective rainfall. As this moist air shifts toward the poles, it cools and precipitates, creating wetter mid-latitude areas but leaving subtropical dry belts were descending air masses extract surface moisture, promoting arid conditions (Climate Central, 2024). Alterations in these drivers due to global warming change precipitation frequency, intensity, and distribution, while also modifying evapotranspiration rates (Nazeri Tahroudi et al., 2025). Such variations manifest as stronger droughts, heavier rainfall, and disrupted seasonal water availability, posing major challenges for global water resource planning.

The hydrological cycle including rainfall, infiltration, runoff, evaporation, transpiration, and aquifer recharge is highly sensitive to climate. Rising atmospheric temperatures intensify the cycle by raising air's saturation vapor pressure, allowing greater moisture retention and thus elevating evaporation (Kleidon, 2013). According to Huntington (2023) research and simulations, each 1 °C temperature rise enhances the global hydrological cycle by about 2–3% in terms of moisture flux. This heightened cycle amplifies flood and drought risks, as more moisture is rapidly circulated yet distributed unevenly across space and time (IPCC, 2023a, 2023b, 2023c, 2023d). Furthermore, warming alters snowmelt timing, reduces snowpack, and accelerates glacier retreat, reshaping runoff patterns, especially in mountainous and polar zones where snow and ice serve as natural reservoirs (Yang et al., 2021). Groundwater systems are likewise affected, with recharge rates and water quality shaped by precipitation shifts and evapotranspiration, threatening long-term water sustainability. To predict how climate change will impact water resources, researchers use diverse modeling techniques combining climate dynamics with hydrological processes. General Circulation Models (GCMs) or Global Climate Models simulate atmosphere–ocean–land interactions to project temperature, rainfall, and radiation under different emission pathways (IPCC, 2023a, 2023b, 2023c, 2023d). Because GCM outputs operate at broad scales, downscaling and integration with hydrological models are required for local and regional assessments (Yang et al., 2021). Hydrological models represent watershed and river basin processes like runoff, groundwater flows, and evapotranspiration. These range from conceptual rainfall-runoff tools to physically based basin models (Mengistu et al., 2025). Integrated climate-hydrology modeling allows assessments of future water availability, flood potential, drought frequency, and shifts in water quality across varying climate scenarios.

Recent progress has introduced artificial intelligence and data-driven approaches to enhance hydrological prediction accuracy and efficiency (Mengistu et al., 2025). These models are refined against historical meteorological and hydrological data to capture feedback and system nonlinearities while minimizing uncertainties. Increasingly, they incorporate socio-economic and land-use variables to account for human influences on climate-driven hydrological changes (Chivambo, 2025). Nonetheless, modeling challenges remain due to uncertainties in emissions, climate sensitivity, downscaling approaches, and incomplete knowledge of hydrological feedback mechanisms (Domínguez-Tuda et al., 2021). Expanding observational systems, remote sensing capacity, and global scientific collaboration are critical to improving forecasts and informing adaptive water governance (Table 2.1).

2.3 Impacts on Water Availability

Climate change remarkably transforms the quality, timing and availability of freshwater supplies through a complex interplay of processes. Increasing global temperatures, shifts in atmospheric circulation, and evolving precipitation patterns collectively impact surface water flows, groundwater recharge, glaciers and snowpack (cryospheric storage), and seasonal runoff variability leading to diverse regional outcomes with some areas experiencing water abundance while others face severe shortages (IPCC, 2023a, 2023b, 2023c, 2023d; USGS, 2025). Some of the primary pathways through which climate change influences water availability include:

2.3.1 Shifts in Precipitation Patterns

Shift in precipitation concentration such as intensity, seasonal timing and frequency in a climate driven modifications have a far-reaching consequence for water supply reliability, flood hazards, and drought frequency. Evidence from observational records

Table 2.1 Climate change and hydrological systems

Features	Description	References
Greenhouse effect and warming	GHGs trap heat, raising global temps by ~ 1.1 °C which is expected to rise further	Friedlingstein et al. (2024)
Climatic drivers	Solar energy and circulation shape rainfall; warming shifts patterns, causing droughts and floods	Nazeri Tahroudi et al. (2025)
Hydrological cycle	+ 1 °C boosts water cycle intensity 2–3%; alters snowmelt, runoff, and glaciers	Kleidon (2013), Huntington (2010), IPCC (2023a, 2023b, 2023c, 2023d)
Groundwater	Recharge and quality affected by rainfall shifts and evapotranspiration	Yang et al. (2021)
Modeling	GCMs + hydrological models project water availability and extremes	(IPCC 2023a, 2023b, 2023c, 2023d), Mengistu et al. (2025)
Technology advances	AI improves prediction using climate, land-use, and socio-economic data	Mengistu et al. (2025), Chivambo (2025)

and climate model ensembles confirms that a warmer atmosphere leads to increased atmospheric moisture capacity, increasing the intensity of heavy rainfall events, even as some regions endure longer and more frequent dry spells (IPCC, 2023a, 2023b, 2023c, 2023d). These changes are spatially heterogeneous, many mid- and high-latitude areas and parts of the tropics are projected to experience increases in mean precipitation, while subtropical regions typically become drier (IPCC, 2023a, 2023b, 2023c, 2023d). Additionally, the timing and duration of wet and dry seasons are shifting across many river basins, complicating water storage management and agricultural scheduling (EPA, 2025). The intensification of brief, intense precipitation episodes lead to higher flood peaks that can overwhelm existing drainage infrastructure and natural floodplains (IPCC, 2023a, 2023b, 2023c, 2023d). At the same time, increased length and severity of dry intervals exacerbate soil moisture deficits and reduce groundwater recharge, further stressing water resources (Davamani et al., 2024). The coexistence of heavier rain and longer droughts undermines the predictability and reliability of freshwater availability essential for agricultural, industrial, and municipal needs. Moreover, interactions between altered precipitation patterns and land-use changes like deforestation and urbanization amplify runoff and reduce infiltration, aggravating local water scarcity (EPA, 2025). For water resource managers, this dynamic creates a challenging balancing act: designing systems robust enough to handle more frequent high-flow extremes while ensuring sufficient water storage and supply during extended low-flow periods.

2.3.2 Glacial Retreat and Snowpack Decline

Glaciers and seasonal snowpacks in mountainous regions act as natural freshwater reservoirs by accumulating precipitation as solid ice and snow during winter, which then melts gradually during warmer months, regulating downstream river flows (USGS, 2025). Across many mountain systems, accelerating warming has caused substantial reductions in seasonal snow cover and widespread glacier mass loss. Initially, the increased meltwater from retreating glaciers may boost summer river flows, somewhat compensating water demands. However, this "peak water" phase is transitional; ultimately, continued glacier decline leads to diminished dry season baseflows that elevate water scarcity risks during critical agricultural and hydropower demand periods (Copernicus TC, 2024). The spatial variability of glacial responses is significant. The southern Hindu Kush-Himalaya and tropical Andes have experienced rapid ice mass loss and altered melt timing, affecting hundreds of millions dependent on these water sources (Copernicus TC, 2024; Johansen et al., 2024). The decline of perennial cryospheric reserves also increases glacial lake formation, heightening downstream risks of glacial lake outburst floods (GLOFs), which threaten human settlements and infrastructure (Copernicus TC, 2024). These ongoing cryosphere changes necessitate regionally tailored water management and hazard mitigation strategies to cope with shifting runoff regimes and flood risks.

2.3.3 Groundwater Recharge Under Modified Rainfall Regimes

Aquifers, which store groundwater, provide a crucial buffer for maintaining water supplies during dry periods and supporting agriculture and ecosystems. Climate change influences groundwater recharge through both direct changes in precipitation characteristics amount, frequency, and intensity and indirect changes mediated by vegetation dynamics, soil moisture, and land-use alterations (Davamani et al., 2024). Increased precipitation intensity can result in more surface runoff and less infiltration where soils are compacted or vegetation cover is sparse, thereby reducing recharge effectiveness. Conversely, more frequent but lighter precipitation events might enhance infiltration in certain contexts (Davamani et al., 2024). Furthermore, rising temperatures increase evapotranspiration rates, which can reduce soil moisture availability and consequently curb groundwater recharge in many regions (Davamani et al., 2024). Studies using climate-hydrology-groundwater coupled models highlight complex, site-specific responses of recharge rates: some aquifers experience long-term declines in recharge and falling water tables, while others show temporary recharge increases linked to shifts in seasonal rainfall or human water management practices like managed aquifer recharge (Davamani et al., 2024). The high uncertainty in recharge projections stems from the interconnected nature of climatic drivers, hydrogeologic conditions, and anthropogenic pressures, emphasizing the need for enhanced groundwater monitoring and integrated modeling frameworks.

2.3.4 Regional Variability in Water Availability Impacts

2.3.4.1 Himalayan Glacier-Fed River Systems

The Hindu Kush-Karakoram-Himalaya region sustains major river systems such as the Indus, Ganges, and Brahmaputra, which support dense populations and intensive agriculture across South Asia. Recent syntheses reveal accelerated glacier retreat in much of this region, with notable spatial heterogeneity: while some northern and western sectors exhibit slower melting (Karakoram anomaly), the southern and eastern Himalayas face pronounced ice mass losses and earlier snowmelt (Copernicus TC, 2024). These changes have reshaped seasonal streamflows, often increasing pre-monsoon discharge in the short term but reducing dry-season baseflows as glaciers shrink. This threatens irrigation reliability and hydropower generation during critical dry months. Compounding these challenges is the increasing erraticism of the monsoon, elevating flood risks and mid-season drought likelihood (Copernicus TC, 2024).

2.3.4.2 Rainfall Variability in Sub-Saharan Africa

Sub-Saharan Africa experiences notable inter-annual and decadal rainfall variability influenced by large-scale climate phenomena such as the El Niño-Southern Oscillation and Indian Ocean variability, as well as local temperature increases (IPCC, 2023a, 2023b, 2023c, 2023d). Parts of southern and eastern Africa have witnessed more frequent, concurrent droughts and heatwaves, while some equatorial regions see increased episodes of intense rainfall. This compound variability causes episodic recharge surges alternating with extended deficits, challenging groundwater and surface water systems responsible for agriculture and food security (IPCC, 2023a, 2023b, 2023c, 2023d). Water infrastructure is often limited, and climate-induced volatility reinforces vulnerability. Building resilience demands investments in storage capacity, groundwater development, early warning systems, and institutional strengthening to enable coordinated water allocation and sustainable use.

2.3.4.3 Andean Glaciated Basins

In the tropical and extratropical Andes, glaciers provide vital low-flow water sources during dry seasons for cities and irrigated valleys. Accelerated glacier retreat is widespread, initially boosting runoff but now contributing to declining late-season flows as ice reservoirs dwindle (Johansen et al., 2018). The most rapid glacier loss is noted in northern and central Andes, where warming rates are highest. This trend exacerbates competition among urban, agricultural, and hydropower demands, necessitating adaptive measures such as reservoir expansion, improved demand management, and upstream watershed protection. Regional inventories underscore the urgency of enhanced cryospheric monitoring and equitable water management planning to confront these emerging stresses (Johansen et al., 2018).

2.4 Impacts on Water Quality

Water quality is intrinsically linked to climate processes, where changing temperatures, varied precipitation patterns, sea-level rise, and fluctuating hydrology alter the chemical, physical, and biological makeup of aquatic environments. Beyond mere water quantity, climate change disrupts the lakes, health of rivers, groundwater, estuaries, and coastal waters by influencing oxygen levels, nutrient cycles, pollutant dispersion, and salinity regimes. These effects are compounded by land use alterations, urban growth, and contamination sources, producing complex cumulative stress that jeopardizes ecosystem function, food production, and human health. This section reviews key mechanisms by which climate change degrades water quality, underpinned by recent global and regional research findings.

2.4.1 Temperature Effects on Water Chemistry

Rising water temperatures represent one of the clearest consequences of global warming for aquatic systems. Increased thermal loads reduce the solubility of dissolved oxygen (DO), which is vital for aerobic aquatic organisms and biochemical processes. A temperature rise of 1 °C currently corresponds to an approximate 2% decrease in oxygen solubility, fostering hypoxic conditions especially in stratified lakes and slow-moving rivers (Jane et al., 2021). Reduced DO harms fish and invertebrates, dampens aerobic microbial degradation, and accelerates nutrient releases like phosphorus from sediments, creating reinforcing cycles that deteriorate water quality further. Elevated temperatures also boost phytoplankton metabolic rates, leading to more frequent, persistent, and intense harmful algal blooms (HABs). Cyanobacteria such as *Microcystis aeruginosa* flourish in warmer, nutrient-rich waters, producing toxins that threaten drinking water safety and public health (Paerl & Otten, 2023). Observations from North America, Europe, and China indicate that HABs are expanding spatially and seasonally, linked to warming temperatures, prolonged thermal stratification, and enhanced nutrient inputs (Chapra et al., 2022). HAB outbreaks not only disrupt ecosystems but also increase water treatment costs and jeopardize food security via contamination of fisheries and aquaculture.

2.4.2 Increased Runoff and Sedimentation Due to Extreme Rainfall

Climate change intensifies the hydrological cycle, causing more frequent and severe heavy rainfall events. These events produce high runoff volumes that transport sediments, nutrients, pathogens, and contaminants from agricultural, urban, and industrial areas into aquatic systems. Increased sediment deposition reduces water clarity, smothers bottom habitats, and disturbs aquatic food chains, while also carrying pollutants like phosphorus and heavy metals (Zhang et al., 2022). Catchments with degraded soils and sparse vegetation are especially vulnerable to sediment surges during storms that overwhelm reservoirs and diminish storage (Nguyen et al., 2021). Urban stormwater compounds this issue by delivering hydrocarbons, microplastics, and heavy metals into waterways, often straining combined sewer systems and increasing fecal pollution risks (Shrestha et al., 2022). Moreover, intensified rainfall speeds up contaminant transport, reducing natural filtration and microbial breakdowns that typically mitigate pollution, thereby necessitating more advanced treatment and resilient watershed management.

2.4.3 Salinization in Coastal Aquifers (Sea Level Rise)

Rising sea levels, resulting from ocean thermal expansion and polar ice melting, threaten coastal freshwater resources by allowing saltwater intrusion into groundwater aquifers. This saltwater encroachment jeopardizes freshwater availability for drinking, agriculture, and ecosystems by altering groundwater flow gradients (Michael et al., 2022). Particularly affected areas include low-lying delta regions such as the Ganges–Brahmaputra-Meghna, Mekong, and Nile, which supply water to millions. For example, salinity has intruded up to 100 km inland in Bangladesh, undermining water security, rice production, and health (Acharjee et al., 2023). Salinization additionally mobilizes toxic trace elements like arsenic in sediments, further degrading groundwater quality. Projected sea level rise between 0.44 and 0.77 m by 2100 under high emission pathways will exacerbate these challenges globally (IPCC, 2023a, 2023b, 2023c, 2023d). Adaptive measures such as managed aquifer recharge, desalination, and freshwater buffer zones will be costly and often beyond the reach of vulnerable coastal communities.

2.4.4 Pollutant Concentration During Droughts

Prolonged droughts that reduce precipitation and stream flow lead to increased pollutant concentrations in surface waters due to diminished dilution capacity. Levels of nutrients, pesticides, and pathogens rise as flows lessen, compounding ecological and health risks (Mosley, 2015). High temperatures during droughts enhance evapo-concentration, further concentrating contaminants in rivers and reservoirs. The 2018 European drought exemplified these effects, where elevated nitrate and pesticide concentrations surpassed drinking water standards due to reduced base-flow and assimilation capacity (Koch et al., 2020). In many developing areas, drought-induced reductions in water treatment infrastructure exacerbate contamination. Increased reliance on marginal sources such as untreated wells or brackish water during droughts also raises exposure to chemical and microbial pollutants. Coupled drought and heatwave stress worsen oxygen depletion, promotes algal blooms, and spread pathogens, creating complex water quality crises.

2.4.5 *Emerging Contaminants*

Climate change intersects with the growing risk posed by emerging contaminants, including pharmaceuticals, microplastics, personal care products, synthetic organics, and threatening aquatic ecosystems and human health. Extreme rainfall events mobilize these substances from wastewater, landfills, and urban runoff, while drought conditions cause them to concentrate in water bodies (Ahmed et al., 2021). Pharmaceutical residues, such as antibiotics and endocrine disruptors, resist conventional treatment and accumulate downstream, with changing flow regimes altering their dilution and degradation (Patel et al., 2023). Microplastics, transported more efficiently during storms and persistent during droughts, have been detected at high levels in rivers like the Thames, Yangtze, and Mississippi (Li et al., 2022). Increasing water temperatures also promote microbial colonization on microplastic surfaces, facilitating pathogen and antibiotic resistance gene dispersal (Zhang et al., 2021). These dynamics intensify ecological and public health risks, demanding improved monitoring, advanced treatment technologies, and stronger regulations.

2.5 Impacts on Hydrological Extremes

Climate change is reshaping the occurrence, magnitude, and persistence of hydrological extremes worldwide, exerting profound stress on water systems. These changes are evident through rising flood events, intensifying droughts, stronger storm surges and cyclones that disrupt coastal freshwater, and greater variability in river discharges.

2.5.1 Increasing Flood Frequency and Magnitude

The enhanced moisture-retention capacity of a warming atmosphere is driving more intense and frequent flood events globally (IPCC, 2023a, 2023b, 2023c, 2023d; WMO, 2025a, 2025b). A strengthened hydrological cycle produces heavy, short-lived rainfall episodes, generating runoff surges and flood peaks that often exceed the design capacity of drainage networks and mitigation infrastructure (Mejía et al., 2023). Europe, for instance, has witnessed some of its most devastating floods in recent decades, with immense economic losses and fatalities (WMO, 2025a, 2025b). In numerous areas, extended wet periods and rising seas magnify flooding threats, particularly across deltas and expanding coastal cities (Diop et al., 2025). Flood impacts extend beyond immediate damage, creating long-term repercussions for water quality, public health, agriculture, and ecosystems. Rapid urbanization further heightens risks, as impervious surfaces accelerate runoff while curtailing infiltration, amplifying both flood peaks and duration (Shrestha et al., 2022). The unpredictability of flood magnitude and timing complicates management showing the need for adaptive designs and improved predictive capacity.

2.5.2 Droughts Intensification and Extension

Conversely, many arid and semi-arid regions are experiencing increasing drought severity and persistence, driven by altered rainfall regimes and higher evapotranspiration rates linked to warming (IPCC, 2023a, 2023b, 2023c, 2023d; Yun et al., 2025). Such droughts are characterized by declining soil moisture and diminished baseflows, straining supplies crucial for domestic, industrial, and agricultural uses (Jisem Journal, 2024). Hydrological droughts sustained deficits in surface and groundwater reserves are becoming more severe, particularly in southern Africa, Australia, and the Mediterranean (IPCC, 2023a, 2023b, 2023c, 2023d). Frequently coinciding with heatwaves, they exacerbate water scarcity, heighten wildfire risks, and trigger agricultural losses (WMO, 2025a, 2025b). The socio-economic consequences are most acute for marginalized populations reliant on natural resources. Effective responses require integrated demand management, greater investment in water reuse and storage, and strengthened institutional coordination.

2.5.3 Storm Surges and Cyclones Threatening Coastal Freshwater

Sea-level rise from glacial melt and thermal expansion increases the magnitude and reach of storm surges linked to tropical cyclones and other extremes (Konikow et al., 2022). Stronger storms, carrying greater energy and water, produce devastating coastal floods that impair freshwater supplies and damage infrastructure (Diop et al., 2025). Storm-driven saline intrusion into aquifers and estuaries further compromises freshwater quality and supply in deltas and islands (Acharjee et al., 2023). Vulnerable regions such as the Ganges–Brahmaputra-Meghna and Mekong deltas face escalating risks, affecting millions of residents. Adaptive measures include protective infrastructure, advanced forecasting and evacuation systems, and ecosystem-based strategies such as mangrove and wetland restoration to buffer surge effects.

2.5.4 River Flow Variability and Its Sectoral Impacts

Altered rainfall patterns, glacier retreat, and shifting snowmelt dynamics are driving greater variability in river discharges, complicating water use across multiple sectors (IPCC, 2023a, 2023b, 2023c, 2023d; Yun et al., 2025). Variations in flow intensity and timing disrupt shipping routes, undermine irrigation reliability during peak agricultural demand, and destabilize hydropower generation (Jisem Journal, 2024). In mountain-fed systems, shrinking glaciers generate volatile downstream flows, while lowland rivers face increasingly erratic cycles of drought and flood (Diop et al., 2025). Major river basins such as the Amazon, Mekong, and Nile exhibit altered discharge regimes that jeopardize water security for agriculture, cities, and industries (WMO, 2025a, 2025b). Addressing these requires integrated basin-scale management informed by climate projections, flexible allocation systems, and investments in adaptable infrastructure.

2.6 Impacts of Climate Change on Water-Dependent Sectors

Climate change is significantly altering the operational environments and risk profiles for vital sectors reliant on freshwater, including agriculture, energy, public health, urban infrastructure, and freshwater ecosystems.

2.6.1 Agriculture

Globally, agriculture remains the largest consumer of freshwater and exhibits extreme sensitivity to climate variations. Elevated temperatures and heightened evapotranspiration rates increase crop water requirements, raising irrigation demands even where rainfall volumes remain unchanged (Lobell et al., 2025). Shifts in rainfall timing and increased heatwave incidence exacerbate risks of crop failure, particularly in rain-fed agricultural systems heavily reliant on predictable seasonal precipitation in regions such as sub-Saharan Africa and South Asia (Lobell et al., 2025). Increased irrigation demands drive intensified groundwater pumping, accelerating aquifer depletion and undermining long-term viability (Davamani et al., 2024). Where irrigation depends on snow or glacier melt, altered melt timing reduces water availability during late growing seasons, heightening competition amongst agricultural, urban, and ecological uses. Furthermore, soil salinization increasingly threatens low-lying irrigated deltas and semi-arid regions where diminished river flows, intensified evaporation, and inadequate drainage concentrate salts in root zones, damaging crop yields and forcing costly land remediation or abandonment (Davamani et al., 2024).

2.6.2 Energy

Water and energy sectors are tightly coupled. Hydropower generation depends heavily on river flows and reservoir storage, which climate change disrupts through altered precipitation, glacier retreat, and snowmelt variations. Modeling studies project substantial declines in hydropower productivity in various regions by mid-century, with flow seasonality shifts creating mismatches between availability and demand peaks. This variability challenges grid stability and energy security. Thermoelectric power plants, including nuclear facilities, face dual climate-driven risks: water scarcity restricts cooling water withdrawals, limiting operational output, and elevated intake water temperatures reduce cooling efficiency and may violate regulatory thermal discharge limits (UNEP Finance Initiative, 2024). Empirical evidence links rising heatwaves and droughts to increasing weather-related outages and operational constraints precisely during periods of peak electricity demand for cooling, exacerbating grid stress (UNEP Finance Initiative, 2024). Enhancing resilience thus entails diversifying cooling technologies (closed cycle, air-cooled), integrating water availability into power system planning, and boosting renewable energy capacity and storage flexibility.

2.6.3 Public Health

Climate variability increases waterborne disease risks through multiple ways. Temperature rises increase pathogen replication rates and extend transmission seasons for vectors like *Vibrio* species. Extreme precipitation and flooding mobilize fecal contamination, overwhelming sanitation infrastructure and polluting drinking water sources. Conversely, droughts reduce dilution volumes, concentrate pollutants and pushing populations towards unsafe water sources (World Health Organization, 2024; NEJM Review, 2023). These dynamics disproportionately endanger low- and middle-income regions with limited sanitation coverage, where climate shocks trigger spikes in diarrheal disease, cholera, and other waterborne illnesses (World

Health Organization, 2024). Climate-resilient interventions emphasize climate-proofing water and sanitation infrastructure, decentralizing point-of-use water treatment, strengthening surveillance and early-warning systems, and integrating water, sanitation, and hygiene (WASH) programs with public health emergency planning. Equitable governance facilitating access during crises remains vital.

2.6.4 Urban Systems

Rapid urban growth concentrates demand in megacities, where climate impacts further strain water systems. Increasing numbers of cities face projected chronic water scarcity by mid-century fueled by population growth, land conversion, and climate change (Nature, 2021). Heatwaves and drought undermine potable water supply, disrupt energy-water feedback loops essential for cooling, and impair wastewater collection and treatment. Extreme storms damage infrastructure and threaten contamination (Nature, 2021). Recent urban analyses reveal many water utilities lack institutional flexibility and diverse supply sources (surface water, groundwater, reuse, desalination) crucial for addressing climate and urban pressures (Nature, 2021). Effective adaptation hinges on demand management (pricing, leakage control), supply diversification including stormwater harvesting and recycling, bolstering decentralized services for informal settlements, and integrating climate risk into urban planning and investment.

2.6.5 Ecosystems

Freshwater ecosystems rank among the most threatened globally, with climate change compounding pressures from land use, pollution, and fragmentation. Global assessments document alarming declines in aquatic biodiversity driven by climate extremes and altered flow regimes (PMC, 2025). Wetlands, crucial for water filtration, flood regulation, and aquatic habitat, are rapidly degrading due to climate impacts such as sea-level rise and shifts in precipitation, diminishing their ecological functions and economic benefits (Reuters, 2025). Fish populations, especially migratory and cold-water species, suffer habitat fragmentation and temperature stress tied to dams, flow modification, and warming, causing fishery collapses affecting millions' livelihoods and food security (PMC, 2025). Conservation imperatives call for barrier removal, restoration of natural flow regimes, expansion of protected areas, and basin-scale land management reforms.

2.7 Regional Perspectives

The impacts of climate change on water resources are strongly region-specific: while global physical drivers such as warming, altered precipitation patterns, and sea-level rise operate universally, their local expressions and socio-ecological outcomes vary significantly based on regional climate characteristics, glacier dependency, land use, and governance strength. Here, we summarize the major water-related challenges identified across different global regions, drawing on the latest scientific assessments.

2.7.1 Africa

Africa faces a disproportionate burden of climate-induced water stress, despite minimal contributions to global greenhouse gas emissions. Rising temperatures and increasing precipitation variability have heightened drought frequency and severity, impacting surface water flows, groundwater recharge, and food security in expansive areas (IPCC, 2023a, 2023b, 2023c, 2023d). Semi-arid zones are experiencing accelerated desertification, driven by diminished rainfall, extensive land degradation, and unsustainable water practices that hinder productivity and amplify adaptation costs (WMO, 2025a, 2025b). Compounding the physical trends are social vulnerabilities such as limited water storage capacity, insufficient groundwater monitoring, and high reliance on rain-fed agriculture, which turn drought into recurrent humanitarian emergencies. Priorities for adaptation include enhancing resilient water storage infrastructure like managed aquifer recharge, restoring degraded landscapes, boosting irrigation efficiency, and improving early-warning systems alongside strengthening transboundary water governance to bolster community resilience (IWA Publishing, 2025).

2.7.2 Asia

Asia's water systems are dominated by monsoon rainfall patterns and high-elevation glacier and snowmelt, both susceptible to climate change. Warming accelerates glacier retreat throughout the Himalaya and adjacent ranges, reshaping runoff timing and magnitude in major river basins such as the Indus, Ganges, Brahmaputra, and Mekong (ESSD, 2025). While meltwater volumes have temporarily increased during this peak water phase, progressive ice losses are diminishing dry-season flows vital for millions downstream (ESSD, 2025). Monsoon patterns are also shifting, with increased extreme rainfall events and altered onset and withdrawal timing disrupting agricultural calendars and reservoir operations (IPCC, 2023a, 2023b, 2023c, 2023d). This combination of less reliable cryospheric water buffering and intensified monsoon volatility necessitates integrated river basin management, enhanced real-time glacier and hydrological monitoring, and demand-side water management to reduce downstream exposure to water stress and hazards.

2.7.3 *Europe*

Europe exhibits increasingly polarized water risks, featuring long-lasting droughts especially across the Mediterranean and central regions alongside greater flood risk from intense, short-duration rainfall in parts of northern and Western Europe (Copernicus Climate Change Service, 2025). The heatwaves and localized floods of 2024–2025 exemplify these trends, generating substantial economic damage and straining water supplies and ecosystems (The Guardian, 2025). Growing urbanization and aging water infrastructure intensify vulnerabilities as low flows reduce hydropower and navigation capacity while focusing pollutants, and extreme storms test drainage systems and reservoir safety. European adaptation primarily focuses on nature-based solutions such as catchment restoration and urban green infrastructure, multi-year water storage, and coordinated transboundary water resource management to handle increasing seasonal variability.

2.7.4 *North America*

Water challenges in North America vary by region. The western US and Mexico experience more frequent and severe droughts and aridification due to reduced winter precipitation and increased evapotranspiration, deeply affecting agriculture, municipal supplies, and wildfire risks (NOAA, 2025; USGS, 2025). Conversely, the Atlantic and Gulf coasts face stronger hurricanes and sea-level rise exacerbating storm surge damages to freshwater infrastructure (NOAA, 2025). The Laurentian Great Lakes display complex variability in ice cover, stratification, and precipitation, influencing lake levels, temperatures, and ecosystem services (USGS, 2025).

2.7.5 *Latin America and the Caribbean*

The Amazon basin shows unprecedented hydrological variability, with extreme droughts and floods disrupting the regional moisture recycling mechanism critical for continental rainfall patterns (NASA Earth Observatory, 2024). Such disruptions threaten agriculture and hydropower systems continent-wide. Recent droughts and decreased river levels (2023–2024) have impaired navigation, increased wildfire risks, and stressed energy infrastructure (NASA Earth Observatory, 2024). Coastal zones in the Caribbean and Latin American deltas face rising inundation risks from sea-level rise compounded by storm surges, compromising freshwater lenses and salinizing soils and aquifers. Policy approaches include ecosystem-based adaptation like mangrove restoration, revising floodplain land use, and expanding monitoring and early-warning capabilities.

2.7.6 Small Island Developing States (SIDS)

Small Island developing states are among the most vulnerable to climate-related water security threats due to small land areas, dependence on shallow coastal aquifers, and limited water storage capacity (IPCC, 2023a, 2023b, 2023c, 2023d; WMO, 2025a, 2025b). Saltwater intrusion into freshwater lenses degrades potable groundwater supplies, while extreme storms frequently damage fragile infrastructure and contaminate surface water catchments, with serious humanitarian and development implications. Adaptation strategies prioritize protection and restoration of coastal buffers such as coral reefs and mangroves, feasible managed aquifer recharge, decentralized rainwater harvesting, and scaling-up financial and technical support for integrated water security.

2.8 Adaptation and Mitigation Strategies

Effectively and efficiently manage the impacts of climate change on water resources, demands integrated approaches that will combine advanced technological solutions, robust policy structure, nature-oriented strategies, and socio-economic initiatives. These approaches seek not only to alleviate the detrimental impacts of climate change but also to bolster adaptive capacity, enhance systemic resilience, and sustainably secure water availability for all users.

2.8.1 Technological Solutions

2.8.1.1 Water Recycling and Reuse

Recycling and reusing water has become central to climate adaptation efforts, particularly in arid and rapidly expanding urban areas. Wastewater that is treated is increasingly leveraged for industrial activities, agricultural irrigation, and groundwater replenishment. This strategy alleviates pressure on conventional freshwater sources and improves water reliability during scarcity periods. Countries like Israel and Singapore showcase successful advanced wastewater treatment schemes producing reclaimed water for diverse uses, significantly advancing water security goals (Qadir et al., 2020). Studies predict that, with adequate safety and treatment standards, wastewater reuse could supply up to 10% of worldwide irrigation demand by 2050 (Jiménez & Asano, 2015; Rasul & Sharma, 2016).

2.8.1.2 Desalination Technologies

Desalination is a vital tool for adapting water-scarce coastal regions. Innovations in reverse osmosis membranes and coupling with renewable energy-powered desalination plants have decreased both costs and environmental impacts (Jones et al., 2019). Countries such as Saudi Arabia and the UAE heavily depend on desalinated water for municipal consumption, while Australia has established large-scale desalination infrastructure to buffer against droughts (Ghaffour et al., 2019). Although concerns remain about brine disposal and energy consumption, pairing desalination plants with solar or wind energy offers sustainable pathways to reduce emissions and improve resilience.

2.8.1.3 Smart Irrigation Systems

Given agriculture's dominance in freshwater withdrawals (~ 70%), improving irrigation efficiency is crucial. Smart irrigation, utilizing soil moisture sensors, satellite monitoring, and AI, optimizes water application according to crop demand and meteorological conditions, minimizing losses through evaporation or seepage (Blatchford et al., 2020). Pilot programs in India and sub-Saharan Africa integrating remote sensing with mobile platforms demonstrate enhanced water efficiency and crop yield stability under variable rainfall regimes (Ravelo et al., 2021).

2.8.2 Nature-Based Solutions

2.8.2.1 Wetland Restoration and Watershed Management

Wetlands provide essential ecosystem services, including water retention, aquifer recharge, pollutant filtration, and flood attenuation. Unfortunately, over 35% of global wetlands have been lost since the 1970s, eroding these natural buffers (Ramsar, 2018). Restoration efforts in places like the Mississippi Basin and Danube floodplains effectively reduce flood vulnerability while boosting water quality and biodiversity (Davidson et al., 2020). Similarly, managing entire watersheds through reforestation and erosion control stabilizes hydrological processes and enhances resilience to climate variability, especially in mountainous and semi-arid landscapes (Grigg, 2016).

2.8.2.2 Reforestation and Soil Moisture Conservation

Increasing vegetative cover via reforestation and afforestation improves water infiltration, reduces surface runoff, and enhances soil water retention. Forests also moderate microclimates, buffering drought effects. For example, China's "Grain for Green" initiative expanded vegetation cover, stabilizing soils and improving water conservation in degraded areas (Feng et al., 2016). At the farm scale, mulching, agroforestry, and conservation tillage conserve soil moisture and boost water use efficiency, supporting crops against climatic extremes (Locatelli et al., 2015).

2.8.3 Policy and Governance

2.8.3.1 Integrated Water Resources Management (IWRM)

Integrated water resources management provides a governance framework harmonizing competing water demands across sectors while incorporating climate resilience into basin management. By aligning socioeconomic and ecological goals, IWRM mitigates conflicts and enhances resource use efficiency (UNESCO, 2019). The Mekong River Commission exemplifies basin-wide planning that balances hydropower, agriculture, and ecosystem health under evolving climate conditions (Johnston & Kummu, 2019).

2.8.3.2 Climate-Resilient Governance Frameworks

Beyond IWRM, resilient governance necessitates embedding adaptive capacities into institutions, regulations, and water infrastructure investments. This involves climate-informed water allocation, flexible permitting, and cross-sector coordination (OECD, 2021). In Europe, the Water Framework Directive progressively incorporates climate resilience measures, including drought management and flood risk planning (EEA, 2020).

2.8.3.3 Transboundary Water Cooperation

Segmenting nearly 40% of the global population in shared river basins underscores the importance of transboundary water collaboration for climate adaptation. Treaties like the Indus Waters Treaty and initiatives such as the Nile Basin Initiative reveal both the complexities and opportunities in managing cross-border resources under climatic stress. While climate change heightens the risk of resource conflicts, it also incentivizes cooperative monitoring, information exchange, and benefit-sharing agreements (Zeitoun et al., 2020). The Senegal River Basin's joint infrastructure projects highlight resilience gains achievable through coordinated flood control and hydropower development (UNECE, 2020).

2.8.4 Socio-Economic Measures

2.8.4.1 Community-based Adaptation (CBA)

Communities, especially in developing contexts, are frontline actors in managing climate-water risks. CBA prioritizes participatory, decentralized approaches leveraging local knowledge for water management interventions, including rainwater harvesting, decentralized treatment, and community-operated irrigation schemes (Archer et al., 2014). East African participatory groundwater monitoring supports better drought preparedness and conflict reduction over scarce water resources (Ngigi et al., 2017).

2.8.4.2 Education and Capacity Building

Sustaining resilience requires enhancing human capacities to anticipate, plan, and address water-related climate challenges. Educational programs promoting water conservation, hygiene, and early-warning systems build adaptive capacity broadly (Adger et al., 2017). Training water professionals in climate risk assessments, advancing technical expertise in nature- and technology-based solutions, and empowering marginalized groups particularly women, integral in household water management are critical components of this effort (UN Women, 2020).

2.9 Conclusion

Climate change and its impact on water resources shows that climate change is transforms water resources worldwide. It affects the hydrological systems, modifying water availability, quality, and extreme event frequency with significant implications for ecosystems and human communities. Water which is an important for climate adaptation becomes evident and essential to protect freshwater resources for supporting food production, energy generation, health, and biodiversity. Regional differences in vulnerability and exposure indicate that universal approaches are inadequate. Rather, comprehensive responses across global, regional, and local levels are vital to develop adaptive capacity and resilience successfully. Tackling these issues associated with water resources demands integrating technological innovations in water recycling, desalination, and intelligent irrigation with nature-based approaches such as watershed management and wetland restoration. Importantly, adequate policy structures including integrated water resources management, transboundary collaboration, and participatory governance are required to balance competing water needs sustainably. Socio-economic approaches that strengthen communities, develop capacity, and promote equitable water access support long-term achievement.

References

Acharjee, T., Rahman, M., & Hasan, M. (2023). Sea level rise and salinity intrusion in coastal Bangladesh: Impacts on water, agriculture, and health. *Environmental Research Letters, 18*(7), Article 074003.

Adger, W. N., et al. (2017). In *Human security and climate change: Policy responses*. Cambridge University Press.

Ahmed, M. B., Zhou, J. L., Ngo, H. H., & Guo, W. (2021). Emerging organic contaminants in wastewater: A critical review on treatment technologies. *Journal of Hazardous Materials, 401*, Article 123292.

Archer, D., Almansi, F., DiGregorio, M., Roberts, D., Sharma, D., & Syam, D. (2014). Moving towards inclusive urban adaptation: Approaches to integrating community-based adaptation to climate change at city and national scale. *Climate and Development, 6*(4), 345–356. https://doi.org/10.1080/17565529.2014.918868

Blatchford, M. L., Mair, A., & O'Connell, E. (2020). Smart irrigation technologies for climate-resilient agriculture. *Agricultural Water Management, 240*, Article 106267. https://doi.org/10.1016/j.agwat.2020.106267

Brekke, L. D., Kiang, J. E., Olsen, J. R., Pulwarty, R. S., Raff, D. A., Turnipseed, D. P., Webb, R. S., & White, K. D. (2009). *Climate change and water resources management: A federal perspective* (Circular 1331). U.S. Geological Survey. https://doi.org/10.3133/cir1331

Chapra, S. C., Zhou, Y., & Xiao, Y. (2022). Climate warming and harmful algal blooms: Impacts on inland waters. *Limnology and Oceanography, 67*(4), 795–814. https://doi.org/10.1002/lno.11960

Chivambo, J. M. F. (2025). Effect of climate change on water resources management. *Environmental Science and Policy, 134*, 102–110.

Copernicus, T. C. (2024). Regional glacier mass loss and hydrological impacts in mountain systems. *The Cryosphere, 18*, 5595–5614.

Copernicus Climate Change Service. (2025). European state of the climate 2024 report. European Environment Agency.

Davamani, V., John, J. E., Poornachandhra, C., Gopalakrishnan, B., Arulmani, S., Parameswari, E., Santhosh, A., Srinivasulu, A., Lal, A., & Naidu, R. (2024). A critical review of climate change impacts on groundwater resources: A focus on the current status, future possibilities, and role of simulation models. *Atmosphere, 15*(1), 122. https://doi.org/10.3390/atmos15010122

Davidson, N. C., Davy, A. M. J., & Finlayson, M. A. (2020). Wetland conservation and restoration as climate adaptation. *Conservation Biology, 34*(5), 1137–1147. https://doi.org/10.1111/cobi.13611

Diop, S. B., Ekolu, J., Tramblay, Y., Dieppois, B., Grimaldi, S., Bodian, A., Blanchet, J., Rameshwaran, P., Salamon, P., & Sultan, B. (2025). Climate change impacts on floods in West Africa: New insight from two large-scale hydrological models. *Natural Hazards and Earth System Sciences, 25*, 3161–3184. https://doi.org/10.5194/nhess-25-3161-2025

Domínguez-Tuda, M., et al. (2021). Global analysis of the hydrologic sensitivity to climate variability. *Journal of Hydrology, 600*, Article 126700.

EEA. (2020). Climate change adaptation and water management in Europe. European Environment Agency.

EPA. (2025). *Climate change impacts on freshwater resources*. Environmental Protection Agency.

ESSD. (2025). Glacier retreat and hydrological implications in high mountain Asia. *Earth System Science Data, 17*, 4293–4320.

Feng, X., et al. (2016). Revegetation in China's Loess Plateau is helping reduce soil erosion and flooding. *Nature Geoscience, 9*(12), 830–835. https://doi.org/10.1038/ngeo2820

Feulner, G. (2015). Global challenges: Climate change. *Global Challenges*.

Friedlingstein, P., Artaxo, P., Gallego-Sala, A. V., Jia, G., Jones, C. D., Kawamiya, M., Loisel, J., Loutre, M.-F., Rehfeld, K., Rovere, A., Smith, C., Séférian, R., van der Wel, N., & Ziegler, E. (2024). Earth system responses to different levels of greenhouse gas emissions mitigation. *Frontiers in Climate, 6*, Article 1480208. https://doi.org/10.3389/fclim.2024.1480208

Ghaffour, N., Missimer, T. M., & Amy, G. L. (2019). Technical review and evaluation of desalination technologies. *Desalination, 467*, 2–14. https://doi.org/10.1016/j.desal.2019.05.003

Grigg, N. S. (2016). Integrated water resources management: Balancing views and improving practice. *Water International, 41*(7), 885–898. https://doi.org/10.1080/02508060.2016.1209928

Huntington, T. G. (2010). Climate warming-induced intensification of the hydrologic cycle: A review. *Advances in Agronomy, 109*, 1–53.

IPCC. (2023a). *Climate change 2023: Regional assessments*. Intergovernmental Panel on Climate Change.

IPCC. (2023). Climate change 2023: Synthesis report. Contribution of working groups I, II and III to the sixth assessment report. IPCC.

IPCC. (2023). Sixth assessment report (AR6): Climate change 2023. Physical Science Basis. Intergovernmental Panel on Climate Change.

IPCC. (2023). *Climate change 2023: The physical science basis*. Cambridge University Press.

Jane, S. F., Hansen, G. J. A., & Carey, C. C. (2021). Warming reduces dissolved oxygen in global lakes. *Nature, 594*, 66–70.

Jiménez, B., & Asano, T. (2015). In *Water reclamation and reuse around the world*. IWA Publishing.

Jisem Journal. (2024). Impacts of climate variability on hydrological extremes. *Journal of Integrated Science for Environment and Policy, 15*.

Johansen, K. S., Alfthan, B., Baker, E., Hesping, M., Schoolmeester, T., & Verbist, K. (2018). *The Andean glacier and water atlas: The impact of glacier retreat on water resources*. (pp. 77). UNESCO Publishing/GRID-Arendal. ISBN 978-92-3-100286-1.

Johnston, R., & Kummu, M. (2019). Water resource models in the Mekong: Progress and limitations. *Journal of Hydrology: Regional Studies, 25*, Article 100620. https://doi.org/10.1016/j.ejrh.2019.100620

Jones, E., Qadir, M., van Vliet, M. T., Smakhtin, V., & Kang, S. M. (2019). The state of desalination and brine production: A global outlook. *Science of the Total Environment, 657*, 1343–1356. https://doi.org/10.1016/j.scitotenv.2018.12.076

Kleidon, A. (2013). A simple explanation for the sensitivity of the hydrologic cycle to surface temperature and solar radiation. *Earth System Dynamics, 4*, 455–465.

Koch, H., Vögele, S., & Wagner, S. (2020). Droughts, water quality and water supply: Impacts of droughts on drinking water supply in Europe. *Sustainability, 12*(18), 7553.

Konikow, M., Di Long, M., Mekonnen, M. M., Schmied, H., Mukherjee, A., MacDonald, A., Reedy, R. C., Shamsudduha, M., Simmons, C. T., Sun, A., Taylor, R. G., Villholth, K. G., Vörösmarty, C. J., & Zheng, C. (2022). Groundwater and climate change. *Nature Reviews Earth and Environment, 3*(7), 472–486. https://doi.org/10.1038/s43017-022-00378-6

Li, J., Song, Y., & Cai, H. (2022). Microplastics in inland freshwater systems: Sources, fate, and ecological risks. *Water Research, 215*, Article 118238.

Lobell, D. B. et al. (2025). Climate variability and risks to global crop production. *ScienceDirect*. https://www.sciencedirect.com/science/article/abs/pii/S2590332225001447

Locatelli, B., Pavageau, C., Pramova, E., & Di Gregorio, M. (2015). Integrating climate change mitigation and adaptation in agriculture and forestry: Opportunities and trade-offs. *Wiley Interdisciplinary Reviews: Climate Change, 6*(6), 585–602. https://doi.org/10.1002/wcc.357

Marinova, S., Bach, V., Link, A., & Finkbeiner, M. (2025). Criticality assessment of global freshwater resource use: Adapting abiotic resource criticality frameworks—Water SCARCE. *Science of the Total Environment, 966*, Article 178676. https://doi.org/10.1016/j.scitotenv.2025.178676

Mejía, S. X. N., et al. (2023). Climate change impacts on hydrometeorological and river flow extremes. *Science of the Total Environment, 850*, Article 157927.

Mengistu, T. D., et al. (2025). Modeling and prediction of climate change impacts on water resources using machine learning and hydrological models. *Environmental Modelling and Software, 158*, Article 105472.

Michael, H. A., Scott, K. C., Yu, X., & Sengupta, A. (2022). Groundwater and climate change. *Nature Reviews Earth and Environment, 3*(7), 472–486. https://doi.org/10.1038/s43017-022-00294-6

Mosley, L. M. (2015). Drought impacts on the water quality of freshwater systems: Review and integration. *Earth-Science Reviews, 140*, 203–214. https://doi.org/10.1016/j.earscirev.2014.11.010

NASA Earth Observatory. (2024). Water-cycle disruption in the Amazon basin.

NASA GPM. (2020). How does climate change affect precipitation? NASA Global Precipitation Measurement.

Nature. (2021). The growing threat of urban water scarcity. *Nature Communications, 12*, 978. https://www.nature.com/articles/s41467-021-25026-3

Nazeri Tahroudi, M. (2025). Comprehensive global assessment of precipitation trend and pattern variability considering their distribution dynamics. *Scientific Reports, 15*, Article 22458. https://doi.org/10.1038/s41598-025-06050-5

Ngigi, S. N., Kungu, J. B., & Thuo, M. (2017). Participatory groundwater management in East Africa. *Water Policy, 19*(6), 1099–1113. https://doi.org/10.2166/wp.2017

Nguyen, H. T., Kondolf, G. M., Park, E., & Kummu, M. (2021). Extreme rainfall and sediment transport in the Mekong River Basin. *Hydrology and Earth System Sciences, 25*(3), 1437–1454. https://doi.org/10.5194/hess-25-1437-2021

NOAA. (2025). Billion-dollar disasters and climate impacts in the United States.

Noh, S. J., et al. (2024). Climate change impact assessment on water resources. *Environmental Impact Assessment Review, 89*, Article 106623.

OECD. (2021). Water and climate change adaptation: Policies to navigate uncharted waters. OECD Publishing.

Paerl, H. W., & Otten, T. G. (2023). Climate change and harmful cyanobacterial blooms: Emerging challenges and solutions. *Science of the Total Environment, 857*, Article 159422.

Patel, M., Kumar, R., & Sharma, S. (2023). Antibiotics and climate change: A dangerous synergy in aquatic environments. *Frontiers in Environmental Science, 11*, 11234.

PMC. (2025). Freshwater biodiversity and climate change vulnerability. *PMC Articles*. https://www.ncbi.nlm.nih.gov/pmc/articles/PMC11718631/

Publishing, I. W. A. (2025). A comprehensive review of climate change adaptation in water resources. *Journal of Water and Climate Change, 16*(5), 1831–1860.

Qadir, M., Wichelns, D., Raschid-Sally, L., & McCornick, P. G. (2020). Agricultural use of marginal-quality water and climate change adaptation. *Irrigation and Drainage, 69*(1), 47–59. https://doi.org/10.1002/ird.2403

Ramsar Convention Secretariat. (2018). Global wetland outlook: State of the World's wetlands and their services to people.

Rasul, G., & Sharma, B. (2016). The nexus between water, energy, and food security in the context of climate change: Towards a water–energy–food nexus approach. *Water Policy, 18*(1), 12–24. https://doi.org/10.2166/wp.2015.162

Ravelo, A. C., Cuesnes, P., de Ruijter, F. J., Hou, Y., Wesseler, J., & van der Velde, M. (2021). Digital tools for climate-smart irrigation in smallholder farming systems. *Frontiers in Sustainable Food Systems, 5*, Article 659883. https://doi.org/10.3389/fsufs.2021.659883

Reuters. (2025). World risks up $39 trillion economic losses to vanishing wetlands: report. https://www.reuters.com/sustainability/climate-energy/world-risks-up-39-trillion-economic-losses-vanishing-wetlands-report-says-2025-07-15/

Review, N. E. J. M. (2023). Climate change and infectious diseases. *New England Journal of Medicine, 388*, 2347–2356.

Ringler, C., Agbonlahor, M., Barron, J., Baye, K., Meenakshi, J. V., Mekonnen, D. K. & Uhlenbrook, S. (2022). The role of water in transforming food systems. *Global Food Security*, *33*, 100639. https://doi.org/10.1016/j.gfs.2022.100639

Royal Society. (2015). Climate change: Evidence and causes.

Shrestha, S., Shukla, S., Shrestha, P., Bhatta, B., & Talchabhadel, R. (2022). Urban stormwater pollution under climate extremes: A global perspective. *Environmental Pollution, 299*, Article 118857. https://doi.org/10.1016/j.envpol.2022.118857

The Guardian. (2025). Europe's summer of extreme weather caused $43 billion of short-term losses.

UN Women. (2020). Gender, water and climate change. UN Women Policy Brief.

UNECE. (2020). Transboundary water cooperation in a changing climate. United Nations Economic Commission for Europe.

UNEP Finance Initiative. (2024). Climate risks in power generation sector. https://www.unepfi.org/wordpress/wp-content/uploads/2024/05/Climate-Risks-in-the-Power-Generation-Sector-1.pdf

UNESCO. (2019). Leaving no one behind in integrated water resources management. United Nations World Water Development Report 2019.

USGS. (2025). Climate change and future water availability in the United States. U.S. Geological Survey Scientific Investigations Report.

WMO. (2025). Africa faces disproportionate climate change burden and adaptation costs. World Meteorological Organization.

WMO. (2025). State of global water resources. World Meteorological Organization Annual Report.

World Bank. (2025). Water overview: Development news, research, data. Retrieved from https://www.worldbank.org/en/topic/water/overview

World Health Organization (WHO). (2019). Climate change and health. Retrieved from https://www.who.int/health-topics/climate-change

World Health Organization. (2024). Climate change and health: Fact sheet. https://www.who.int/news-room/fact-sheets/detail/climate-change-and-health

Yang, D., Yang, Y., & Xia, J. (2021). Hydrological cycle and water resources in a changing world: A review. *Geography and Sustainability, 2*(2), 115–122. https://doi.org/10.1016/j.geosus.2021.05.003

Zeitoun, M., Goulden, M., & Tickner, D. (2020). Why transboundary water cooperation matters for climate change adaptation. *Nature Climate Change, 10*(5), 345–353. https://doi.org/10.1038/s41558-020-0755-8

Zhang, H., Wang, J., Zhou, B., Zhou, Y., Dai, Z., Zhou, Q., Chriestie, P., Luo, Y., & Ji, R. (2021). Microplastics and the plastisphere: Climate change interactions. *Environmental Science and Technology, 55*(5), 3179–3190. https://doi.org/10.1021/acs.est.0c07731

Zhang, Y., Liu, S., Xu, M., Liu, Y., & Zhao, W. (2022). Sediment and nutrient transport under changing precipitation regimes: Implications for water quality. *Science of the Total Environment, 820*, Article 153282. https://doi.org/10.1016/j.scitotenv.2022.153282

Chapter 3
Water–Energy Nexus and Sustainability Issues

Mariam Iyabo Adeoba, Olalekan Joseph Ogunniyi, Damilola Rebecca Oladejo, and Iyanuoluwa Busola Feyijin

Abstract This Chapter addresses the linkages between water and energy systems in Nigeria from the perspectives of socio-economic development, climate variability, and sustainability. Despite the vast resources, Nigeria grapples with a lack of power and clean water on account of decayed infrastructure. Within the framework of water–energy nexus, this review focuses on interconnections identified through the literature between water for energy (e.g. hydropower generation, thermal plant cooling) and energy for water services (irrigation pumps, urban water supply, wastewater treatment). The synthesis reports that the gas-based power-generating plants form 80% of Nigeria's electricity, whilst hydropower plants constitute only about 20%. By contrast, there is restricted access to clean water and electricity, as only 14% of the population has clean water, with about 45% lacking electricity. The ubiquitous environmental pollution in the Niger Delta, the acute water scarcity of northern Nigeria and fragmented governance inhibiting integrated planning were identified as some of the major sustainability challenges from available literature. This chapter highlights the importance of coordinating investments in data-driven basin management, renewable energy and resilient infrastructure for achieving SDGs 6, 7 and 13. One of the strategic ways for Nigeria to achieve an inclusive growth, climate resilience and long-term resource security is by strengthening the water-energy nexus.

Keywords Water–energy nexus · Hydropower · Rural electrification · Water security · Sustainability · Nigeria · Climate resilience

M. I. Adeoba
UNISA Biomedical Research Group, Department of Mechanical, Bioresources and Biomedical Engineering, College of Science, Engineering and Technology (CSET), University of South Africa, Johannesburg, South Africa

O. J. Ogunniyi (✉)
Department of Industrial and Engineering Management, College of Science, Engineering and Technology, University of South Africa, Johannesburg, South Africa
e-mail: lekjonline@gmail.com

D. R. Oladejo
Department of Biochemistry, Federal University, Oye-Ekiti, Nigeria

I. B. Feyijin
Department of Environmental Health Sciences, University of Ibadan, Ibadan, Nigeria

M. I. Adeoba et al. (eds.), *Water Resources Management in Nigeria*, World Water Resources 10, https://doi.org/10.1007/978-3-032-22813-0_3

3.1 Introduction

The intricate relationships between energy and water supply systems are referred to as the "water–energy nexus". A nexus framework offers an integrative perspective for examining how changes in one sector impact another (Mooren et al., 2025; van den Heuvel et al., 2020). In reality, water is necessary for all forms of energy production, including cooling, hydropower, and bioenergy. Likewise, energy is needed for every step of water supply and treatment, including pumping, purification, and desalination (Shokri & Sanavi Fard, 2022). This interdependence means that water and energy sectors cannot be considered in isolation: stress in one often propagates to the other. Sustainable management, therefore, requires ensuring reliable and affordable water and energy services that support development without compromising future resource security. For example, Nigeria's energy policy explicitly calls for "clean, affordable, secure, and sustainable" energy, and recent analyses of Nigerian water governance stress the need for integrated, climate-resilient policies to ensure universal access to water (Adeshina et al., 2024; Irene et al., 2025; Ojo et al., 2025).

In this chapter, the nexus framework was adopted, which recognises these dual challenges: how Nigeria can secure water and energy for development in ways that are economically and environmentally sustainable. Figure 3.1 summarises the conceptual framework guiding this chapter, linking water resources, energy systems, socioeconomic drivers and governance to sustainability outcomes in Nigeria. Bidirectional dependencies between Water Resources and Energy Systems are shown alongside Socioeconomic Factors that drive demand. Governance & policy mediate these interactions and shape Sustainability Issues. Arrows indicate principal influence pathways, including water for energy and feedback from energy to sustainability outcomes.

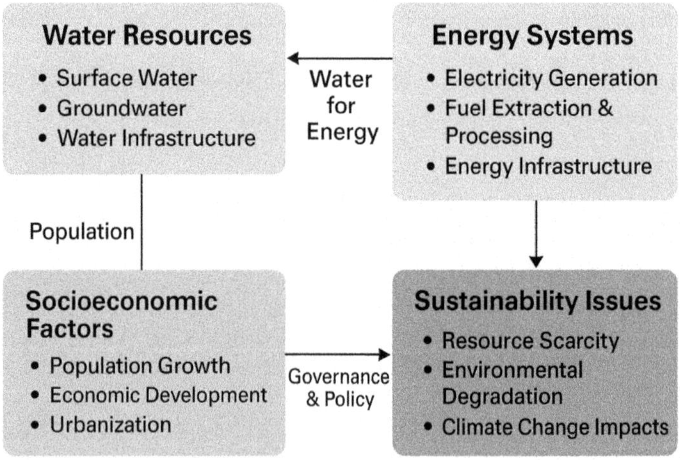

Fig. 3.1 Conceptual framework of the water–energy nexus in Nigeria

Nigeria's water and energy challenges underscore the need for a nexus perspective. Although Nigeria is traversed by major rivers (e.g. the Niger and Benue) and has considerable hydro potential, it faces acute water insecurity. Per-capita renewable water resources are modest (about 1011 m^3/year) and highly uneven, and actual water consumption in many communities is extremely low (often < 20 L per person per day, versus the WHO-recommended 100 L/day) (Irene et al., 2025). As a result, many Nigerians lack a reliable public water supply. Surveys confirm that households often rely on expensive and unsustainable resources, such as tanker trucks, private wells, bottled water, or hand pumps, to meet their basic needs. Concerns regarding equity and public health are raised by the fact that women and children frequently travel great distances to fetch water from seasonal springs or buy it from vendors in rural and peri-urban areas. Climate variability exacerbates these stresses: severe flooding can be devastating, and droughts are becoming more frequent in some areas of northern Nigeria. For example, the 2012 floods in Nigeria displaced over 2.1 million (about 7 million) and caused a massive economic loss (Abubakar et al., 2024; Okafor, 2020). The main sustainability concern of ensuring access and resilience to shocks is illustrated by the pressure on Nigeria's water systems due to increasing demand, unpredictable rainfall and ageing infrastructure (Irene et al., 2025).

The Nigerian energy sector faces similar problems. Despite huge hydrocarbon and renewable resources, access to and reliability of electricity are very limited. More than 86 million Nigerians still lack access to electricity, the highest number of any country without a grid. The sharp urban–rural divide is reflected in the fact that electrification rates in rural areas are less than one third of the level in urban areas, while they are around 89% in urban areas (Williams et al., 2024). Nigeria's installed capacity of around 13 GW meets a peak demand of more than 20 GW, but only about 5–6 GW is produced. According to Ghanbarzadeh et al. (2025), this example demonstrates that the grid area is also affected due to chronically insufficient production. Low voltage and frequent power outages are problems for both homes and businesses. For example, companies that depend on diesel generators lose billions of dollars annually since heavy fuel–oil-powered turbines are required most of the time, and the national grid is less dependable than many of its competitors. These unfavorable circumstances are a result of a lack of infrastructure and investment rather than the availability of resources. With a population of just over 220 million people, Nigeria's population is still growing rapidly, increasing demand for electricity at a time when regulatory and financial barriers are hindering the development of additional generating capacity. This has implications for sustainability because 85% of urban households rely on backup generators, which increase air pollution and carbon emissions and hinder economic growth by supplying an intermittent supply of energy (Golobish et al., 2023; Nitte, 2023). The government has even set a 2030 goal for universal access to renewable energy; thus, a just transition will be necessary to guarantee both increased capacity and less environmental impact (Yusuf & Akande, 2023). The relationship between energy and water requirements has to be better understood to address these sustainability challenges, and some compromises can be found in this area.

Water and energy issues are closely linked to infrastructure constraints and climate change. As mentioned earlier, the changing rainfall patterns in Nigeria have rendered water availability more unpredictable. Scientific analysis shows that rainfall during the rainy season has increased slightly in the north of Nigeria (2–4 mm per year) and decreased slightly in the south-east (about 5 mm per year) (Ibebuchi & Abu, 2023). These changes have different impacts on the storage of hydropower, irrigation potential and groundwater recharge. For instance, water-intensive irrigation and processing are usually required to meet energy and food needs, Sustainable Development Goals (SDGs) 2 and 6, while efforts to combat water pollution and increase water consumption are frequently required to meet water needs (SDG 7). If pursued in isolation, the focus on one objective can exacerbate the other, which is a classic sustainability problem highlighted by the Nexus study. Addressing frequent blackouts in Nigeria is a key issue for sustainable development, as they cause significant economic losses, especially in rural areas (Yusuf & Akande, 2023). Integrated water management, for example. In summary, coordinated planning that acknowledges the interdependence of the two and the larger context of sustainability is necessary to ensure water and energy security in Nigeria.

This chapter aims to review the interlinkage between water and energy systems in Nigeria, identify key sustainability challenges from the literature, and discuss integrated strategies for managing these resources.

3.2 Nigeria's Water Resources

3.2.1 Spatial and Seasonal Rainfall Variability

Sharp regional rainfall differences characterise Nigeria's climate. Annual precipitation varies from the wet south (over 2300 to 3000 mm in coastal areas) to the semi-arid north (often less than 500 to 650 mm) (Morrison et al., 2023). For example, the far south state of Cross River receives, on average, around 2680 mm of rain per year, while Borno in the north-east receives only 631 mm (Morrison et al., 2023). The wettest areas are the coastal Niger Delta and rainforest areas, while the driest areas are in the Sahelian north. Nigeria has a rainy season from April to October and a dry season from November to March on a seasonal basis (Ibebuchi & Abu, 2023). Most regions have only one peak of summer rainfall (e.g., the northern peak occurs in July–August), but the far south exhibits a bimodal pattern with peaks in June and July, as well as in September, with a break in August. During the dry season, particularly in December to February, the rainfall is minimal (in January, often < 50 mm) (Morrison et al., 2023). This pronounced spatial and seasonal variability, characterised by abundant rains in the south during the wet season versus aridity in the north during the dry season, is shaping water availability in Nigeria.

3.2.2 Surface and Groundwater Availability by Region

Nigeria has extensive water resources, but their distribution is uneven. Surface water (rivers, lakes, reservoirs, wetlands) covers only about 2% of Nigeria's land area (Ibrahim et al., 2024a, 2024b). The two largest rivers, the Niger and Benue, meet in central Nigeria and drain most of the country. Many smaller rivers (Gongola, Sokoto, Kaduna, etc.) form four major drainage systems: the Niger Basin (including Benue and tributaries), the Lake Chad Basin in the northeast, and separate Atlantic basins east and west of the Niger Delta (Adeniran et al., 2021; Ohunene et al., 2025). Most natural surface water is concentrated in the south (the Niger Delta's creeks, lagoons, and humid coastal rivers), whereas the drier north relies on dammed reservoirs and seasonal streams. Nigeria has built numerous dams in the northern inland basins for irrigation and hydropower (e.g. Kainji and Tiga dams), reflecting the north's need to store water from a short rainy season (Ibrahim et al., 2024a, 2024b; Irene et al., 2025).

Groundwater is an essential resource everywhere and is present under all regions, but aquifer productivity varies with geology and climate. Groundwater potential is generally highest in the river floodplains and southern coastal aquifers, and lower in the crystalline rock regions of the northwest and central uplands. According to recent groundwater modelling, no part of Nigeria has an "inferior" groundwater potential, but 18% of the land (especially areas underlain by hard rock) falls into the low potential category. Approximately 44% of the country has moderate aquifer potential, while the most productive zones (classified as very high) cover only around 2% (mainly the Niger Delta, coastal Lagos, and patches along major rivers). Northern Nigeria's groundwater potential is constrained by low rainfall and high evaporation, yet in some basins (e.g. Chad basin), thick sedimentary aquifers offer significant storage. Overall, southern Nigeria enjoys both plentiful rainfall and permeable soils, which aid recharge, whereas parts of the north rely on deeper aquifers and river-fed reservoirs to meet their water demand (Ibrahim et al., 2024a, 2024b).

3.2.3 Water Access Disparities: Urban Versus Rural and Regional

But for a country with such apparent water wealth, access to safe drinking water in Nigeria is extremely uneven, whether between urban and rural areas or regions. At least 27% of Nigerians reportedly drink water from untreated surface sources, such as streams, rivers, ponds, or rainwater, especially in rural areas with no access to piped water (Iduseri et al., 2022; Nyika & Dinka, 2023). While urban areas have more functional water infrastructure, service gaps are common: according to the 2022 survey, just 15% of households access piped water; most get their water from hand pumps or through a delivery system (Ferdowsi et al., 2024). These statistics underscore the infrastructure gap: In many households, even in urban settings, families have turned to self-supply (via private wells) or costly transportation of water due to a lack of access to city water systems.

And at the regional level, it's the north, rather than the south, which has more water problems. For instance, research indicates that on the water stress and water inequality indicators in our study, the North-Central region has higher scores relative to other parts of Nigeria. Adding to the challenge are higher population growth rates and lower investment in water infrastructure, resulting in more frequent service interruptions throughout the northern states and greater dependence on groundwater. Southern states, on the other hand, have more pollution and older urban infrastructure. Cities generally have better water coverage than rural areas, although many urban Nigerians still do not have access to safe water every day. Daily water intake estimates in some Nigerian communities are less than 20 L (much lower than the WHO-recommended 100 L), which represents poor access to water (Irene et al., 2025; Sakketa, 2023). This disparity in water access, rich and poor, North and South, urban to rural areas, is a crucial factor for public health and regional development. Indeed, 63.7% of Nigerians experienced frequent or constant water scarcity, and the absence of clean water leads to the prevalence of water-borne-related diseases in rural/peri-urban regions (Irene et al., 2025). Otherwise, bridging these breaks will mean large-scale pipeline investments, improved governance and community-led solutions in underserved areas.

3.2.4 Trends in Water Storage and Groundwater Recharge

(Adetokunbo et al., 2025) utilised Gravity Recovery and Climate Experiment (GRACE) satellite data to reveal significant trends in Nigeria's total water stock over the last two decades. Surprisingly, the country has shown a statistically significant upward trend in land storage between 2002 and 2024. The increase has accelerated in recent years, from around 0.68 cm per year before 2018 to 2.87 cm per year after 2018. An analysis of change points to a sharp rise around July 2019, indicating a new regime of increased storage (Adetokunbo et al., 2025; Ibrahim et al., 2024a, 2024b). This positive trend is counterintuitive in the context of repeated droughts, and scientists attribute it partly to anthropogenic interventions (addition of irrigation, reservoirs, and better water management) that have kept more water on the ground, particularly in the arid north. Spatially, northern Nigeria is experiencing more substantial increases in TWS than central or southern areas; many northern regions are now experiencing water retention increases of more than 2.5 years compared to parts of the south (Adetokunbo et al., 2025). This north–south contrast in storage changes is consistent with the development of dams and irrigation in the north and the likely increase in groundwater recharge in some southern basins. It is essential to note that these increases in water storage occurred without corresponding increases in rainfall; annual precipitation in the north did not increase significantly, and in some areas of the south, it decreased slightly (Ibebuchi & Abu, 2023). As a result, improved water retention (e.g., through storage and recharge of aquifers) and reduced extraction can increase yields. While this is an encouraging indicator for water security, GRACE satellite data also warns that this increase could pose a flood risk if not managed, as more water is now stored in soil, which could exacerbate flooding in years of extreme rainfall. Overall, the analysis carried out with GRACE highlights the dynamic nature of water storage in Nigeria, the impact of human activities on the water cycle and the need for sustainable management as long as climate variability.

3.2.5 Major River Basins and Dam Infrastructure

Nigeria's river network is administered through eight hydrological areas and 12 River Basin Development Authorities (RBDAs), which reflect the major river basins. The Niger River Basin (which includes the Benue) covers the central and western parts of the country and is by far the largest, draining 63% of Nigeria (Omokaro et al., 2024). The Niger River enters Nigeria in the northwest, while the Benue flows from Cameroon in the east; they join at Lokoja to form a broad river that exits through the Niger Delta (Irene et al., 2025). Up north, the Lake Chad Basin is an endorheic drainage basin in the northeast, fed by rivers such as the Hadejia–Komadugu system. Along the Gulf of Guinea coast, Nigeria has smaller westward-flowing rivers (Ogun, Osun, Benin, etc., in the southwestern Littoral basin) and eastward-flowing rivers (Cross, Imo, Qua Iboe in the southeastern Littoral basin) (Odeloui et al., 2022; Olowoyeye & Kanwar, 2023).

Throughout these basins, Nigeria has constructed numerous dams for water supply, irrigation, and hydropower. Most large dams are concentrated in the Sudan–Sahel zone of the north, where surface water is seasonally scarce (Ibrahim et al., 2024a, 2024b). For instance, the Kainji Dam on the Niger River creates Nigeria's largest reservoir (Kainji Lake) and is a key hydroelectric facility (Irene et al., 2025). Upstream on the Niger and its tributaries lie other major dams (Jebba, Shiroro on the Kaduna River, etc.), which together form a cascade for power generation and dry-season flow regulation. In the northwest, the Sokoto-Rima basin is home to the Goronyo and Bakolori dams, which serve irrigation purposes. By contrast, the humid South has fewer and smaller dams; instead, natural wetlands and swamps (e.g., in the Niger Delta) moderate flows (Eteh et al., 2024; Ibrahim et al., 2024a, 2024b). Nigeria's total reservoir storage is modest in relation to its needs, and many dams face siltation and operational challenges. Figure 3.2 (panel B above) indicates the distribution of major dam sites. The spatial mismatch is evident: the populous south gets most of its water from free-flowing rivers and aquifers, whereas the arid north depends on engineered storage and transfers. This has motivated inter-basin water transfer schemes (e.g. proposals to channel water from the Niger or Congo basins to Lake Chad) to alleviate northern water shortages. The efficient management of these river basins and dam infrastructures remains of paramount importance. Nigeria is suffering from regular flooding (when dams overflow or are poorly coordinated) and drought (when water levels drop), which highlights the need for integrated basin-wide planning (Isukuru et al., 2024).

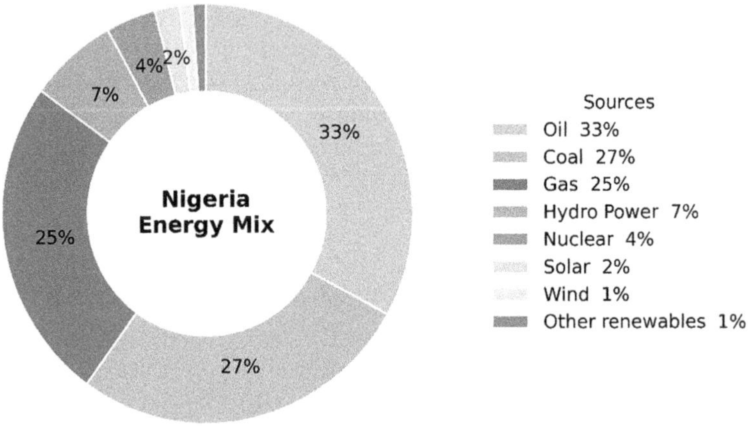

Fig. 3.2 Global primary energy mix by source (2024). Adopted from Adeshina et al. (2024)

3.2.6 Water Quality and Pollution Hotspots

Water quality in all Nigerian waters is a serious concern. Rapid industrialisation, oil extraction and urban sprawl have polluted many rivers and groundwater sources. In the Niger Delta, decades of oil production have resulted in chronic pollution of the water. There has been a documented increase in oil spill incidents in the Delta region since the start of oil production (Peter et al., 2022). On average, hundreds of oil spills occur every year because of pipeline leaks, theft, sabotage and operational malfunctions. For instance, in 2015, 39 oil spill events were reported in Bayelsa State, affecting at least 5 local government areas (South Ijaw, Nembe, Ogbia, Ekeremor and Yenagoa) (Apata et al., 2019; Chijioke et al., 2021). Many spills are only partially cleaned up, leaving a legacy of hydrocarbon pollution in streams, soil and groundwater. Studies have shown that groundwater in the Niger Delta often contains high levels of hydrocarbons and related toxins, which pose a serious health risk for local populations. In addition to oil, the Delta is also affected by gas flares (acid rain) and agricultural runoff, making it one of the most polluted regions in the country (Apata et al., 2019; Peter et al., 2022).

In urban and industrial areas of Nigeria, water quality is impaired by untreated wastewater and discharges. Less than 10% of the industries in Nigeria are reported to treat their wastewater before discharge. As a result, heavy metals and chemicals (e.g. lead, chromium, iron and pesticides) are present at hazardous levels in many inland waterways (Darma et al., 2024; Kayode et al., 2018). Rivers flowing through cities like Lagos, Kano, and Port Harcourt exhibit high biochemical oxygen demand and pathogen counts, primarily due to the discharge of sewage and industrial waste. Nationwide, about 70 million Nigerians lack access to safe water and rely on sources that may be microbiologically or chemically contaminated (Irene et al., 2025). The consequences are evident in the frequent outbreaks of cholera, dysentery and other water-borne diseases, which affect children. Ground water in metropolitan areas is also at risk; groundwater wells may be contaminated by leachate from landfill sites, oil tank farms and septic systems (e.g. observed PCB and benzene contamination in some wells) (Aralu et al., 2025; Ediagbonya et al., 2025).

Geographically, pollution "hotspots" include: the Niger Delta (oil and hydrocarbons), the Lagos-Ogun industrial corridor (industrial effluents, sewage), the Kano and Kaduna areas (tannery and textile waste, heavy metals), and mining districts in the north (acid mine drainage, metals) (Anyanwu et al., 2023; Ewim et al., 2023). Coastal waters and lagoons are also affected by urban runoff and oil transport. In short, although Nigeria is rich in water, water quality management lags, leading to localized degradation that is threatening ecosystems and human health. Strengthening environmental enforcement, expanding water and sanitation infrastructure, and cleaning up polluted areas (in the Niger Delta) are all of the key actions that recent studies have recommended (Ewim et al., 2023).

3.3 Nigeria's Energy Sector

3.3.1 Electricity Generation and Energy Mix

The energy sector in Nigeria is overwhelmingly dependent on fossil fuels, with thermal power (mainly natural gas) accounting for the largest percentage of electricity generation capacity. The country has a total installed generation capacity of around 12.5 trillion kWh, but due to infrastructure and operational bottlenecks, only 3.5 to 5.0 trillion kWh are usually available for daily distribution (Ishaku & Quadri, 2025). This sharp difference between installed and actual production can be attributed to factors such as poor maintenance, problems with gas supply and transmission bottlenecks. As a result, per capita energy consumption remains very low (on order of 150,000 kWh annually), reflecting a chronic lack of supply in the face of increasing demand from the nation (Ishaku & Quadri, 2025).

Figure 3.2 provides a global primary-energy baseline that can be used to contrast Nigeria's electricity fuel mix, discussed below. Oil (33%), coal (27%), and natural gas (25%) together supply the bulk of global primary energy, while hydropower (7%), nuclear (4%), solar (2%), wind (1%), and other renewables (1%) are smaller contributors (Adeshina et al., 2024).

The energy mix for on-grid electricity is dominated by natural gas. Nigeria is richly endowed with gas reserves, and gas-fired power plants form the backbone of generation. By contrast, non-hydro renewables (such as solar or wind) still contribute only a marginal share of on-grid electricity. Large-scale solar or wind farms are virtually absent, though solar photovoltaic (PV) capacity has begun to grow through off-grid and embedded generation projects. On the demand side, recurrent power deficits have led to a widespread reliance on diesel and petrol generators. Nigerians are estimated to spend around 1.56 trillion naira (about $3.8 billion) per year on fuel for backup generators (Pelz et al., 2023). Indeed, Nigeria is the biggest market in Africa for small-scale private generators, which help to fill the gap in supply, but at a much higher price (often as high as \$0.30–\$0.50 per kWh) than electricity from the grid. The widespread use of electricity generators also suggests that the nominal energy mix underestimates the role of liquid fuels in energy supply. Although official electricity generation is 80% natural gas and 20% hydroelectric, a significant part of the electricity consumed is produced by oil (diesel and petrol) in households and businesses. This pervasive dependence on off-grid fossil fuels has serious economic and environmental impacts, including high energy costs for consumers and increased greenhouse gas and pollutant emissions (Ishaku & Quadri, 2025).

3.3.2 Hydropower Resources and Development

Hydroelectric power has long been part of Nigeria's energy strategy and is the country's largest source of renewable electricity. Large dams: notably Kainji (760,000 MW), Jebba (578,000 MW) and Shiroro (600,000 MW) on the Niger River, have a combined installed capacity of around 2,3,3 trillion kWh, but their actual output is often below potential due to problems with maintenance and water management (John et al., 2025). Until 2023, these three plants represented the major hydropower supply in Nigeria. The Zungeru hydroelectric power plant (700.0 MW), the second largest in Nigeria, was fully commissioned at the end of 2023 and is expected to supply about 10% of the country's electricity needs (Ingram, 2023). With the coming online of Zungeru, Nigeria's total hydropower capacity is approximately 2.8 trillion kWh, which is still only about one-fifth of the total installed generation capacity.

The theoretical hydropower potential in Nigeria is indeed high, on the order of 14,000 MW, concentrated in large areas along the Niger and Benue River basins and their tributaries (John et al., 2025). Other medium- and large-scale projects are planned to exploit this potential. The most ambitious of these is the long-delayed Mambilla Dam (3,050,000 MW) on the River Donga, which is expected to be the largest dam in Nigeria upon completion (Adeshina et al., 2024). Other planned developments include projects in Lokoja (750 MW) and Makurdi (1000 MW) in the Benue region. In recent years, several medium-sized projects have advanced, including the Kashimbila Dam (40.5 MW), which provides both electricity and flood protection, and the Gurara II (360 MW), which increases the capacity of the existing dam (Nkwoh, 2023). However, many of these plans are facing financial and logistical obstacles.

Small hydropower (SHP) resources also represent a significant opportunity for supplying electricity to rural areas and for the energy-to-energy nexus. Studies have identified 278 promising small hydroelectric sites in Nigeria with an estimated combined potential of around 3500 kW (Adeshina et al., 2024). These are typically river basins or mini dams (ranging from tens of kW to several MW) scattered across rural areas. However, only 60.0 MW (< 1.7%) of the small hydro potential has been developed so far, reflecting the predominance of large dams in past hydroelectric investment. The federal government, through the river basin development authorities, has had a policy since 1979 to promote small hydroelectric power in irrigation and water supply systems; however, implementation has been slow (Adeshina et al., 2024). Accelerating the development of large and small hydroelectric projects could bring several co-benefits to Nigeria, including providing renewable energy, improving grid stability, hydro as a baseload and peak source, and offering water management benefits, such as irrigation and flood control. This potential will require overcoming financing, resettlement and environmental impact management challenges and integrating these projects into the wider Nigerian water strategy.

3.3.3 Rural Electrification and Off-Grid Renewables

Nigeria faces a significant gap in access to electricity between urban and rural areas. Until 2020, only 55.4% of the population had access to electricity, and access in rural areas was estimated to be only 24.6% (compared to around 83.9% in urban areas) (Pelz et al., 2023). This means that some 85 million Nigerians, mostly in the countryside, are without electricity (Williams et al., 2024). Even for those classified as 'electrified', supplies are often erratic or insufficient to meet household needs. Bridging this access gap is a major development priority, with implications for the health, education, and livelihoods of communities which are not connected to the electricity grid.

Rural electrification efforts are increasingly shifting towards decentralised renewable energy solutions. Nigeria has set ambitious targets in its National Electricity Plan and related programmes. For example, the 2016 policy envisaged the deployment of 10,000 solar mini grids by 2023 to serve remote communities and to achieve 75% rural electrification (Oke et al., 2024). Although implementation has fallen short of the target, the country has made considerable progress in recent years. As of 2023, Nigeria have around 67 solar mini-grids in operation, together with a number of solar thermal power plants, providing an estimated 52.5 million kilowatts of off-grid renewable energy capacity. These systems had already helped some 1.5 million households (over 5.5 million people) that previously had little or no access to reliable electricity (Oke et al., 2024).

International support has enhanced off-grid development in Nigeria. The World Bank and the African Development Bank have invested hundreds of millions of dollars to catalyze private sector engagement in Nigeria's national grid and clean solar energy projects. This has led to an increase in renewable off-grid connections: for example, energy efficiency performance grants under the National Energy Efficiency Programme and the roll-out of solar energy systems that reached more than 5 million households in 2022 (Oke et al., 2024). The government's Solar Power Nigeria initiative (part of the Economic Sustainability Plan) also aims to equip 5 million households with solar energy systems, with a focus on the rural and urban poor. These efforts not only provide basic lighting and telephone chargers, but also electricity for productive uses such as pumping water, processing, clinics and schools, thus fostering the development of rural areas.

Despite these gains, challenges remain. Stand-alone solar systems face several challenges, including high initial investment costs, reliance on imported items and the requirement that they be serviced continuously in isolated areas. Nevertheless, the renewable power market in Nigeria has been expanding quite rapidly, largely due to decreasing solar photovoltaic (PV) costs and new business models including on-site PV and micro-grid systems. Ramping up the off-grid electricity is important to bridge Nigeria's energy access gap since it is economically and logistically impossible for all the villages in Nigeria to be connected to the national grid. Leveraging its plentiful solar resources (Nigeria has 5–7 kWh/m^2day of solar irradiation in large parts), Nigeria can reach universal access to electricity by 2030 with both grid and off-grid solutions (Adeshina et al., 2024). Additionally, distributed RE systems decrease reliance on fossil-based energy producers, reducing the economic pressure from rural households and eliminating indoor air pollution and carbon production.

3.3.4 Energy Security and Systemic Challenges

Nigeria's energy industry continues to struggle with energy security issues, including inconsistent supply, inadequate production, and affordability concerns. In recent years, the national grid has experienced multiple annual breakdowns and total system failures (Golobish et al., 2023). The daily spotless blackouts and grid failures are signs of a stressed system running on extremely thin margins. One of the underlying causes is that available generation capacity is often much lower than peak demand and sometimes even below baseload demand, which leads to the imposition of forced load shedding. As mentioned above, only about 40% of the installed capacity is online due to factors such as maintenance backlog, shortage of gas supply to power stations and transmission bottlenecks (Ishaku & Quadri, 2025). This capacity shortcoming means that the network often cannot cope with the load, resulting in widespread blackouts (Golobish et al., 2023). The Transmission Company of Nigeria (TCN) has a rolling capacity of around 7.5 GW, which is a constraint. Even if more production were available, the old transmission network cannot reliably release power above this threshold, which underlines the need to modernise the network (Ishaku & Quadri, 2025).

Nigeria's electricity supply is among the poorest in the world. The average Nigerian household with electricity on the grid receives only 6.6 h of electricity per day, and many experience dozens of blackouts per month. The World Bank Business Outlook 2021 found that businesses suffer an average of 32.8 blackouts per month, with an estimated 11% loss of sales per year (Pelz et al., 2023). Such unreliable power has led to almost universal use of backup generators by those who can afford to always keep them on. Surveys show that more than 40% of households and up to 80% of businesses own or share generators as an alternative source of energy. This imposes a significant economic burden: Nigerian generator users spend an average of 48% of their monthly income on fueling and maintaining these generators (Ishaku & Quadri, 2025). Nigerians are effectively paying twice for electricity on the grid (whether they receive it, because of the fixed charges for connection) and then for the costly alternative of generating it themselves. High energy costs also translate into energy poverty, as poorer households cannot afford to run generators for extended periods of time, leaving many with only a few hours of electricity a day or no electricity at all.

Tariffs are placing more pressure on energy affordability. Electricity rates in Nigeria have also historically been subsidised and held below cost recovery levels, which has meant that public utilities haven't been able to earn enough money or invest enough back into the infrastructure. However, many residents think that poor service does not deserve tariffs, and some areas are still poorly collected. The result is a downward spiral of falling services and unsound finances. In the context of energy security, this is not only about the lack of physical energy but also about the reliability and predictability of a regular supply.

These systemic challenges are currently being targeted by several endeavours. All three programs aim to increase the country's energy production, and a new joint project with Siemens, the Presidential Energy Initiative, is targeting the re-establishment and progressive expansion of the electrical grid's capacity to 25,000 kW. On the other hand, there is a need to introduce new generation capacity (gas and solar as well as hydroelectric and possibly coal production), modernise transmission lines, substations, and distribution networks (Okibe & Mokuye, 2018). For example, off-grid systems can provide essential services in the event of a grid failure, and solar and wind can complement hydropower seasonally. But enabling further variable renewable penetration will require a modernised framework, as well as investments in grid flexibility and storage.

In summary, the Nigerian energy situation is a country faced with two problems: widening access and enhancing supply to meet its much-desired demand, while ensuring that the system or service provider is more sustainable and dependable. The present situation is characterised by a lack of capacity (high population density due to the central grid, high dependence on one fuel source(gas) for power generation and underutilization of hydro resources. It will require a sustained commitment in terms of finance and politics to tackle these problems. These and other recent developments, such as solar mini-grids and new hydroelectric power plants, are encouraging us to believe that more progress will be made in the years ahead. Nigeria needs to effectively utilise its numerous natural resources, including sunlight, water, gas, and wind, to secure its future. This entails ensuring that water and other sustainability goals are fulfilled, improving current infrastructure, and bridging the gap in rural energy access. Table 3.1 summarises key indicators of Nigeria's electricity sector (2020–2025) and highlights the structural constraints that undermine energy security, notably the large gap between installed capacity and average operational dispatch, high aggregate technical and commercial losses, and substantial sections of the population remaining without reliable access.

Table 3.1 Key indicators of Nigeria's electricity sector (2020–2025)

Indicator	Value (Latest estimate)	Source
Total installed generation capacity	12.5 GW (80% thermal, 20% hydro)	Ishaku and Quadri (2025)
Operational generation (Dispatch)	3.5–5.0 GW on average available	Ishaku and Quadri, (2025)
Peak demand (suppressed)	7 GW (estimated, unmet due to capacity constraints)	Golobish et al., (2023)
Per capita electricity use	150 kWh per year	Pelz et al., (2023)
National electrification rate	55% (83.9% urban, 24.6% rural as of 2020)	Pelz et al., (2023)
Population without electricity	85 million (primarily rural)	Williams et al., (2024)
Hydropower installed capacity	2.8 GW (large hydro); 60 MW (small hydro)	John et al., (2025)
Hydropower potential (untapped)	11–14 GW large; 3.4 GW small (undeveloped)	John et al., (2025)
Largest hydropower plants	Kainji (760 MW), Shiroro (600 MW), Jebba (578 MW), Zungeru (700 MW, new)	Ingram (2023)
Off-grid renewable capacity	52 MW (solar mini-grids and SHS combined)	Oke et al., (2024)
Households using off-grid solar	1 million (5 million people reached)	Oke et al., (2024)
Number of mini-grids installed	67 (solar mini-grids operational as of 2023)	Oke et al., (2024)
Backup generator usage	> 40% households (up to 80% of businesses) own generators	Adeniyi and Ige (2024)
Average grid outages	33 outages per month (per firm, 2021)	Pelz et al., (2023)
Average grid supply (Household)	6–7 h per day	Pelz et al., (2023)
Consumer spending on GenFuel	₦1.56 trillion/year (> $3.5 billion) on generator fuel	Pelz et al., (2023)
Transmission capacity	7.5 GW nationwide (wheeling limit)	Ishaku and Quadri (2025)
Aggregate technical and commercial losses	46% (est., indicating high losses in transmission and distribution)	Golobish et al., (2023); Ishaku and Quadri (2025)

3.4 Water–Energy Nexus Interdependencies

3.4.1 Water–Energy Nexus in Nigeria

Water and energy systems in Nigeria are tightly linked. Large amounts of water are required for power generation (especially hydropower and thermal cooling), while the water sector depends on energy (electricity or fuel) for abstraction, treatment, and distribution. Approximately 214×10^9 m^3 of surface water and 87×10^9 m^3 of groundwater in Nigeria are renewable annually. The uses of freshwater ($12{,}472 \times 10^6$ m^3 in 2010) are divided among agricultural (approximately 44%), domestic (40%), and industrial (16%) uses (Ibrahim et al., 2022). Hydroelectric plants (Kainji, Jebba, Shiroro) account for 1.94 GW of installed capacity and have historically supplied around 20% of Nigeria's grid power. By contrast, natural gas thermal plants now dominate generation (80–82% gas vs. 18–20% hydro) (Machina & Sharma, 2017). Interdependencies are produced by this dual dependence. For instance, droughts can lower river flows and hydropower production, and power outages can stop water pumping for urban supply or irrigation.

3.4.2 Irrigation and Pumping

Agricultural irrigation in Nigeria is highly energy dependent. Many irrigation schemes rely on diesel engines or electric pumps to lift water to fields. For example, a study of five smallholder irrigation schemes in North-Central Nigeria found that farmers used, on average, 360–391 L of diesel per hectare per cropping season (Shanono et al., 2022). This high energy use (at rising fuel costs) limits irrigation expansion. National projects (e.g. Fadama flood-plain development) often include diesel or solar pumps to boost output. Overall, irrigation accounts for nearly half of Nigeria's freshwater withdrawals; therefore, diesel fuel consumption in agriculture is a significant component of the nexus.

Irrigation in Nigeria primarily relies on surface and groundwater sources, which sets up the energy demand for pumping discussed below. Simplified schematic showing surface water and groundwater as the principal sources feeding agricultural operations via water use in irrigation. The diagram provides context for the subsequent discussion of irrigation pumping and its energy implications, in a sector that accounts for 44% of national freshwater withdrawals (Adeniran et al., 2021; Ibrahim et al., 2024a, 2024b).

3.4.3 Urban and Rural Water Supply

Cities also consume energy to provide water. Urban water treatment and distribution networks rely on electric pumps and motors; unreliability of the grid forces many water utilities to use backup generators. For instance, Lagos and Abuja's water corporations supplement their power supply with diesel-driven generators to maintain the supply (often unquantified in the literature). In rural areas, borehole pumps are typically powered by diesel or gasoline. A transition to solar-powered pumps is emerging as costs fall, highlighting a potential synergy (solar irrigation) to decouple water pumping from fossil fuel. Similarly, wastewater treatment (where it exists) requires power, tying sanitation to the electricity sector (Egwom et al., 2025; Nnaji et al., 2024).

3.4.4 Illustrative Case Studies

3.4.4.1 Kainji–Jebba–Shiroro Hydropower Cascade

The interconnected dams on the Niger and Kaduna rivers serve as an example of the water nexus. Together, they generate the Kainji Lake, which sustains regional irrigation and fisheries, and provides about 20% of the nation's electricity. The seasonal trade-off is a result of operators having to lower their output during dry spells due to decreased flows. Kainji's capacity is now 760 MW due to a recent concession (Adeshina et al., 2024; Faleke et al., 2023).

3.4.4.2 Bakolori and Dadin Kowa Irrigation Projects

During the rainy season, floodwaters are stored in large dams, such as Dadin Kowa (Gombe State) and Bakolori (Sokoto State), for use in irrigation during the dry season. Well pumping and canal operation are necessary for these projects. These projects require canals and well pumping to be operated. At Bakolori (capacity 450×10^6 m^3), most of the pumps are powered by diesel engines, resulting in high fuel costs and lower consumption. Food production may be limited by water scarcity, which may compel farmers to reduce irrigation or leave land fallow (Adeshina et al., 2024; Shanono et al., 2022).

3.4.4.3 Urban Water Supply (Lagos State)

Lagos' demand far exceeds the current supply capacity, so treated water is pumped over long distances. To ensure a steady supply in the event of grid outages, the Lagos State Water Corporation operates large pumping stations and has dedicated diesel generators (8–10 MW). Despite the paucity of data, this case shows that water infrastructure frequently requires steady electricity generation and substantial investment to meet service objectives, and that the availability of water in cities depends on a reliable power supply (Babajide & Brito, 2021; Macharia et al., 2020).

3.5 Challenges and Interdependencies

Nigeria's water and energy sectors face intertwined problems. Seasonality is key: drought in the Sahel can cut water flows by half, cut hydroelectric power production sharply, and force gas-fired power plants to fill the gap, increasing fuel consumption and costs. On the other hand, frequent power cuts disrupt water supplies and irrigation. Many dams serve multiple purposes and therefore compromise. Approximately 79% of the large dams in Nigeria are for domestic or industrial water supply, while only 4% are for hydropower. For example, Kainji and Jebba were originally designed for electricity production, but their dams are also used for irrigation and flood control, which makes the balancing of water releases difficult (Adeniran et al., 2021; Adeshina et al., 2024).

Energy and water expansion can sometimes conflict with each other. Building new hydropower projects, such as the under-construction Zungeru Dam with a capacity of 700 MW, may inundate farmland or alter flows to downstream irrigation systems (Adeshina et al., 2024). Likewise, expanding irrigation (pumping more water) increases fuel or electricity demand, potentially straining the grid. There are also synergies: solar-powered borehole pumps in northern Nigeria are emerging to reduce diesel fuel consumption, and small off-grid hydropower systems could power rural water systems. Overall, integrated planning is needed; for example, it has been estimated that Nigeria has over 323 dams (with a total storage capacity of approximately 30×10^9 m^3), with only a fraction optimised for both energy and water (Bello et al., 2022; Kaleem & Sabi, 2021). Recognising these interdependencies and utilising modern data and modelling is critical for sustainable water resource and energy management in Nigeria.

3.5.1 Sustainability Challenges at the Nexus

Nigeria's water–energy nexus is strained by a convergence of environmental and socioeconomic stressors that undermine progress on SDGs 6 (clean water), 7 (clean energy) and 13 (climate action). The nation has abundant groundwater and two large rivers, the Niger and Benue, but due to unequal distribution and deteriorated infrastructure, many people lack access to dependable services (Irene et al., 2025). Only about 14% of Nigerians have safe managed water, and roughly 45% have no electricity (Pelz et al., 2023). Nigeria's national policies (e.g., the National

Table 3.2 Regional climate hazards and impacts on water–energy systems in Nigeria (Irene et al., 2025)

Region	Climate hazards	Water system impacts	Energy system impacts
North (NW, NE)	Intense droughts, high heat	Depleted rivers/aquifers; severe scarcity for irrigation and drinking water	Reduced hydro reservoir levels (e.g. Kainji/Jebba), higher demand for pumped water
Middle belt (NC)	High rainfall variability; occasional floods	Seasonal flooding of rivers/basins, erratic water supply	Damage to substations and distribution, erratic generation scheduling
South (SW, SE)	Heavy rains, coastal flooding	Flooded/contaminated urban water systems; saltwater intrusion in coastal aquifers	Overland floods damage power lines/pipelines; hydro plants on southern rivers risk inundation
Niger delta (SS)	Sea-level rise, extreme floods	Salinization and pollution (oil spills) of freshwater; inundation of wells	Flooding and erosion threaten oil/gas pipelines and generators; frequent outages in the delta power grid

Water Resources Policy 2016, Master Plan 2013, and Renewable Energy Policies) acknowledge integrated water–energy planning; however, implementation gaps persist. Federal and state agencies often have overlapping mandates, impeding a cohesive nexus strategy. As one review notes, technical, financial, environmental, social and institutional deficits are all "predominant challenges in achieving water infrastructure sustainability" in Nigeria, reflecting the multi-sectoral nature of the problem (Adeoti et al., 2023).

The north faces severe drought and water scarcity, stressing irrigation pumps and reducing hydropower output; the Middle Belt and South experience erratic rain and floods that pollute water supplies and damage pipelines or grid assets; while the oil-rich Niger Delta suffers coastal flooding and contamination that impair both water quality and energy (oil/gas) infrastructure. Table 3.2 summarizes the significant climate hazards by region and their impacts on Nigeria's water and energy systems.

3.5.2 Infrastructure Challenges

Nigeria's water and power infrastructure is frequently old and dilapidated. Siltation, mechanical wear and lower inflows associated with changing weather patterns have led to decreased generation at the key hydropower dams Kainji, Jebba and Shiroro, which were constructed in 1968, 1985, and 1990, respectively. Likewise, countless municipal water pipes, treatment plants and rural pumps have gone past their design life without being replaced or maintained sufficiently. A recent study points out that these failures are 'polydimensional' in nature, comprising technical, financial and institutional factors, among others, and without an overhauling of the core structures of water infrastructure, it would continue to be unsustainable (Adeoti et al.,

2023). Nigeria loses electricity and water from its networks frequently: a survey showed that there were 32.8 monthly blackouts per household on average, and many urban water systems function at below capacity levels or even totally collapse during flooding (Pelz et al., 2023). This under-performance of infrastructure compounds the inter-linkage: for instance, ageing infrastructure is more energy-intensive (cooling or purification) and power-intensive (pumping) than required, and often failures in one will cross over to the other (e.g. energy losses which impact power demand and delivery). The overall impact is that service provision, even where physical resources exist, remains unreliable, inarguably a prime source of non-sustainability (Adeoti et al., 2023; Pelz et al., 2023).

3.5.3 Access Disparities: Urban Versus Rural

The nexus issue is particularly acute in Nigeria's impoverished, outlying regions. Only 70% of Nigeria's roughly 200 million people have access to basic water, and 47% of the population lacks electricity, with the disparity between rural and urban areas widening. Other sources of electricity were found in 25% of rural households, whereas 84% of urban households had grid-connected electricity, compared to approximately 10% and 33% in sub-Saharan Africa as a whole (Edefo, 2025; Pelz et al., 2023). Likewise, 87% of the urban population (as opposed to 60% of the rural population) has improved access to drinking water (2018). The discrepancies of these factors result in great inequity of care. Wealthy urban users often make up for the deficit by tapping private wells, generators, and bottled water, while rural users rely on occasional wells and community groundwater sources. This creates twin disadvantages of inequality: poor communities already have less reliable and lower-quality water, exacerbating health and economic disparities.

Access disparities are compounded by geographical factors: areas in the south with plentiful oil supplies or major population centres tend to have better coverage than the empty northern and central zones. With some exceptions, everyone in Abuja and Lagos has electricity, while some northern states receive only a few hours a day. Few who are connected receive dependable amounts of service, and few reliable contributors obtain access to copious services (Pelz et al., 2023). All of this reflects an imbalance in the nexus, where it is the rural and the marginal that are most threatened when the governance system and infrastructure fail.

3.5.4 Institutional and Governance Challenges

The first is the long-standing issue of fragmented government. Water and energy responsibilities are distributed among numerous governmental levels and agencies under Nigeria's federal structure. The Federal Ministry of Water Resources, river basin authorities, and many parastatal agencies share responsibility for water management. Parts of the electrical grid are controlled by state and federal public utilities. The complexity reveals inefficiencies and a lack of coherence. Vague responsibilities and poor policy coherence have been repeatedly highlighted in reviews; for example, a systematic review identified multiple regulatory and policy gaps. 6 Political instability, ambiguous operational policy directives and incomplete projects that can undermine the proper functioning of water infrastructure (Adeoti et al., 2023). In fact, this means long delays or outright cancellations of essential projects, like transmission lines. Lack of funding is aggravated by long-term underinvestment in many systems; low budgetary allocation and poor cost recovery contribute to shortfalls.

In summary, several interacting challenges faced by the water and energy nexus in Nigeria due to climate variability; resources are not only becoming stressed but effectively being delivered strained through aging and undersized infrastructures, as well as resource governance impeded by fragmentation. They hinder the achievement of SDGs 6, 7, and 13 by perpetuating inequities and inefficiencies (Irene et al., 2025; Pelz et al., 2023). According to Nigeria's energy and energy transition pathway, continued investment in "strong infrastructural development" is a necessary component to addressing the above challenges. As showcased by the examples cited above, Nigeria is at risk of failing to fulfil its national and sustainable development agendas in terms of universal access to energy and water.

3.6 Conclusion

The availability of unequal natural resources, aged infrastructure, and inadequate institutional coordination all influence Nigeria's water nexus. The only way to combat poverty, hunger, and inequality, improve human life, and achieve justice is with strength and energy. However, as numerous studies demonstrate, not all parts of the nation have access to clean energy and water. For example, rural areas have been without electricity or water for decades, while many urban people rely on expensive private generators and wells for their households. Examples of the profound interdependence between the sectors include irrigation, thermal power plants, and dams; a disruption in one of these would have an effect on the others as well. This relationship is growing as climate variability increases the frequency of floods and droughts, and consequently, insecurity.

To meet these challenges, a holistic approach is required, which integrates water and energy planning into an informed model of sustainability. There also needs to be consistent policies for water regulators and river basin authorities to release water in the most efficient manner possible for residential use, agriculture and hydropower. Existing priority areas focus on aiding environmental protection, renewing transmission lines and dams and developing off-grid renewable energy. Investments in data systems, water efficiency, and community energy efficiency solutions will be crucial to building long-term resilience. Nigeria can harness its substantial solar, hydro, and groundwater potential to advance development and progress towards SDGs 6 (clean water), 7 (affordable clean energy), and 13 (climate action) by applying the nexus principle.

References

Abubakar, M. L., Abdussalam, A. F., Ahmed, M. S., & Wada, A. I. (2024). Spatiotemporal variability of rainfall and drought characterization in Kaduna, Nigeria. *Discover Environment, 2*(1), 72. https://doi.org/10.1007/s44274-024-00112-7

Adeniran, A., Daniell, K. A., & Pittock, J. (2021). Water infrastructure development in Nigeria: Trend, size, and purpose. *Water, 13*(17), 2416. https://doi.org/10.3390/w13172416

Adeniyi, A., & Ige, V. (2024). Maximizing Nigeria's renewable energy potentials for sustainable growth and development. *Andersen.* https://ng.andersen.com/maximizing-nigerias-renewable-energy-potentials-for-sustainable-growth-and-development/#:~:text=Maximizing%20Nigeria%27s%20Renewable%20Energy%20Potentials,of

Adeoti, O. S., Kandasamy, J., & Vigneswaran, S. (2023). Water infrastructure sustainability in Nigeria: A systematic review of challenges and sustainable solutions. *Water Policy, 25*(11), 1094–1111. https://doi.org/10.2166/wp.2023.173

Adeshina, M. A., Ogunleye, A. M., Suleiman, H. O., Yakub, A. O., Same, N. N., Suleiman, Z. A., & Huh, J.-S. (2024). From potential to power: Advancing Nigeria's energy sector through renewable integration and policy reform. *Sustainability, 16*(20), 8803. https://doi.org/10.3390/su16208803

Adetokunbo, P., Eluyemi, A. A., Aguda, S., Jegede, E., Omoseyin, T., Sanuade, O. A., & Adesina, R. B. (2025). Decadal trends and periodicity of terrestrial water storage in Nigeria-GRACE satellite observations. *Open Journal of Geology, 15*(07), 343–357. https://doi.org/10.4236/ojg.2025.157017

Anyanwu, I. N., Beggel, S., Sikoki, F. D., Okuku, E. O., Unyimadu, J.-P., & Geist, J. (2023). Pollution of the Niger delta with total petroleum hydrocarbons, heavy metals and nutrients in relation to seasonal dynamics. *Scientific Reports, 13*(1), 14079. https://doi.org/10.1038/s41598-023-40995-9

Apata, M. D., Okujagu, D. C., & Beka, F. T. (2019). Geospatialization of spilling facility from Spdc 2015 oil spill reports of the Niger delta region of Nigeria. *Journal of Applied Geospatial Information, 3*(1), 160–165. https://doi.org/10.30871/jagi.v3i1.1005

Aralu, C. C., Agbo, K. E., Nweke, N. D., Nwoke, S. U., Okah, A. T., Abugu, H. O., Egbueri, J. C., Agbasi, J. C., Ezugwu, A. L., Omeka, M. E., & Ucheana, I. A. (2025). Levels, sources and toxicity assessment of PCBs in surface and groundwater in Nigeria: A systematic review. *Journal of Hazardous Materials Advances, 17*, Article 100527. https://doi.org/10.1016/j.hazadv.2024.100527

Babajide, A., & Brito, M. C. (2021). Solar PV systems to eliminate or reduce the use of diesel generators at no additional cost: A case study of Lagos, Nigeria. *Renewable Energy, 172*, 209–218. https://doi.org/10.1016/j.renene.2021.02.088

Bello, U., Yawas, D. S., & Oyedeji, A. N. (2022). Assessment of perceived performance of solar-powered borehole projects in Nigeria. *Management Science and Business Decisions, 2*(2), 32–43. https://doi.org/10.52812/msbd.47

Chijioke, N., Audu-Bako, S., & Uwakwe, I. (2021). Strategies for managing oil spill related conflicts in the Niger delta: A case study of Bayelsa state. *Advances in Social Sciences Research Journal, 8*(4), 465–483. https://doi.org/10.14738/assrj.84.9976

Darma, A. I., Zakari, S. A., Sani, A., & Mamman, H. S. (2024). Soil pollution and heavy metal contamination: Insight from effluent discharged from Challawa industrial area, Nigeria. *Fudma Journal of Sciences, 8*(6), 355–361. https://doi.org/10.33003/fjs-2024-0806-2749

Edefo, J. W. (2025). Water and sanitation access in Nigeria. *Water Conservation Science and Engineering, 10*(2), 72. https://doi.org/10.1007/s41101-025-00402-y

Ediagbonya, T. F., Olaniyan, O. I., Orisawayi, A. O., Oziegbe, F. E., & Warekromor, H. T. (2025). Risk assessment of polychlorinated biphenyls (PCBs) in surface and groundwater in three states in Nigeria. *Water Practice and Technology, 20*(7), 1557–1574. https://doi.org/10.2166/wpt.2025.088

Egwom, O. J., Godwin, A. I., Ogar, M. O., & John, G. (2025). Digital solution for seamless water supply management in Wukari metropolis. *Journal of Science Research and Reviews, 2*(1), 62–75. https://doi.org/10.70882/josrar.2025.v2i1.29

Eteh, D. R., Egobueze, F. E., Paaru, M., Otutu, A., & Osondu, I. (2024). The impact of dam management and rainfall patterns on flooding in the Niger delta: Using sentinel-1 SAR data. *Discover Water, 4*(1), 123. https://doi.org/10.1007/s43832-024-00185-8

Ewim, D. R. E., Orikpete, O. F., Scott, T. O., Onyebuchi, C. N., Onukogu, A. O., Uzougbo, C. G., & Onunka, C. (2023). Survey of wastewater issues due to oil spills and pollution in the Niger delta area of Nigeria: A secondary data analysis. *Bulletin of the National Research Centre, 47*(1), 116. https://doi.org/10.1186/s42269-023-01090-1

Faleke, S. A., Nwabeze, G. O., & Buhari, H. L. (2023). Characteristics of capture and culture fishery production in Kainji lake basin, Nigeria. *Journal of Agricultural Extension, 27*(3), 35–40. https://doi.org/10.4314/jae.v27i3.4

Ferdowsi, A., Piadeh, F., Behzadian, K., Mousavi, S.-F., & Ehteram, M. (2024). Urban water infrastructure: A critical review on climate change impacts and adaptation strategies. *Urban Climate, 58*, Article 102132. https://doi.org/10.1016/j.uclim.2024.102132

Ghanbarzadeh, T., Habibi, D., & Aziz, A. (2025). Addressing reliability challenges in generation capacity planning under high penetration of renewable energy resources and storage solutions: A review. *Renewable and Sustainable Energy Reviews, 212*, Article 115461. https://doi.org/10.1016/j.rser.2025.115461

Golobish, S., Yeganyan, R., Tan, N., Cannone, C., & Howells, M. (2023). *The burden of the broken grid: Modelling improved power-sector reliability to support low carbon development in Nigeria.* https://doi.org/10.21203/rs.3.rs-2753971/v1

Ibebuchi, C. C., & Abu, I.-O. (2023). Rainfall variability patterns in Nigeria during the rainy season. *Scientific Reports, 13*(1), 7888. https://doi.org/10.1038/s41598-023-34970-7

Ibrahim, E. S., Chiroma, J. T., Abubakar, M. A., Ojih, S. A., Waziri, E. S., Daffi, R. E., & Yenne, E. (2024a). Spatial modelling of present and future groundwater potentials in Nigeria; towards a sustainable water demand and supply. *Sustainable Water Resources Management, 10*(4), 137. https://doi.org/10.1007/s40899-024-01107-1

Ibrahim, M. M., Ibrahim, M., & Aisha, M. M. (2024b). Groundwater storage variability in West Africa using gravity recovery and climate experiment (GRACE) and global land data assimilation system (GLDAS) data. *EQA -International Journal of Environmental Quality, 64*, 87–102. https://doi.org/10.6092/issn.2281-4485/20167

Ibrahim, U. A., Alkali, A. N., Sanyaolu, B. O., & Usman, B. (2022). Nigerian water resources management—An overview. *Arid Zone Journal of Engineering, Technology and Environment, 18*(1), 23–30.

Iduseri, E. O., Abbas, I. I., Izunobi, J. U., Ogedegbe, S. O., & Ogbonna, D. O. (2022). Access to safe drinking water in developing countries: A comparative analysis of the urban and rural areas of Zaria, Kaduna State, Nigeria. *Journal of Applied Sciences and Environmental Management, 25*(11), 1869–1878. https://doi.org/10.4314/jasem.v25i11.4

Ingram, E. (2023). *Units commissioned for the 700 MW Zungeru hydropower plant in Nigeria.* Factor this. https://www.renewableenergyworld.com/hydro-power/technology-equipment/units-commissioned-for-the-700-mw-zungeru-hydropower-plant-in-nigeria/#:~:text=Located%20on%20the%20Kadina%20River,support%20irrigation%20for%20the%20region

Irene, J., Irene, B. N., & Daniels, C. (2025). Not a drop to drink: Addressing Nigeria's deepening freshwater crisis. *Water, 17*(12), 1731. https://doi.org/10.3390/w17121731

Ishaku, H. P., & Quadri, Q. Y. (2025). Modeling renewable energy integration for Nigeria's 2030 electricity target using EnergyPLAN. *Journal of Energy Systems, 9*(2), 159–171. https://doi.org/10.30521/jes.1598604

Isukuru, E. J., Opha, J. O., Isaiah, O. W., Orovwighose, B., & Emmanuel, S. S. (2024). Nigeria's water crisis: Abundant water, polluted reality. *Cleaner Water, 2*, Article 100026. https://doi.org/10.1016/j.clwat.2024.100026

John, N. K., Onwuagbu, C. C., & Chigozie, C. J. (2025). Potential of integrated energy solution in Nigeria: Opportunities and challenges for sustainable development-multi facet assessment model. *Discover Sustainability, 6*(1), 239. https://doi.org/10.1007/s43621-025-00915-5

Kaleem, O., & Sabi, A.-F.B.S. (2021). Overview of aquaculture systems in Egypt and Nigeria, prospects, potentials, and constraints. *Aquaculture and Fisheries, 6*(6), 535–547. https://doi.org/10.1016/j.aaf.2020.07.017

Kayode, O., Luethi, C., & Rene, E. (2018). Management recommendations for improving decentralized wastewater treatment by the food and beverage industries in Nigeria. *Environments, 5*(3), 41. https://doi.org/10.3390/environments5030041

Macharia, P., Kreuzinger, N., & Kitaka, N. (2020). Applying the water-energy nexus for water supply—A diagnostic review on energy use for water provision in Africa. *Water, 12*(9), 2560. https://doi.org/10.3390/w12092560

Machina, M., & Sharma, S. (2017). Assessment of climate change impact on hydropower generation: A case study of Nigeria. *International Journal of Engineering Technology Science and Research, 4*, 2394–3386.

Mooren, C. E., Munaretto, S., La Jeunesse, I., Sievers, E., Hegger, D. L. T., Driessen, P. P. J., Hüesker, F., Cirelli, C., Canovas, I., Mounir, K., & Madrigal, J. G. (2025). Water–energy–food–ecosystem nexus: How to frame and how to govern. *Sustainability Science, 20*(6), 2313–2334. https://doi.org/10.1007/s11625-025-01691-x

Morrison, A. I., Emmanuel, N. O., & Ighovie, E. F. E. S. (2023). Analysis of rainfall distributions and variations in Nigeria. *Himalayan Journal of Agriculture, 04*(02), 1–6. https://doi.org/10.47310/hja.2023.v04i02.012

Nitte, I. S. (2023). Global energy transition and its implications on energy security in Nigeria: A critical review. *EPRA International Journal of Climate and Resource Economic Review*, 1–9.

Nkwoh, B. (2023). *Electricity, power systems and renewable energy*. The International Trade Administration, U.S. Department of Commerce. https://www.trade.gov/country-commercial-guides/electricity-power-systems-and-renewable-energy#:~:text=6%2C024%20MW%2C%20including%3A

Nnaji, C. C., Udokpoh, U. U., & Ifeakor, A. R. (2024). Assessing the efficiencies of domestic water pumps and distribution systems for household water supply in Enugu State, Nigeria. *Indian Journal of Engineering, 21*(55), 1–25. https://doi.org/10.54905/disssi.v21i55.e5ije1680

Nyika, J., & Dinka, M. O. (2023). *Water challenges in rural Sub-Saharan Africa* (pp. 39–55). https://doi.org/10.1007/978-3-031-26271-5_3

Odeloui, D., Nlend, B., Huneau, F., Celle, H., Garel, E., Alassane, A., Boukari, M., & Sambienou, G. (2022). Insight into groundwater resources along the coast of Benin (West Africa) through geochemistry and isotope hydrology; recommendations for Improved Management. *Water, 14*(14), 2154. https://doi.org/10.3390/w14142154

Ohunene, I., Adewuyi, T., Garba, A., Mbow, C., Agbaje, G., Rufus, O., & Monsor, O. (2025). Assessing nature-based flood mitigation measures at the general area of the confluence of rivers Benue and Niger in Kogi state, Nigeria. *American Journal of Remote Sensing, 13*(1), 13–31. https://doi.org/10.11648/j.ajrs.20251301.12

Ojo, S., Albrecht, E., Igboji, C., Ibrahim, B., & Enoguanbor, E. (2025). Potentials of water-energy-food security nexus approach as an option for implementation of climate change policies: A perspective for Nigeria. *African Journal of Climate Change and Resource Sustainability, 4*(2), 150–170. https://doi.org/10.37284/ajccrs.4.2.3639

Okafor, J. C. (2020). Flood, livelihood displacement, and poverty in Nigeria: Plights of flood victims, 2012–2018. In: *African handbook of climate change adaptation* (pp. 1–12). Springer.

Oke, I. A., Ahiakwo, C. O., Amadi, H. N., & Ojuka, O. E. (2024). Assessment of off-grid power plants in Nigeria: Location, capacity, performance and current status. *Journal of Energy Research and Reviews, 16*(3), 38–51. https://doi.org/10.9734/jenrr/2024/v16i3341
Okibe, H. B., & Mokuye, S. (2018). *Issues and challenges in the implementation of electric power sector recovery programme (EPSRP) and eligible customer registration policy (ECRP) in Nigeria* (vol. 3, pp. 42–54).
Olowoyeye, O. S., & Kanwar, R. S. (2023). Water and food sustainability in the Riparian Countries of Lake Chad in Africa. *Sustainability, 15*(13), 10009. https://doi.org/10.3390/su151310009
Omokaro, G. O., Idama, V., Airueghian, E. O., & Michael, I. (2024). Water resources, pollution, integrated management and practices in Nigeria—An overview. *American Journal of Environmental Economics, 3*(1), 5–18. https://doi.org/10.54536/ajee.v3i1.2593
Pelz, S., Chinichian, N., Neyrand, C., & Blechinger, P. (2023). Electricity supply quality and use among rural and peri-urban households and small firms in Nigeria. *Scientific Data, 10*(1), 273. https://doi.org/10.1038/s41597-023-02185-0
Peter, C. C., Ezeogu, C. M., & Brisibe, O. (2022). Analysis of oil pollution intensity in Niger-delta, Nigeria using normalized difference vegetation index. *Journal of Geography, Environment and Earth Science International*, 1–10. https://doi.org/10.9734/jgeesi/2022/v26i11641
Sakketa, T. G. (2023). Urbanisation and rural development in sub-Saharan Africa: A review of pathways and impacts. *Research in Globalization, 6*, Article 100133. https://doi.org/10.1016/j.resglo.2023.100133
Shanono, N. J., Nasidi, N. M., & Yahaya Usman, N. (2022). Appraising the current state of irrigation schemes in Northern Nigeria using sustainability pillars. *Environmental Research and Technology, 5*(4), 305–314. https://doi.org/10.35208/ert.1097795
Shokri, A., & Sanavi Fard, M. (2022). A sustainable approach in water desalination with the integration of renewable energy sources: Environmental engineering challenges and perspectives. *Environmental Advances, 9*, 100281. https://doi.org/10.1016/j.envadv.2022.100281
van den Heuvel, L., Blicharska, M., Masia, S., Sušnik, J., & Teutschbein, C. (2020). Ecosystem services in the Swedish water-energy-food-land-climate nexus: Anthropogenic pressures and physical interactions. *Ecosystem Services, 44*, Article 101141.
Williams, N., Raji, T., & Ekoh, C. (2024). Illuminating Nigeria: Blurring the lines between the grid and off-grid electricity. *Georgetown Journal of International Affairs*. https://gjia.georgetown.edu/2024/08/07/illuminating-nigeria-blurring-the-lines-between-the-grid-and-off-grid-electricity/#:~:text=With%20less%20than%20six%20years,29%20Nigeria%20Electricity%20System
Yusuf, A., & Akande, O. K. (2023). *Drivers, enablers, barriers and technologies (debt) for low-energy public housing delivery in Nigeria.*

Chapter 4
Legal Aspects of Water Resources Management in Nigeria: Challenges, Governance, and Sustainable Practices

Olawunmi Opeyemi Obisesan

Abstract Water is an essential natural resource that supports life, economic growth, and environmental sustainability. In Nigeria, the constitutional provisions, statutory legislations, and customary law scatter the legislative structure for the governance of water resources. This chapter critically analyses the legal and institutional regimes governing ownership, control, allocation, and protection of water resources within the Nigerian federation. It emphasizes the mandates of federal institutions like the National Water Resources Institute (NWRI) and the Nigeria Integrated Water Resources Management Commission (NIWRMC), State Water Agencies, Local Government infrastructure, and community-based organizations. The chapter also considers the contribution of customary leaders, citizens' organizations, and civil society organizations toward participatory and sustainable water management. Key issues that have been documented are jurisdictional clashes, poor enforcement, under-financing, and poor stakeholder consultation. This chapter submits that harmonization of the current legal tools, strengthening intergovernmental coordination, and domestication of integrated water resources management (IWRM) concepts are necessary to ensure equitable, efficient, and environmentally friendly use of Nigeria's water resources.

Keywords Customary water rights · Environmental governance · Integrated water resources management · Nigeria · Water law

4.1 Introduction

Water is an essential resource that supports Nigeria's social and economic growth, but water access is unequal and poor in Nigeria (Martins et al., 2025). Nigeria, with its varied hydrological regimes and more than 200 million inhabitants, is experiencing

O. O. Obisesan (✉)
Department of Clinical Legal Education, School of Law, Kampala International University, Kampala, Uganda
e-mail: olawunmi.obisesan@kiu.ac.ug; ogundipe0359@pg.babcock.edu.ng

M. I. Adeoba et al. (eds.), *Water Resources Management in Nigeria*, World Water Resources 10, https://doi.org/10.1007/978-3-032-22813-0_4

increasing resource supply challenges, sustainability, and efficiency in regulation (Omokaro et al., 2024). Nigeria's legal and institutional regime for the management of the water resource has experienced periodic restructuring, with fragmentation and enforcement continuously being encountered. The legal dimension of governing the waters involves the set of laws, policies, institutional arrangements, as well as judicial pronouncements that guide ownership, use, control, and conservation of waters (Barua & Tongper, 2019). It dictates the distribution of waters between contesting uses, delineates the rights as well as obligations of the people, the government bureaucracies, the procedures to resolve conflicts, and the use of waters sustainably (Singh & Roy, 2024). Water law, therefore, acts as a regulative as well as distributive tool that balances private rights with the public interest as well as environmental sustainability (Singh & Roy, 2024).

In Nigeria, the governance of waters occurs within a multifaceted federal constitutional setup, with legislative and governing power fragmented between the federal, state, and local governments. The 1999 Constitution of the Federal Republic of Nigeria (as amended) remains the primary legal document that regulates such power delegation (Samuel et al. 2021). However, the Constitution is bereft of a definable or express provision on the ownership or governance of waters. Instead, power is inferred from scattered provisions, more actually in the Second Schedule, and corollary enactments such as the Land Use Act 1978, the Water Resources Act 2004, as well as the predicate legislation that the states and the federal governments enact. The patchy legal framework is a regular source of jurisdictional overlaps, institutional confrontations, and uncertain delineations between the extent of the power of the federation and the states over waters (Ngene et al., 2021).

The Constitution as well as the Water Resources Act gives the federal government jurisdiction over surface waters and groundwater crossing state boundaries or of international character. This is a demonstration of the desire to give preference to centralized control over transboundary as well as interstate waters, such as the Niger and Benue Rivers, international watersheds, such as the Niger Basin and Lake Chad. However, intra-state waters, particularly those within the state borders used mainly locally to take care of sanitation, agriculture, as well as home supplies, are within the jurisdiction of state governments. Local governments, though bound within the context of the Constitution, play ancillary functions in the provision of sanitation as well as the provision of water supply at the local level (Samuel et al., 2021).

Despite these multi-level systems, institutional coordination remains a persistent vulnerability. The state and the state agencies usually implement their programs independently, with overlapping mandates with inadequate intergovernmental coordination (Ngene et al., 2021). The growing number of agencies, such as the Federal Ministry of Water Resources, River Basin Development Agencies, as well as State Water Agencies, produces a fragmented policy environment that discourages the adoption of integrated water resource management (IWRM), a best practice universally stipulated by the United Nations as well as international conventions. Secondly, the environmental as well as human rights paradigm of the waters of Nigeria must be framed within the context of environmental as well as human rights jurisprudence. Section 20 of the Constitution of Nigeria obliges the state to "protect as well as improve the environment, as well as safeguard the waters, air, land, forest, as well as wildlife of Nigeria." Despite being located within the non-justiciable Fundamental Objectives as well as Directive Principles of State Policy, the provision enshrines the norm of environmental conservation as well as the use of water in a non-destructive manner. Moreover, the international as well as the regional human rights obligations towards the waters of Nigeria under the International Covenant on Economic, Social as well as Culture Rights (ICESCR), the African Charter on Human as well as Peoples' Rights, as well as the UN General Comment No. 15 on the Right to Water, expand the legal argument to include the right of the people to access to waters that are clean, safe, yet affordable as a corollary to the right to life as well as human dignity (Ngene et al., 2021). In practice, the challenges of water management in Nigeria are made worse by climate change, pollution, uncontrolled exploitation, weak enforcement regimes, and deterioration of infrastructure. Water bodies are threatened due to pollution with industrial effluents, oil spillages, farming runoff, as well as indiscriminate waste disposal. The absence of credible legal frameworks to control pollution as well as the monitoring of the quality of water makes environmental degradation and the risk to public health even harder to control. Besides, unjustifiable access to water, particularly in the rural setting as well as the peri-urban areas leaves one wondering about a role to be played by legal as well as policy frameworks in the enforcement of social justice as well as equitable distribution of resources (Martins et al., 2025).

Recent policy developments, notably the proposed National Water Resources Bill (2016–2022), have reignited debates about the appropriate constitutional balance between federal and state powers in water governance. Proponents argue that the Bill would harmonize Nigeria's numerous and outdated water laws and promote integrated management of the nation's water resources. Critics, however, contend that it would centralize control in the federal government, undermine state autonomy, and violate the principle of federalism enshrined in the Constitution (Ukpai, 2022).

Accordingly, the examination of the legal framework of the governance of the waters of Nigeria should include both the provisions of the constitution, statutory acts, judicial dicta, as well as international obligations (Ladan, 2013). The examination should also cover the examination of institutional cooperation, continuity of policy, as well as complementarity between development imperatives as well as environmental conservation. Ultimately, the effective governance of the waters in the state of Nigeria depends on the rule of law, efficient policy making, as well as collaborative federalism—components to the effective governance of the waters of the state sustainably, equitably, as well as in the public interest (Ladan, 2013).

This chapter analyzes the Nigerian legal framework for the regulation of water resources, paying particular focus to constitutional and legislative systems, the functioning of predominant institutions, legislative trends of recent years, challenges to regulation, and reform suggestions.

4.2 Constitutional and Legislative Framework

4.2.1 Constitutional Overview

The 1999 Constitution of Nigeria (as amended) distributes legislative power over water resources between the federal government and the states, water resources being a "concurrent" issue. Though the federal government retains exclusive rights over interstate waters and watercourses affecting more than one state, (Constitution of the Federal Republic of Nigeria, 1999a, Sch. 2, Pt. 1, Item 64) states own water within them. The Constitution establishes basic principles regarding the protection of the environment, such as water resources. (Okorie, 2019; Constitution of the Federal Republic of Nigeria, 1999a, Sch. 2, Pt. 1, Item 64), Such provisions are placed under the non-justiciable chapter of Fundamental Objectives and Directive Principles of State Policy. Law and development (Ikpeze, 2014).

Although the Constitution identifies water as a public good, it does not establish an independent right to water, and there is no enforceable requirement stated for water supply and sanitation.

4.2.2 Statutory Instruments and Key Legislation

The key federal legislation is the Water Resources Act, Cap W2, Laws of the Federation 2004 (originally enacted in 1993). The Act places management and control of surface water and groundwater, except for solely intrastate water, solely in the Federal Government. The Act gives authority to the Minister of Water Resources to regulate use, prevent pollution, grant licenses, and encourage coordination between government agencies. Section 4.2 guards statutory and customary domestic and agricultural use of water, describing the state-indigenous legal orders point of contact. Other constitutional legislation includes.

4.2.2.1 River Basins Development Authorities Act, Cap R9 LFN 2004

It establishes the River Basin Development Authorities (RBDAs) as institutions responsible for managing hydrological catchments and shaping resource allocation: The River Basin Development Authorities have not made much positive contribution to Nigeria's water resources management, and several studies have recorded poor performance against significant investment.

RBDAs, which were established between 1973 and 1976, affected some developments in irrigation farming but did not deliver the anticipated economic development. Research supported the inadequacy of planned basins, with negative environmental effects and poor economic performance. Subsequent studies validate persistent issues: a 2019 case study of 12 RBDAs established that they "failed to spur agricultural and rural development" as they were disorganized and non-basin wide. A comparison study of 2020 determined that failure of institutional interfaces between state institutions and RBDAs was responsible for being ill-organized. Case studies pin down very low project utilization, and it has only developed 3.4% of the land acquired. The repeated finding for decades is that with so much water resources in Nigeria, RBDAs have failed to attain their developmental targets (Gana et al., 2019).

4.2.2.2 Environmental Impact Assessment Act, Cap E12 LFN 2004

It oversees the environmental impact assessment of projects that have the potential to impact water quality: Different assessments identify systemic problems in EIA application, such as "inadequate capacity of EIA approval authorities, deficiencies in screening and scoping, poor EIA quality, inadequate public participation and weak monitoring". There are recurrent observations from research works that "EIA is typically conducted much later once the project proponents are already attached to a design concept" and is marred by "shortage of post-project monitoring." In tandem with this, Nigeria too has severe water pollution, especially in the Niger Delta due to oil activities, (Omokaro et al., 2024) and is ineffective in integrated water management (Ngene et al., 2021). The sources, however, do not say anything about whether these water management failures are because of EIA Act shortcomings.

4.2.2.3 National Inland Waterways Act (Cap N47 LFN 2004)

Nigerian water law is dispersed throughout several documents and instruments but are not sufficient to address the issues of control, ownership, administration, and protection of water resources (Ngene et al., 2021).

Research indicates that there are urgent management issues of inland waterways, such as policy inconsistency, private sector deficiencies, and agency rivalries among and between agencies (Chukwuma et al., 2017; Ngene et al., 2021; Nwankwoala et al., 2014). Institutional weaknesses and duplication of agency roles are noted by authors as the major impediments to efficient management of water resources.

4.2.2.4 The Nigerian Minerals and Mining Act (Cap N162 LFN 2004)

This act provides legal provisions for the use and regulation of water in mining operations. However, evidence suggests that these provisions have not been enforced and implemented effectively, resulting in the degradation of water resources.

Conclusion from the analyzed relevant law in the mining sector of Nigeria is that the government's response to the problems in the use of water in mining is inadequate (Igbayiloye & Niyi-Gafar, 2025), even with legal protection. It validated that "Nigeria is not lacking in laws to govern the mining sector" but highlighted the necessity to "strengthen the regulatory institutions to facilitate compliance" (Akinsulore, 2024). The study criticized the Nigerian Minerals and Mines Act community development stipulations. For water authorities at the state and local government levels, enabling law and decrees govern them in terms of managing water supply and sanitation but with customary law still exerting a strong influence, particularly in rural and communal settings. Nigeria has dual water governance through formal state-level institutions and ongoing customary practices but with nothing much being documented on the role of customary law.

Various studies affirm the existence of formal structures of water governance. State water agencies (SWAs) are "regulated agencies supplying water to urban dwellers" in formal arrangements. Local government is composed of "local governments and other community-based organizations in participatory decision-making" with "multiple provision regimes" defining water and sanitation governance (Samuel et al., 2021).

But proof of the existence of customary law is extremely remote. According to a research report, "cultural, socio-economic, physical-environmental and temporal factors" shape water practice, which implies customary systems continue to exist. The 2016 National Water Policy "fails to engage local communities in water and sanitation management", possibly allowing customary practice to make up for shortcomings in regulation (Sanjo et al., 2024).

Whereas comparative Zimbabwe evidence informs us that "traditional informal practices still dominate among rural communities" even in the face of formal water law (Shoko & Naidu, 2018) evidence of the impact of customary law on Nigerian water law is something for further research. Cross River State, for instance, has passed custom-tailored water legislation aimed at enhancing sanitation and supply (*Cross River State Water Supply and Sanitation Law* No. 5, 2025).

4.3 Institutional Framework and Stakeholders

4.3.1 Federal Institutions

4.3.1.1 Federal Ministry of Water Resources

The Federal Ministry of Water Resources (FMWR) has a mandate for directing national policy formulation, resource planning, and coordination of regulation. It oversees water allocation and quality frameworks, in collaboration with sectoral agencies and states.

The general evidence finds Nigeria's water management system grappling with severe institutional problems. Various sources reference problems in overlapping mandates, coordination issues, and poor policy implementation (Ngene et al., 2021; Nwankwoala et al., 2014). The sources all mention that River Basin Development Authorities (RBDAs) have the core working function in water resources development and management (Okeola & Balogun, 2017). Historical review indicates more stringent federal coordination, with initial propositions to create a National Water Resources Board (Isukuru et al., 2024), although up-to-date sources do not debate whether FMWR is an efficient coordinator under the necessity of such a requirement.

4.3.1.2 River Basin Development Authorities (RBDAs)

Twelve RBDAs oversee strategic hydrological basins throughout the nation. Their legislative mandate is tripartite, extending to irrigation, rural water supply, and environmental protection. But they are routinely prosecuted for dilution of resources, lack of independence, and area overlaps with other agencies (Gana et al., 2019). River Basin Development Authorities in Nigeria have been a significant part of water resource management since 1976 but have shown consistently poor performance and limited impact with huge investments.

Independent evaluations repeatedly record poor performance. RBDAs were established to harness water resources and stimulate agricultural development via large-scale irrigation, but their contribution "is hardly felt" on food production potential (Balogun, 2022). Research shows that RBDAs have "failed to accelerate agricultural and rural development" (Gana et al., 2019) and are at the peak of underutilization levels—e.g., the Cross River Basin Development Authority developed only 3.4% of land acquired (Balogun, 2022). Critical shortcomings are weak institutional structure, poor integration between hydrological basins and regions, poor coordination, and poor consideration of local context.

4.3.1.3 National Water Resources Institute (NWRI) and Nigeria Integrated Water Resources Management Commission (NIWRMC)

National Water Resources Institute (NWRI)

NWRI provides capacity building by undertaking research and technical assistance, while the NIWRMC was established to enforce integrated water resources management policy and regulatory legislation across the country. The National Water Resources Institute (NWRI) and Nigeria Integrated Water Resources Management Commission (NIWRMC) are flagship institutions of the Federal Ministry of Water Resources with the responsibility of enhancing Nigeria's water governance and sustainability (Martins et al., 2025).

The National Water Resources Institute (NWRI) was founded in 1979 (*National Water Resources Institute Act*, Cap N83 LFN 2004) and saddled with the following tasks: Trains and develops water professionals, conducts research in hydrology, groundwater, irrigation, and water quality, Compiles and coordinates hydrological data, assists government in water policy and planning, Promotes water conservation and public awareness.

These roles have witnessed successes in improved technical capability in the water sector, hydrological data and research planning, national water policies and sustainable practice, and regional cooperation through UNESCO and ECOWAS links.

Nigeria Integrated Water Resources Management Commission (NIWRMC)

This was Created in 2007. They regulate and license water abstraction and wastewater discharge, enforce Integrated Water Resources Management (IWRM), resolve water-use disputes between stakeholders, develop policies, regulations, and standards for sustainable use, keep a National Water Resources Database, and encourage environmental sustainability in water development projects (Martins et al., 2024).

These have led to improved water governance and institutional coordination, Curbed groundwater overuse through regulation, improved stakeholder participation and conflict resolution, and enhanced data-driven national water planning.

The roles of both the National Water Resources Institute (NWRI) and the Nigeria Integrated Water Resources Management Commission (NIWRMC) complement each other in that NWRI offers training, data, and research while NIWRMC exercises regulation, coordination, and licensing. They complement each other and provide for sustainable and integrated water management in Nigeria.

4.3.2 State and Local Agencies

States govern potable water supply via statutory bodies and ministries of state. Rural water supply is governed by local authorities and must conduct sanitation programs as well as manage small-scale systems. Coordination between different levels of government is generally poor, exacerbated by inadequate financing as well as ill-defined delegation of regulatory powers (Samuel et al., 2021).

The role and influence of State and Local Agencies for water resources management in Nigeria supplements the NWRI and NIWRMC.

State and Local Government Agencies in Water Resources Management in Nigeria.

4.3.2.1 State Water Agencies (SWAs)

Every Nigerian state has a State Water Agency (SWA) or Water Corporation, created by the state legislature for managing and regulating water supply, sanitation, and distribution within urban and semi-urban areas (Sanjo et al., 2024). Examples include the Lagos Water Corporation, Kano State Water Board, and Kaduna State Water Corporation (Bello et al., 2021).

They are guided by their own State Ministries of Water Resources or Public Utilities and complement the federal endeavors coordinated by the Federal Ministry of Water Resources.

The Roles of State Water Agencies

1. Water Supply and Distribution
   - Design, build, and manage water treatment facilities and distribution pipes.
   - deliver safe water for domestic, industrial, and commercial use.
2. Infrastructure Development and Maintenance
   - provides pipeline, reservoir, and treatment plant maintenance.
   - Rehabilitate old water systems to maximize service delivery.
3. State-Level Policy and Regulation
   - Institute state-level water policies from national standards.
   - Establish standards for customer service, billing, and water quality.
4. Private Sector Partnerships and Involvement
   - Private firm and donor agency partnership through Public–Private Partnerships (PPP).
   - Collect funds for the development of infrastructure.
5. Monitoring and reporting

Collect data on water production, consumption, and losses for planning and accountability.

Influence of State Water Agencies

- Enhanced Urban Water Access: Enhanced treated water coverage in urban areas.
- Decentralized Administration: Increased state-level autonomy and sensitivity to local needs.

However, SWAs usually experience funding deficits, outdated infrastructure, and inefficiency that manifests in low rural coverage of services, among others.

4.3.2.2 Local Government Authorities (LGAs)

Local Governments, instituted by the 1999 Constitution (Sect. 4.7), are primarily responsible for rural water supply and sanitation.

Their role is normally carried out by the Department of Water, Sanitation, and Hygiene (WASH) or Rural Water Supply and Sanitation Agencies (RUWASSA), sometimes supported by foreign development partners like UNICEF and the World Bank.

The Functions of Local Agencies

1. Rural Water Supply
   - Increase boreholes, hand pumps, and small water schemes in rural communities.
   - Make rural water facilities sustainable.
2. Sanitation and Hygiene Promotion
   - Coordinate community sensitization on sanitation, hygiene, and water conservation.
3. Community Participation
   - Organize Water User Associations (WUAs) and community-based organizations (CBOs) for local ownership and maintenance of facilities.
4. Monitoring and Reporting
   - Monitor functionality and coverage of boreholes in rural areas.
   - Support national surveys like the National Water, Sanitation and Hygiene (WASH) Survey.

The Impact of Local Agencies

- Rural Access Improved: Increased access to water by millions of people in rural areas.
- Empowerment of Communities: Fostered local ownership and sustainable use of water systems.

They face challenges that include, but are not limited to poor funding, weak institutional capacity, and low technical manpower which limits long-term sustainability (Sanjo et al., 2024).

4.3.2.3 Coordination and Integration

Federal Level: NIWRMC, NWRI, and FMWR establish policies and regulate national frameworks.

State Level: SWAs implement policies and operate large-scale infrastructure.

Local Level: RUWASSAs and LGAs implement rural and community projects.

All these levels constitute an integrated system of water governance, although coordination problems result from overlapping mandates and low resources.

4.3.3 Other Stakeholders

Water resource management in Nigeria is not solely the government's affair. There is a very diverse range of non-state actors such as traditional leaders, citizens' associations, and non-governmental organizations (NGOs), who are also key players in ensuring community engagement, accountability, and sustainability in water management (Ugbah et al., 2017) Traditional leaders, civil society organizations, and NGOs play a key role in water governance, particularly in rural areas, where indigenous conflict resolution and collective ownership remain the foundation (Awuah & Bijimi, 2024). Private sector participation is on the rise but is hampered by regulatory intricacy as well as market entry barriers (Martins et al., 2025). Traditional leaders, civil society groups, and NGOs provide an inclusive definition of Nigeria's participatory and multi-level water governance system (Antai et al., 2025a, 2025b).

4.3.3.1 Traditional Authorities and Institutions

Trad system (situations like chiefs, village heads, community elders, and traditional rulers are the de facto managers of water in most rural and peri-urban Nigeria.

They are legitimized and empowered by communal confidence and local customs, frequently preceding formal government institutions.

Roles of Traditional Authorities and Institutions:

1. Local Governance and Conflict Resolution
   - Manage conflicts around access to water points, grazing corridors, and irrigation channels.
   - Facilitate peaceful coexistence between herders, farmers, and communities.
2. Historical Regulation of Water Use
   - Enforce local traditions of equitable use of rivers, wells, and irrigation channels.

- Regulate time and volume of domestic and agricultural water use.

3. Water Works Organization
   - Authorize community contribution (labor or money) towards the construction of water infrastructure.
   - Act as intermediary agencies between government authorities and locals.
4. Conservation of Environment and Culture
   - Enforce customary standards that conserve sacred water bodies and prevent pollution.

Impact of the Traditional Authorities and Institutions

- Increased local compliance with traditional water regulations.
- Improved prevention of conflict and social cohesion.
- Supported government programs by providing local acceptance and cooperation.

4.3.3.2 Citizen Organizations and Community-Based Organizations (CBOs)

Citizen organizations such as Water User Associations (WUAs), community water committees, and civil society coalitions are increasingly playing a role in water service delivery monitoring and holding authorities accountable.

Roles of Citizen Organizations and Community-Based Organizations (CBOs)

1. Community Ownership and Participation
   - Operate local water schemes and perform regular maintenance of boreholes and hand pumps.
   - Collect user fees and organize repairs.
2. Advocacy and Accountability
   - Engage in policy dialogue, budget tracking, and performance monitoring of water utilities.
   - Call for transparency and fair access, particularly to poor and vulnerable people.
3. Awareness and Education
   - Ensure hygiene promotion activities and empower homes on safe handling and use of water.
4. Partnership with Government and Donors

- Join in rural water scheme planning and implementation.

Impact of Citizen Organizations and Community-Based Organizations (CBOs)

- Enhanced sustainability of community water schemes with local ownership.
- Enhanced public voice and control over water management.
- Ensured women and youth-inclusive decision-making.

4.3.3.3 International Partners and Non-Governmental Organizations (NGOs)

NGOs at international and national levels have been actively involved in Nigeria's water sector for a long time. The major ones are the WaterAid Nigeria, Action Against Hunger, UNICEF, and the Global Water Partnership (GWP).

They supplement government action through finances, technical assistance, and influence on policies.

Roles of International Partners and Non-governmental Organizations (NGOs)

1. Implementation of Projects and Capacity Building
   - Build and renovate boreholes, wells, and sanitation facilities.
   - Train local government officials and technicians in water quality management.
2. Policy and Institutional Support
   - facilitate the development of national and state water policy.
   - Offer technical support for Integrated Water Resources Management (IWRM).
3. Advocacy and Rights-Based Approaches
   - Campaign for the recognition of access to safe water as a human right under the UN General Assembly Resolution (2010).
   - Advocate gender-inclusive and gender-sensitive approaches to water service delivery.
4. Research and Innovation
   - Create policy briefs, tracking reports, and impact studies to guide reform.

Impact of International Partners and Non-governmental Organizations (NGOs)

- Improved access to clean water for marginalized groups.

Table 4.1 Synergy between non-state and government actors in Nigerian water resources management

Stakeholder	Main contribution	Collaboration level
Customary authorities	Conflict resolution, local regulation	Work with LGAs, RUWASSA, and RBDAs
Citizen groups/CBOs	Operation, maintenance, monitoring	Partner with state water agencies and NGOs
NGOs/international partners	Funding, capacity building, advocacy	Collaborate with FMWR, NIWRMC, and state ministries

Note Together, these actors strengthen bottom-up participation, policy legitimacy, and sustainability in Nigeria's water resources management framework (Isukuru et al., 2024)

- Improved institutional arrangements and mechanisms of accountability.
- Improved the rights-based discourse on Nigeria's water governance (Table 4.1).

4.4 Legal Issues and Barriers

4.4.1 Fragmented Mandates and Overlapping Jurisdiction

Nigeria's multi-level water governance generates fragmented mandates and overlapping mandates, as the federal, state, and local governments tend to duplicate efforts. It does not have a specialized water regulation commission at the federal level that accounts for every element of the management, despite efforts to have the Nigeria Water Resources Regulatory Commission through recent (Omokaro et al., 2024).

4.4.2 Appropriations, Deficits, and Lack of Enforcement

Funding deficiencies are a persistent impediment to effective water resource management and infrastructure expansion. Enforcement of pollution control mandated by federal statutes remains lax, with penalties insufficient to deter offenders and statutory provisions frequently ignored. Institutionally, agencies such as the RBDAs and state water boards are historically under-resourced and subject to politics (Okeola & Balogun, 2017; Omokaro et al., 2024).

4.4.3 Customary and Statutory Rights and Issues of Land Tenure

Statutory formal rights tend to clash against customary occupancy, especially in peri-urban and rural regions. Land ownership controversies are often intertwined with water usage claims, and statutory licensing regulations occasionally challenge age-long, customary water access regime (Okeola & Balogun, 2017).

4.4.4 Licensing and Barriers to Regulation

The Water Resources Act 2004 places a requirement for licensing water use other than for irrigation or household use, a scheme that has been criticized as unwieldy and expensive and that it deters private finance. Various levels of government have authorization regimes of their own, which further induce confusion and inefficiency (Ngene et al., 2021).

4.5 Recent Policy and Legislative Developments

4.5.1 National Water Resources Bill 2020

The National Water Resources Bill 2020 is a landmark legislative initiative to harmonize the federal water laws, reducing the number of institutions and instituting new regulatory authorities to oversee and license. It sets out to establish:

- Nigeria Hydrological Services Agency (NIHSA) for technical monitoring.
- National Water Resources Institute (NWRI) for research and educational purposes.
- River Basin Development Authorities (RBDAs.
- Nigeria Water Resources Regulatory Commission (NWRRC) for economic and technical regulation, licensing, enforcement, and standard setting.

States may implement catchment control agencies of their own for intrastate rivers, and catchment control is promoted for stakeholder engagement between levels of government.

The Bill supports current policies in fostering aligned water allocation, catchment control, and a national water supply, sanitation, and hygiene fund, but it has been contended that its progressivity, enforceability, and defense of customary law rights must be strengthened. (Antai et al., 2025a, 2025b).

4.5.2 National Water Resources Policy and International Standards

The National Water Policy (2016) and the National Water Sanitation Policy (2004) articulated government intentions to fulfill Sustainable Development Goal 6—universal access to safe water and sanitation. While not legislation, they enumerate priorities for funding, cooperation, and technical standards.

Global best practice champions integrated water resources management (IWRM), cross-sectoral planning, and public–private partnership. Nigeria's NIWRMC brings the country closer to IWRM alignment and international standards.

4.5.3 State and Local Legislative Innovations

Other states enact their own water and sanitation legislation to boost coverage and regulation, among others, Cross River State Water Supply and Sanitation Law No. 5 of 2025, aimed to harmonize both rural and urban areas' service delivery.

4.6 Socio-Legal Perspective of Access and Equity

4.6.1 Disparity in Access to Water

Access to safe water in Nigeria is extremely unequal, and rural residents, slum inhabitants of cities, and minority groups are severely restricted. Barriers to access include geographical remoteness, economic marginalization, and political abandonment.

4.6.2 Gender, Poverty, and Vulnerable Groups

Women, children, and other vulnerable groups are disproportionately hit by water insecurity since they dedicate a disproportionate time and energy to ensuring water for household use. Socio-legal responses have to be responsive to the groups in law, policy, and program.

4.6.3 Regulations and Private Sector Role

Muddled licensing regimes, uncertainty in regulation, and expensive permits restrict participation and investment in water infrastructure by the private sector. Simplification of regulation through reform stimulates entry and sustainable service provision of public–private partnerships.

4.6.4 Customary Law and Participation of Community Stakeholders

The traditional and local means of resolving conflict and the local customary use of water remain paramount to how allocations are determined and disputes are resolved in most communities. Successful law reform must entail local customary and stakeholder engagement throughout. (Ugbah et al., 2017).

4.7 Conclusion

Nigeria's water resources management legal framework is ambitious in design but inconsistent in application. Overlapping competencies, non-enforcement, and funding and stakeholder shortcomings prevent the realization of equitable water access and sustainable resource harnessing. Recent policy and legislation evolutions, e.g., the National Water Resources Bill 2020 and IWRM implementation, show improved guidance, but must address areas of enforceability deficit, customary safeguards, and regulation effectiveness. Building a harmonized, inclusive, and legally enabled water resources management system will be key to ensuring long-term water security, public health, and national progress.

4.8 Recommendation

4.8.1 Integrated Water Resources Management (IWRM)

Adoption of IWRM principles of cross-sectoral planning, catchment-based resource management, and multi-stakeholder engagement will enhance system resilience, efficiency, and legal certainty.

4.8.2 Institutional and Regulatory Enhancement

Nigeria must create a free-standing, well-funded regulator commission that has end-to-end authority across all levels, the federal, the state, and the local. It will harmonize licensing, enforce compliance, and establish standards.

4.8.3 Enhancing Justiciability and Accountability

Constitutional environmental duties must be made justiciable, obliging courts to adjudicate and enforce state responsibilities and citizens' rights to water and sanitation. Sanctions for pollution and rule-violating breaches ought to be enhanced, and public interest litigation eased.

4.8.4 Participation of Stakeholders and Safeguarding Customary Rights

Legislative reform has to affirm and validate customary water rights and engage local authorities and civil society players in planning, conflict resolution, and regulation.

4.8.5 Private Sector Funding, Engagement, and Equitable Distribution

Long-term financing requires innovative funding methods, such as public–private partnerships, donor finance, and ring-fenced governmental spending. Licensing regimes have to be streamlined and simplified, and sector-specific incentives for private finance.

4.8.6 Adaptation to Climate Change and Sustainability

Legislative procedures need to internalize climate flexibility, water conservation incentives, and disaster protection to enable water systems to become resistant to drought, floods, and contamination (Constitution of the Federal Republic of Nigeria, 1999a, b; Emmanuel, 2012; Lakai, 2020; Nwoba, 2022; Oke, 2017; Oyeronke, 2025; Yusuf, 2022).

References

Akinsulore, A. O. (2024). Instream flow rights for protecting river ecology in Nigeria's bitumen zone: Learning from Uganda's experience. *REPLIK: Journal of Environmental Law and Policy.*

Antai, G. O., Aidonojie, P. A., Onyejegbu, D. C., & Obisesan, O. O. (2025a). International environmental crimes: An appraisal of the impact of enforcement efforts and policy response on illegal wildlife trafficking. *African Journal of Sociological and Psychological Studies*, *5*(1).

Antai, G. O., Obisesan, O. O., Umo, M. E., & Okonji, C. I. (2025b). Analyses of the applicability of international arbitration in investor-state dispute settlement in East Africa. *NIU Journal of Humanities, 10*(1), 249–259.

Awuah, K. G. B., & Bijimi, C. K. (2024). Discovering the core stakeholders in the Nigerian urban water supply system. *Sustainable Water Resources Management, 10*(1), 8.

Balogun, E. (2022). Harnessing Nigeria's huge agricultural potentials for food security: A roadmap. *Journal of Economics and Policy Analysis.*

Barua, A., & Tongper, K. (2019). Water conflict, law and governance. In: J. Holden (Ed.), *Water resources: An integrated approach* (2nd ed., pp. 177–190). Routledge.

Bello, N. I., Imam, M. Z., Adamu, H., & Abubakar, A. S. (2021). Overview of domestic water supply in Kano State, Nigeria. *International Journal of Geography and Geography Education, 44*, 1–11.

Chukwuma, O. M., Obeta, C. O., & Obeta, M. C. (2017). Patterns, problems and options for improved domestic water supplies to the rural communities of Enugu State, Nigeria. *Journal of Agricultural Extension and Rural Development, 9*(2), 52–58.

Constitution of the Federal Republic of Nigeria. (1999a). Sch. 2, Pt. 1, item 64.

Constitution of the Federal Republic of Nigeria. (1999b). (as amended), s. 20.

Emmanuel, M. A. (2012). Water meanings, sanitation practices and hygiene behaviours in the cultural mirror: A perspective from Nigeria. *Journal of Water, Sanitation and Hygiene for Development, 2*(3), 168–181.

Gana, B. A., Abdulkadir, I. F., Musa, H., & Garba, T. (2019). A conceptual framework for organisation of river basin development and management in Nigeria. *European Journal of Engineering Research and Science, 4*(6), 34–40.

Igbayiloye, O. B., & Niyi-Gafar, O. L. (2025). Water resource protection in Africa's mining sector: A Nigerian perspective. *Turf Law Journal, 5*, 1–15.

Ikpeze, N. G. (2014). Safe disposal of municipal wastes in Nigeria: Perspectives on a rights-based approach. *Journal of Sustainable Development Law and Policy, 3*(1), 72–86.

Isukuru, E. J., Opha, J. O., Isaiah, O. W., Orovwighose, B., & Emmanuel, S. S. (2024). Nigeria's water crisis: Abundant water, polluted access, and policy challenges. Cleaner Water.

Ladan, M. T. (2013). Review of the regulatory and policy frameworks of the Nigerian water sector and international legal response for sustainable management and protection of water resources.

Lakai, L. I. (2020). Public water and waste management in Nigeria: Legal framework, obstacles and challenges. *KAS African Law Study Library/librairie Africaine D'études Juridiques, 7*(4), 631–641.

Martins, G., Ehinmentan, B., Mafimisebi, P., America, A., & Gbadero, G. (2025). Sustainable water resource management in Nigeria: Challenges, integrated water resource management implementation, and national development. *International Journal of Trendy Research in Engineering and Technology.*

Martins, G., Ehinmentan, B., Mafimisebi, P., America, A., & Gbadero, G. (2024). Sustainable water resource management in Nigeria: Challenges, integrated water resource management implementation, and national development. *International Journal of Trendy Research in Engineering and Technology, 9*(1), 8–16.

Ngene, B. U., Nwafor, C. O., Bamigboye, G. O., Ogbiye, A. S., Ogundare, J. O., & Akpan, V. E. (2021). Assessment of water resources development and exploitation in Nigeria: A review of integrated water resources management approach. *Heliyon, 7*(1), Article e05955.

Nwankwoala, H. O., Ngah, S. A., & Ogbogu, S. A. (2014). Groundwater resources of the Niger delta: Quality implications and management considerations. *International Journal of Water Resources and Environmental Engineering, 6*(5), 155–163.

Nwoba, S. (2022). Water policy reform in the Nigeria water governance system: Assessment of water resources management based on OECD principles on water governance. *Water Policy, 24*(10), 1704–1722.

Oke, T. (2017). Ownership and control of natural resources under the Nigerian constitution 1999 and its implications for environmental law and practice. *International Law Research, 6*(1), 162.

Okeola, O. G., & Balogun, O. S. (2017). Challenges and contradictions in Nigeria's water resources policy development: A critical review. *AFRREV STECH: An International Journal of Science and Technology, 6*(1), 1–19.

Okorie, O. (2019). Re-examining the protection of the environment under the 1999 constitution and international instruments. *Developing Country Studies, 9*(9), 1–11.

Omokaro, G. O., Idama, V., Airueghian, E., & Michael, I. (2024). Water resources, pollution, integrated management and practices in Nigeria—An overview. *American Journal of Environmental Economics, 3*(1), 5–18.

Oyeronke, O. O. (2025). Flooding in Nigeria: An assessment of the laws and institutions. *Beijing Law Review, 16*(2), 908–926.

Samuel, K. J., Agbola, S. B., & Olojede, O. A. (2021). Local governance and the crisis of water and sanitation provision in medium-sized urban centres: Evidence from three cities in Nigeria. *Local Economy, 36*(2), 164–177.

Sanjo, K. A., Chinade, A. A., & Abdulsamad, A. (2024). Water governance: Challenges and prospects for the implementation of sustainable development goal six in Nigeria. *African Journal of Environmental Sciences and Renewable Energy, 14*(1), 86–109.

Shoko, E., & Naidu, M. (2018). Peace-based informal practices around shared communal water resources in Tyrone Village of Mhondoro-Ngezi, Zimbabwe. *International Journal of African Renaissance Studies, 13*, 77–91.

Singh, A. K., & Roy, S. K. (2024). A study of the concept and legal principles of water management. *Journal of Advances and Scholarly Researches in Allied Education, 21*(2), 62–71. https://doi.org/10.29070/mgn02g48

Ugbah, R., Meldrum, A., & Ehiwario, K. (2017). Water access and community engagement: Creating the right environment for maximizing the benefits of community engagement processes and increasing water participation in Nigeria. *WIT Transactions on Ecology and the Environment, 220*, 129–139.

Ukpai, S. N. (2022). Water policy reform in the Nigeria water governance system: Assessment of water resources management based on OECD principles on water governance. *Water Policy, 24*(10), 1704–1722.

Yusuf, T. (2022). Local governance and the crisis of water and sanitation provision in medium-sized urban centres: Evidence from three cities in Nigeria. *H2Open Journal, 7*(1), 1–11.

Part II
Water Access and Sanitation

Chapter 5
Taps, Toilets, and a Thirsting Nation: Unveiling Nigeria's Water and Sanitation Crisis

Enovwo E. Odjegba, Mariam Iyabo Adeoba, and Ayodetimi Omoniyi

Abstract Nigeria is facing a grave water and sanitation crisis, with an alarming paradox: amid plenty of water resources, Nigeria has over 71 million people lacking access to safe drinking water and over 130 million without decent sanitation. This chapter delves into the intricate causes surrounding this crisis, examining water governance issues and their indirect effects on public health that have continuously perpetuated a culture of inadequate service provision across Africa's most populous country. Waterborne diseases including cholera, typhoid fever and dysentery afflict population at risk groups, with a heavy public health burden on Nigeria as the disease results cause an economic loss of 1.3% of Nigeria's GDP every year. These health problems are made worse by cultural norms, behavioral characteristics, and inadequate infrastructure, particularly for women and children who bear the primary burden of collecting and handling water. Institutional and governance shortcomings emerge as crucial roadblocks to progress. Coordination problems, project delays, and budget overruns have resulted from the fragmented institutional structure, which is characterized by unclear roles for different levels of government. Historical patterns in the implementation of water projects show persistent issues that endanger the longevity of infrastructure, including a lack of finance, a lack of technical expertise, procurement concerns, and inadequate maintenance systems. Notwithstanding these challenges, the chapter emphasizes emerging options that hold potential for change, such as community-driven projects, technological breakthroughs, and reform initiatives. Scalable frameworks for sustainable service delivery are offered by successful

E. E. Odjegba (✉)
Department of Water Resources Management and Agrometeorology, Federal University of Agriculture, Abeokuta, Nigeria
e-mail: odjegbaee@funaab.edu.ng

M. I. Adeoba
Department of Mechanical, Bioresources and Biomedical Engineering, University of South Africa, Pretoria, South Africa
e-mail: mariamsalami@yahoo.co.uk

A. Omoniyi
Centre of Excellence in Agricultural Development and Sustainable Environment (CEADESE), Federal University of Agriculture, Abeokuta, Nigeria
e-mail: omoniyiam@funaab.edu.ng

M. I. Adeoba et al. (eds.), *Water Resources Management in Nigeria*, World Water Resources 10, https://doi.org/10.1007/978-3-032-22813-0_5

projects like solar-powered water systems and Community-Led Total Sanitation programs. The study finds that attaining universal access to water and sanitation necessitates holistic strategies that merge enhanced governance, inventive financing methods, community involvement, and climate-adaptive infrastructure advancement. Despite significant challenges, Nigeria's human resources and natural wealth lay the groundwork for the systemic transformations needed to guarantee every Nigerian has access to safe water and sufficient sanitation services vital for health, dignity, and progress.

Keywords Water · Sanitation · Waterborne diseases · Public health · Water governance

5.1 Introduction

As the Sustainable Development Goals 2030 targets draws near, Nigeria as a signatory still lags in her efforts at meeting targets in the areas of water, sanitation, health and food security, to mention a few (Shehu & Nazim, 2022). Nigeria's expansive urban centres have limited or complete lack of access to potable water due to poor functioning or non-function public utilities, while its rural regions rely on contaminated streams and hand-dug wells.

In their National Outcome Routine Mapping (WASHNORM) report for 2021, FMWR/NBS/UNICEF (2022) asserted that about 87% of Nigeria lack access to safe drinking water, with Southwestern (29%) and Northeastern (2%) Nigeria recording the highest and lowest percentage access respectively. The disparity is similar for urban and rural populations which are set at 27% and 2% respectively. Access to safe drinking water is also linked to poverty as poor families are 17 times less likely to have access to safe water when compared to rich households. The effect of poverty equally extends to sanitation as poor families sometimes lack access to toilet facilities or available facilities are not adequate or gender segregated, resulting in open defecation.

To surmount this crisis, Nigeria must adopt an integrated and people-oriented approach that would strengthen institutional capacity, improve funding mechanisms, and prioritise equity (DAI, 2021; UNICEF, 2022). In addition, behavioral change and community engagement must take the centre stage (ISWSC, 2023; News Agency of Nigeria, 2025). Climate-resilient strategies have to be embedded in water resources planning to address environmental stressors (Adesogan & Sasanya, 2025; Akinwale & Udochukwu, 2025). In the absence of urgent and coordinated action, the goal of achieving universal access to water and sanitation will remain elusive (Okorafor et al., 2024; UNICEF, 2023). While the journey ahead may be challenging, achieving access to water and sanitation is not impossible. This chapter, therefore, explores the water and sanitation crisis in Nigeria by providing information on the status of the crisis, the health implications of water insecurity and poor sanitation, infrastructural decay, impacts of climate change and the way forward. The objective of this chapter is to critically examine Nigeria's water and sanitation crisis by exploring its underlying causes, systemic governance challenges, and far-reaching public health implications. It aims to provide a comprehensive analysis of the current state of water access and sanitation services, identify infrastructural and institutional gaps, and assess the socio-economic and cultural factors that perpetuate inequities. Furthermore, the chapter seeks to highlight emerging solutions, including community-driven initiatives, technological innovations, and climate-resilient strategies, while offering actionable recommendations for achieving universal access to safe water and sanitation in alignment with the Sustainable Development Goals.

5.2 Current State of Water Access in Nigeria

The water supply situation in Nigeria is a mixed one of progress and persistent challenges. According to the current report of the Nigerian Bureau of Statistics and other monitoring agencies, access to basic water across the Nigerian population stands at about 70%, however its figures do not reflect the reality of different regions of the country and urban and rural populations of almost 200 million people (UNICEF Nigeria, 2023; Water Supply and Sanitation Collaborative Council, 2024). An overview of Nigeria's water supply chain, including key stages and institutional actors, is presented in Fig. 5.1.

In big cities, such as Lagos, Abuja, Kano, and Port Harcourt, water supply systems are typically operated by the state water corporations/agencies (Adebayo et al., 2024a,b). These are the utility nodes of treatment and distribution of utilities, serving varying proportions of urban residents. For example, Lagos State has made massive investments in water infrastructure, like the Lagos water corporation reforms and the socialisation of a public-private partnership, resulting in improved water service delivery to several million residents in the state (Agboola et al., 2023a,b). However, according to reports, even in well-served urban areas, there are still challenges of old water infrastructure, non-revenue water losses and inadequate coverage of peri-urban settlements (Ogundimu et al., 2024).

The rural water supply sector is more problematic since the responsibilities of delivering the services are shared by several agencies and different levels of government (Bashir et al., 2024). The rural population often depends on boreholes, hand-dug wells, and small-scale water systems that could be operated by the community-based organisations, local governmental bodies, or supported by development partners (Adeleke et al., 2023a,b). Improvement and sustainability of the rural water systems is also significant, since most of the systems are malfunctioning due to poor maintenance, unavailability of spares and lack of community ownership (Danladi et al., 2024a,b).

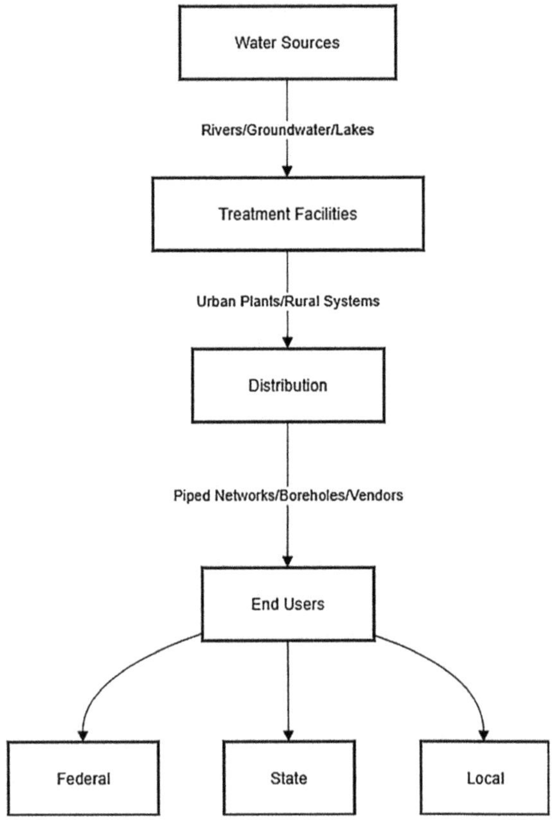

Fig. 5.1 Overview of Nigeria's water supply chain depicting the stages from water sources to treatment, distribution, and end-user access, along with key institutional actors and service delivery pathways

5.2.1 Urban Water Supply Systems

In the biggest cities, formal water utilities comprising of treatment plants, distribution, and customer services systems are available, although coverage and reliability can be quite weak (Adebayo et al., 2024a,b). Other emergent riots have been witnessed through the restructuring of the Lagos State Water Corporation, i.e., corporatization, performance contracts, and privatized partnerships, which have enhanced the delivery of water supply to parts of the metropolis (Idowu et al., 2023a,b).

Abuja, as the federal capital, enjoyed the special investments by the federal government in water facilities. The eight treatment plants and distribution systems organised by the Federal Capital Territory Water Board serve the centre and the satellite towns (Balogun & Islam, 2025). The current trends of rapid population growth and expansion of urban areas have strained the ability of existing systems, causing the water to be supplied irregularly, and the use of water vendors has become more profound in some places (Nwachukwu et al., 2024).

Table 5.1 State-level water project implementation performance in Nigeria showing the number of active projects, completion rates, average delays, budget overruns, and key success factors across selected states adopted from Adebayo et al. (2024a,b); Lagos Water Corporation (2023)

State/utility	Population served	Coverage (%)	Water production (MLD)	Non-revenue water (%)	Tariff (₦/m^3)
Lagos	12.5 million	68	540	35	158
FCT Abuja	2.8 million	72	180	28	142
Kano	1.9 million	45	95	42	89
Rivers	1.2 million	52	78	38	125
Oyo	1.8 million	41	85	45	95
Kaduna	1.1 million	48	72	40	102

Some of the secondary cities, including Kano, Ibadan, Port Harcourt and Enugu, have state water agencies with different levels of proficiency and coverage (Kaine et al., 2024; Ogbonna et al., 2023). These networks tend to experience the burdens of old pipelines, insufficient budgets to build and rehabilitate and all the related copious non-revenue water attributed to the leakages and illegal connections. Performance indicators for selected urban water utilities, including coverage, production capacity, and non-revenue water, are summarized in Table 5.1.

The informal water sector is of vital importance to the urban water supply, both in settlements of the low-income bracket and in settlements where public services are weak. Trucks, carts, or motorcycles are used in water vending as a water vending practice acquires vital services yet at increased unit prices compared to piped water (Eboh et al., 2023a,b). Sachet water manufacturing and distribution have increased exponentially, offering inexpensive drinking water alternatives, but creating issues of control and quality and environmental implications based on packaging waste abundance (Amadi et al., 2023).

5.2.2 *Rural Water Supply Dynamics*

Rural water supply in Nigeria comprises a large variety of technologies, organizational structures, and service delivery levels, which are determined by local situation, community choices and accessible resources (Bashir, et al., 2024). Hand-dug wells are the most widespread water point in numerous localities that were frequently developed by communities with the help of government agencies, non-governmental organizations or development partners (Garba et al., 2024a,b). Such systems are usually found among individual homes or even small neighborhoods and do not need a lot of technical knowledge to use and maintain.

The use of boreholes has become common in the rural set up, especially where groundwater can be found at suitable depths (Danladi et al.,2024a,b). Solar and hand-pump systems are advantageous because they are environmentally friendly and will not rely on electricity or fuel (Adesina et al., 2024a,b). Nonetheless, the sustainability of boreholes is based on issues such as the geological context, the capacity of the communities to manage project, accessibility to spare parts and technical assistance to carry out repairs.

An intermediate area falling between the rural point sources and urban networks is the small-town water systems. They often characterized systems with limitations of people with 2000 to 20,000 population, often covered by small plants, tank elevated storage systems, and limited distribution systems (Bashir et al., 2024). The management differs, with the commonest being a local government control, where the community has some inputs in the operation and maintaining portions.

Water committees, user associations, and cooperative arrangements will offer a means of collection of user fees, maintenance coordination, and a means of intermediary with external support agencies. Such arrangements are bound to succeed only based on social cohesion, quality of leadership, and continuous technical and financial assistance by government or other development partners (Adeleke et al., 2023a,b).

5.2.3 Water Quality and Treatment Challenges

The quality of water is another vital aspect of the Nigerian water supply situation, which has huge relevance to the health sector and development prospects (Agunwamba et al., 2024). Lakes and rivers are among surface water sources that are prone to contamination due to the impact of agricultural runners, industrial effluents, domestic sewage, and garbage dumps (Oladoja, et al., 2024). In spite of the fact that rivers Niger and Benue are sources of the major portion of water, their condition is affected by different grades of pollution which are the most notable close to such areas of population concentration or industrial production as in the first case, and near the largest cities, as in the case of the Benue River (Nigerian Hydrological Services Agency, 2023).

Shallow wells are potentially at risk of surface contamination, whereas deeper or longer boreholes are susceptible to naturally occurring minerals such as iron, fluoride, and salts that lead to change in taste and safety (Adamu et al., 2023; Aliyu, et al., 2024). There are high fluoride levels that may demand special treatment in some northern regions and the problem of saltwater intrusion in some coastal areas (Udoh et al., 2024a,b).

Conventional treatment plants used in most major cities allow use of coagulation, sedimentation, filtration, and disinfection treatment, although aging systems and lack of maintenance impact negative performance. Smaller systems can employ simpler treatment techniques or simpler and basic disinfection systems. Environmental-related household practices involving water treatment and safe storage protocol differ significantly between communities and socioeconomic strata (Okafor et al., 2023). Although point-of-use filters, boiling, and chemical disinfection are used by some households in urban centres, others in rural places do not have access to such forms of treatment.

5.3 Economic Dimensions of Water Supply

Financing mechanisms, cost recovery methods, affordability considerations, and wider economic implications of water services are other areas seen as economical in relation to the water supply situation in Nigeria. Water infrastructure is mostly financed by the public sector as the national and state governments use budget funds to support capital projects and subsidize operations (Federal Ministry of Water Resources, 2023). Nevertheless, in many cases, funds are inadequately allocated to the sector vis-a-vis the needs that each sector requires because of budget limitations and conflict of priorities.

Participation by the private sector in water supply is an activity that has developed through different models including management contracts, concession, and public private partnerships (Ajayi et al., 2024). Full privatization has not been adopted, although some states have used private operators to carry out selected roles (e.g. billing, revenue collection, and providing technical services). There is a considerable amount of small and privately owned contributions of the total water supply by small-scale water vendors and borehole drillers (Eboh et al., 2023a,b).

Water tariff policies endeavour to strike an equilibrium between the cost recovery and affordable aspects, but the practice varies greatly across various utilities and services (Adebayo et al., 2024a,b). Many urban water utilities use block tariff structures, based on which the low-rate subsidy is given to low consumption groups and high rates are given to high consumers. Nevertheless, tariff collection efficiency is a problem in most systems, restricting resources to expand and improve the systems.

5.4 Health on the Line: Waterborne Diseases and Public Health

Waterborne diseases remain a critical public health issue in Nigeria, significantly contributing to illness and death, particularly among children and the elderly. These diseases, caused by pathogenic microorganisms such as bacteria, protozoa, and viruses, are primarily spread through contaminated water sources, resulting in illnesses like cholera, typhoid fever, and hepatitis A and E (Okesanya et al., 2024). The major transmission routes and contamination pathways are illustrated in Fig. 5.2. Despite progress in improving childhood survival rates, Nigeria bears much of the global disease burden, especially in maternal and child health (Asuquo et al., 2025).

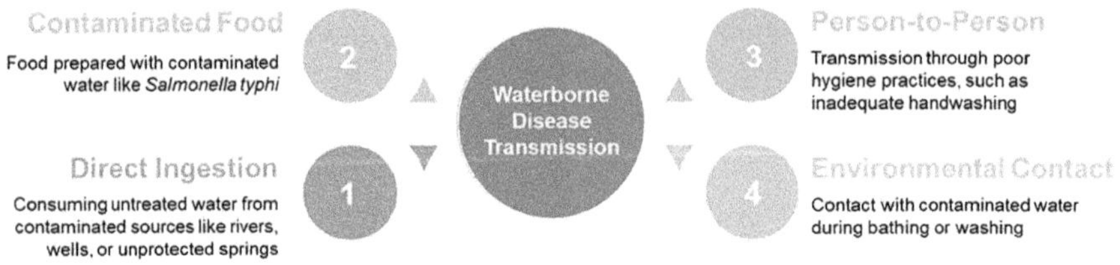

Fig. 5.2 Transmission pathways of waterborne diseases in Nigeria showing the fecal-oral route and key contamination sources such as open defecation, inadequate sewage systems, and flooding, which facilitate pathogen entry into water supplies

The most prevalent illnesses include dysentery, diarrhea, cholera, typhoid fever, and schistosomiasis, all of which contribute significantly to the national disease burden (Asuquo et al., 2025; Okpasuo et al., 2019). Waterborne diseases in Nigeria mainly spread through the fecal-oral route, with polluted water sources acting as the main medium for pathogen transmission. Rivers, lakes, and wells are key sources of contamination due to fecal waste from open defecation, inadequate sewage systems, and flooding. Heavy rains and flooding, common in Nigeria, worsen contamination by washing fecal matter into water bodies (Liu et al., 2024). Poor water treatment practices keep the cycle of infection ongoing, as pathogens like HAV can survive in water for weeks (Liu et al., 2024). Effective measures such as point-of-use water treatment (like boiling or chlorination) greatly decrease the incidence of disease (Pooi & Ng, 2018).

5.4.1 *Ripple Effects: How Unsafe Water Undermines Health, Equity, and Development*

Waterborne illnesses are a pervasive threat in Nigeria, affecting over 75% of the population across all age groups (Abdulkadir, 2018; Asuquo et al., 2025). Children, particularly those under five, bear the brunt of diseases like cholera and dysentery, which claim countless lives annually. Young Nigerians face significant health vulnerabilities, with repeated exposure to waterborne diseases often resulting in malnutrition, delayed physical development, and impaired cognitive function—all of which limit their long-term potential. The situation is even more dire in rural areas, where only about 31% of the population has access to clean drinking water. This lack of access intensifies regional health disparities and leaves rural communities disproportionately affected.

Elderly individuals and those with compromised immune systems are especially vulnerable to complications from waterborne illnesses, placing additional strain on Nigeria's already overstretched healthcare system. Medical facilities in rural zones often lack adequate resources, such as hospital beds, essential supplies, and trained personnel, forcing them to prioritize water-related illnesses over other critical services like maternal health and chronic disease care. Many patients are compelled to travel long distances to urban hospitals, delaying treatment and increasing the likelihood of fatal outcomes. Meanwhile, healthcare workers face overwhelming workloads, leading to burnout and a decline in the quality of care, along with a heightened risk of hospital-acquired infections.

The impact of unsafe water reaches beyond immediate health concerns, reinforcing cycles of poverty and social inequality. Children who survive these illnesses frequently experience long-term setbacks, including missed school days and diminished academic performance, which limit their prospects and perpetuate generational poverty. In rural communities, where sanitation infrastructure is often lacking, these challenges are even more severe, trapping families in persistent socioeconomic hardship.

Adults working in physically demanding sectors like agriculture suffer productivity losses due to prolonged illness, particularly from diseases such as typhoid fever. Fatigue and absenteeism reduce their economic output, with subsistence farmers bearing the brunt of these effects. Women, who are typically responsible for collecting water and caring for sick family members, lose valuable time that could otherwise be spent on education, income-generating activities, or civic engagement. This dynamic further entrenches gender disparities, restricting women's ability to contribute meaningfully to household welfare and societal development. The elderly, with weakened immune systems, face heightened risks of severe illness and death from waterborne diseases. Their loss disrupts family support networks, leaving households without critical emotional and economic anchors. In low-income communities, frequent illness fosters social stigma, as families focus on survival rather than collective well-being, weakening community cohesion.

The economic toll of unsafe water is staggering, costing Nigeria about 1.3% of its GDP annually WHO, (n.d). Direct costs (diagnostics, medications, and hospitalizations) place a heavy burden on households, particularly in rural areas with limited healthcare access. For example, treating severe cases of typhoid can consume a significant portion of a family's income, pushing them deeper into poverty. Indirect costs, such as lost wages from illness or caregiving, further strain low-income households, with some spending up to 20% of their income on water and healthcare (Abdulkadir, 2018). These costs divert resources from critical sectors like education, infrastructure, and poverty alleviation, undermining Nigeria's progress toward the Sustainable Development Goals, particularly those related to health (SDG 3) and clean water (SDG 6). Frequent cholera outbreaks, worsened by events like the 2024 floods (UN OCHA, 2024), strain public budgets and deter foreign investment and tourism, limiting economic growth. Rural clinics, often lacking basic diagnostic tools and essential medicines, struggle to manage outbreaks, further hampering development efforts.

5.4.2 Behavioural and Cultural Factors Influencing Public Health in Nigeria

Traditional practices, varying levels of public awareness, and community engagement can either hinder or advance efforts to improve public health. In many rural communities, particularly in northern Nigeria, open defecation is widely practiced, with nearly a quarter of the population (about 46 million people) engaging in it due to cultural acceptance or convenience (WHO, 2023). In some Hausa communities, for instance, cultural norms discourage latrine use due to beliefs that defecation in enclosed spaces is impure or associated with spiritual taboos (Akpabio, 2012). This leads to fecal contamination of water sources like rivers and wells, fueling outbreaks of cholera and dysentery.

In southern Nigeria, cultural reverence for rivers complicates safe water use. Many communities view rivers as sacred, using them for both spiritual rituals and domestic needs, despite contamination risks. This practice contributes to high rates of diseases like hepatitis A, particularly in areas like Cross River State. Gender norms further exacerbate the issue, as women, who collect water in 80% of rural households, often prioritize nearby sources over safer ones to save time, increasing exposure to contaminated water. In urban slums, resistance to shared sanitation facilities due to privacy concerns perpetuates open defecation. Addressing these deeply rooted practices requires culturally sensitive interventions that respect local traditions while promoting safe water and sanitation habits.

Although some effective campaigns have leveraged local media and community leaders to drive behaviour change. For example, radio campaigns in northern Nigeria have boosted handwashing rates by 15% via engaging audiences in local languages (Oyerinde & Jacobs, 2022). However, many campaigns fall short due to a lack of cultural relevance or inadequate reach. Generic messaging, such as English-language billboards, often fails to resonate with rural populations who have low literacy or limited access to urban media. A national cholera prevention campaign, for instance, had minimal impact in northern states because it overlooked local beliefs about water sources (Akpabio, 2012). One-off campaigns also struggle to sustain change. A handwashing initiative in Lagos saw initial success but faded without ongoing reinforcement, as communities reverted to old habits. To be effective, campaigns must be tailored to local contexts, delivered in local languages, and supported by continuous engagement and monitoring to ensure lasting changes in behaviour that reduce disease prevalence.

These successes highlight the power of participatory approaches in driving health improvements (Alabi et al., 2024). However, initiatives lacking community buy-in often fail. In northern Nigeria, development projects introduced without community involvement, such as the installation of standardized latrines often encounter resistance. This pushback typically stems from cultural disconnects or the absence of local engagement in the planning process. Another notable example is a borehole initiative in Borno State that failed within a year due to a lack of training for residents on how to maintain the system, ultimately resulting in water contamination (Akpabio, 2012).

For interventions to be sustainable, they must be inclusively designed and executed with active community participation to ensure they reflect local values, needs, and capacities.

Despite these challenges, there is a clear opportunity for transformation. By adopting comprehensive policies, empowering communities, and fostering innovative collaborations, Nigeria can disrupt the cycle of disease and inequality. A few key steps that could serve as a foundation for long-term public health improvements include:

1. Increasing Investment in WASH
2. Strengthening Governance and Collaboration
3. Adopt Climate-Resilient Strategies
4. Regulate Water Vendors and Waste Management
5. Sustain Culturally Sensitive Education.

5.5 Infrastructure Interrupted: Challenges in Water Project Implementation

The Nigerian water infrastructure development has been characterised by grandiose plans, huge investments and endemic delays in implementation that have typified the sector over the past several decades (Bureau of Public Procurement, 2024). Since independence in 1960, various governments have launched a combination of water supply projects that have ranged in success, with large-scale urban treatment plants, rural borehole projects, and uneven success rates, functionality and sustainability (National Assembly, 2023). The water infrastructure sector of the country is plagued by decayed projects, half-baked projects, and underutilised systems below their capacity due to system-wide failure in project design, costing and funding, construction and operationalisation, and maintenance (Water Resources Development Consortium, 2023).

The National Water Supply Policy sanitation framework establishes the general policy framework to guide the development of water infrastructure by calling on organized synergies among the various levels of government and the sectors (Federal Ministry of Budget and National Planning, 2023). Although the same policy intention can be translated into functional infrastructure, the realisation has been a problem as implementation gaps have manifested due to innumerable technical, financial, institutional, and socio-political factors (Olajide et al., 2024). All these challenges have led to Nigeria's inability to effectively meet its international standards of water supply provision and failure to provide millions of its citizens with suitable water services.

The task of implementing water projects in Nigeria is carried out by various stakeholders, which include state governments, federal ministries, local government areas, development partners and individuals in the private sector and community organisations (Infrastructure Concession Regulatory Commission, 2023). Such a complicated institutional landscape provides the conditions under which coordination failures, duplication of efforts, conflicting priorities, as well as gaps in accountability may occur, thereby derailing the project's success. The absence of standardisation of the procedures, quality control, and performance tracking systems has also contributed to the lack of changes in improving implementation outcomes (Danladi et al., 2024a,b).

Water projects that failed or have not been completed have economic costs that range beyond wasted financial levy to the missed chances of successful economic impact, trust of the government programs by the citizens, and the continued usage of the alternative water sources by the citizens and the businesses that need them (Jega et al., 2024). It is imperative to understand the patterns and drivers of implementation difficulties to have a better way of improving implementation efforts and transforming the performance of the sector in Nigeria.

5.5.1 Historical Patterns of Project Implementation

The history of water project implementation in Nigeria shows that the same source of challenges has been present in numerous political regimes and eras of development (Water Resources Development Consortium 2023). During the oil boom in the 1970s, many ambitious projects to develop water infrastructure were begun with large government financing, which, however, had substantial delays, cost overruns and quality problems that jeopardised the longer-term viability of the project. The increasing number of projects, which saw the construction of the Greater Lagos Water Supply Scheme and other state capital water systems, all supported the possible problems of dealing with such large-scale infrastructure development in a dynamic economic and political region (Idowu et al., 2023a,b).

The structural adjustment of the 1980s and 1990s imposed further limitations because of the low levels of government expenditure and depreciation of the currency that impacted on the possibility of importing the required equipment and keeping the existing projects underway (Maina et al., 2023). Some projects that were launched during this term were either shelved or took a very long time before they were able to materialise due to a lack of finance to continue with them. The movement to donor-funded operations during this period came with new types of implementation models and gave rise to the challenge of harmonising governmental agencies and international development agencies (World Bank 2024).

The shift to democratic rule in 1999 saw a renewed directive on water infrastructure development, and subsequent administrations embarked on high-profile water projects as a developmental agenda (Ezenwa et al., 2023a,b). The desire on the part of new regimes to do away with the projects of the previous administration has, however, formed a trend of instability in these facilities as well as wastage of investment. Such political discontinuity has been especially problematic with long-term infrastructure projects that need a long-lasting commitment spanning electoral cycles. A summary of major water infrastructure projects, including their planned capacities, budgets, completion status, and key challenges, is presented in Table 5.2.

Table 5.2 Major water infrastructure projects status analysis

Project name	State/location	Planned capacity	Budget (₦ Billion)	Completion status (%)	Key challenges
Gurara water project phase II	FCT/Niger	200 MLD	85.6	65	Funding gaps, contractor disputes
Kano regional water supply	Kano	150 MLD	72.3	45	Land acquisition, technical issues
Ogun-Oshun River basin	Multi-state	300 MLD	124.8	30	Inter-state coordination, funding
Lagos fourth mainland bridge water	Lagos	180 MLD	96.4	78	Environmental clearance delays
Port Harcourt water expansion	Rivers	120 MLD	58.9	52	Security concerns, cost escalation
Kaduna water supply upgrade	Kaduna	100 MLD	45.7	88	Equipment procurement delays

Source Adapted from National Assembly (2023); River Basin Development Authorities (2023)

Significance of the implementation issues has been noted over the past decades, with attempts to enhance project management systems being made, and institutional capacities being boosted, and more realistic approaches to planning being adopted (Ogundimu et al., 2023). The aftermath of partially completed and unsuccessful projects has remained prevalent and influential in terms of affecting sector credibility and making spread allocations decisions.

5.5.2 *Financing and Funding Challenges*

The issue of financial constraints is one of the most enduring problems with the implementation of water projects in Nigeria that seems to affect both small and large projects, irrespective of the modalities through which they are executed (Jega et al., 2024). The allocation of funds to water infrastructure, by government, has been historically undersigned relative to those needed in the sector, meaning that limited available government resources have often led to the erosion of an appropriate amount of funding to be allocated to both capital projects and operational needs (Federal Ministry of Budget and National Planning, 2023). The structure of the budget allocation is slanted towards initiation of new projects rather than completion of older projects leading to the underutilized facilities.

The development partners/international financial institutions have significant amounts of external financing to be used in water infrastructure development, though they come with a lot of strings and requirements on how, when, and in what ways to implement the finances, which have not always coincided with local conditions and capacity (World Bank, 2024). The costs of projects have been impacted by the exchange rate and inflation rates because projects whose components have larger imported elements, such as treatment equipment and special materials, have a (negative) effect on the project budgets in real terms (Maina et al., 2023).

The lack of specialised financial systems for water infrastructure, in the form of a development bank or water bond, has reduced access to long-term financing at low interest rates (Agboola et al., 2023a,b). The lending to water projects by commercial banks is usually limited by perceptions of water sector risk, the uncertainty in water-related revenue streams and the difficulty in coming up with collateral by entities that are in the public sector. This imperfection has been the cause of other issues, such as project delays and cuts, that have implications on the viability of the projects.

An ongoing issue has been cost escalation during the implementation of projects, caused by factors such as inflation, change in the scope of the projects undertaken, unexpected site conditions, and prolonged implementation periods (Ajayi et al., 2023). Cost growth in many projects has ranged between 50 and 100%above the initial estimates and this brings in more sources of finance and some tough decisions to be made regarding changes in the project in terms of modification or completion.

Water projects are weak in terms of revenue generation and costs recovery methods, and most utilities are not earning enough to cover their operational costs as well as servicing the debts (Eboh et al., 2023a,b). This is a vicious cycle that causes a low financial performance to restrict the ability to access more funds to expand the system and improve, furthering the provision of insufficient services and hotels on government subsidies.

5.5.3 *Technical and Engineering Challenges*

Technical issues in water project construction include several factors of system design, equipment choices, building quality, and transferring technology that influence the project performance (Tijani et al., 2023a,b). The under-provision of feasibility studies and other technical assessments, which never totally identify conditions at a site, the availability of water resources, the use and demands surrounding water, and the technical demands thereof, has been behind the suffering of many water projects by the Nigerians (Nigerian Society of Engineers, 2024). The effect has been the overdesign of systems, under design of systems or systems that do not match the local conditions and the capacity of operations.

It is foundational that equipment selection and procurement activities have been influenced by several factors such as cost aspects, supplier relations, maintenance, and the local technical capacity (Garba, et al., 2024a,b). The desire to pursue least-cost procurement methods has, on occasion, led to equipment choices that turn out to be unsuitable for the prevailing local operating conditions or are hard to maintain due to a lack of technical skills. Consistency in the type and specification of equipment across projects has been minimal, and this makes maintenance and spares management difficult.

Quality control and supervision during construction have also been noted as a major weak point in most water projects, with poor on-site management resulting in poor workmanship, utilisation of substandard materials and failure to follow designs (Igbinosa et al., 2024a,b). This situation has been the lack of available local contractors and construction supervisors, which has led to poor quality of construction, coupled with corruption in the procurement and contract management (Abdullahi et al., 2023).

Components of water projects, such as technology transfer and capacity building, have been ill-planned and have not been implemented properly, such that operating agencies lack the technical know-how on how to manage and operate complex systems effectively (Amadi et al., 2024). This has especially been a challenge for high technology treatment technologies and automated control systems that only need specialised skills in operating and maintaining them.

There is the challenge of coordination that has accompanied the integration of water supply systems and other infrastructure like power supply, telecommunications, and transport, which impact the project implementation timelines and cost (Yakubu et al., 2023). Power availability and access to the network, especially reliability, has been a perennial issue with challenges to water treatment and distribution systems, necessitating the need to acquire backup generating capacity via expensive decisions that tend to boost the cost of ownership or use.

5.5.4 Institutional and Governance Issues

The instability in institutions and governance has contributed immensely towards impeding the implementation of water projects in Nigeria, as it has reversed any form of unhindered planning, coordination, and implementation of these projects (Hassan et al., 2023a,b). The lack of integration in the institutional structure of managing the water sector, where the various functions are allocated across many federal, state, and local government bodies, creates the possibility of friction points and inconsistent management strategies in project development and implementation (Olajide et al., 2024).

Capacity limitations in implementing agencies pose a basic challenge since most institutions do not have enough qualified technical, financial and project management personnel (Ogundimu et al., 2023). Excessive turnover rates among the employees, poor training opportunities, and unsatisfactory compensation packages have led to institutional inefficiencies that have caused damage in the manner of implementation and sustainability. There has been a trend in appointments of technical personnel to key offices on a political basis at the expense of qualified and experienced managers.

The procurement system and procedures have been some of the main causes of implementation delays and poor quality, where cumbersome bureaucracies, a limited supply of competitors and poor vendor qualification systems have led to poor contractor selection and performance (Bureau of Public Procurement, 2024). Corruption of procurement activities has led to contracts being awarded to unsound contractors, increased costs of the projects and poor implementation results that do not serve the intentions of the project.

Many of the implementing agencies still have poor project management systems and practices, lack the use of modern project management tools, have insufficient systems of monitoring and evaluation and poor documentation of projects (Danladi et al., 2024a,b). This has led to cost over—runs, schedule failure, and failure to identify and resolve implementation issues timely. There is absence of effective coordination mechanisms between different levels of government and sector agencies and this has led to overlapping activities, clashes in project designs and wasted opportunities of economies of scale and synergistic advantages (Ahmed et al., 2024). The interstate water projects have been notably affected by their lack of ability to coordinate in reaching agreement on cost sharing arrangements, the manner of management and a distribution of the benefits.

5.5.5 Regional and State-Level Implementation Challenges

There are challenges regarding implementation across the various regions and states in Nigeria due to a variation in institutional capacity and political commitment, as well as availability of resources and socio-economic conditions (Ahmed et al., 2024). There are specific challenges that are associated with security concerns making their way to project sites, the inability to attract competent contractors and consultants, and the scarce local technical capacity on how the projects are implemented and supervised (Bashir et al., 2023).

Middle Belt region has faced the challenge of inter-communal conflicts, which poses as a challenge to activities in the project's areas, problems of land acquisition, and accessibility problems during rainy seasons that reduce construction activities. Another issue that has been a challenge to the States in this region is the coordination of a series of local government areas in regional water projects (Chukwu et al., 2023a,b).

Southern states typically enjoy increased access to technical expertise and contractors, but they must also contend with challenging topography in some regions, issues of environmental clearance of some projects based on their impact on sensitive an ecosystem and possible community opposition to projects having the perceived impact on traditional sources of water or fishing industries (Kaine et al., 2023).

Institutional capacity, political stability and governance quality varies across the states and as such, it has had great impacts on successfully implementing projects (Ogbonna et al., 2024; Lagos State Ministry of Works and Infrastructure, 2024). Project implementation results have generally been better in states with stronger

Table 5.3 State-level water project implementation performance

State	Active projects	Completion rate (%)	Average delay (months)	Budget overrun (%)	Key success factors
Lagos	28	72	8	15	Strong institutions, PPP experience
FCT	15	68	12	22	Federal support, technical capacity
Kano	22	45	18	35	Security challenges, funding gaps
Rivers	18	58	15	28	Revenue base, security concerns
Kaduna	20	52	20	42	Institutional reforms underway
Ogun	16	61	14	31	Proximity to Lagos, technical support

Source: Adapted from National Assembly (2023); Infrastructure Concession Regulatory Commission (2023)

institutions, better structures of governance and with relatively stable political environments, whereas states with poorly developed institutions have had problems with project delays, cost escalations and quality problems. Implementation performance indicators for selected states, including active projects, completion rates, delays, and budget overruns, are presented in Table 5.3.

Federal system of government has further complicated interstate water projects because there are problems agreeing on the funding of the project, sharing benefits and calculating management compliances (River Basin Development Authorities, 2023). The areas of legal and regulatory regimes of interstate cooperation in the field of water are under development, which creates a delay in the implementation of multi-state projects and the emergence of disagreements.

5.5.6 Environmental and Social Compliance Issues

Environmental and social safeguard requirements have also grown in significance as factors to consider in the implementation of water projects, with most of them facing delays because of deficient environmental impact assessment, community consultation procedures and resettlement planning (Ibrahim et al., 2024a,b). The increased focus on environmental and social responsibility by donor agencies and the government has brought along with-it new demands that most implementation agencies are ill-equipped to respond.

Water infrastructure projects on large scale have especially faced a challenge in the land acquisition and resettlement with unclear land titles, inadequate compensation to affected communities as well as community relations during the construction that is usually troublesome (Chukwu et al., 2023a,b). The issue of legal rights to land and compensation has caused lengthy delays in several projects and insufficient consultation with the local community has resulted in active opposition and protests interrupting projects.

Environmental clearance takes long, and most of the project teams lack adequate personnel to ensure the environmental impact assessment and mitigation plan is done (Adeleke, et al., 2023a,b). The scarcity of skilled environmental advisors and the limitation of operational capacity in the environmental regulatory agencies have been deemed to cause delays in the clearance approvals. Various cultural and heritage aspects have influenced certain water projects, especially those relating to historical sites or areas of historical value to the people. The archaeological surveys and the need to secure heritage clearance have generated time and cost to the projects as well as in other times forcing a change design (Suleiman et al., 2024a,b).

Gender and social inclusion implications during design and implementation of projects have attracted more consideration and require that some consultations must be done with the women groups, actions taken on considering the impact of the project on men and women and specific actions taken to ensure men and women equally access the benefits of the project (Adesina et al., 2024). These requirements have not only been key determiners of the sustainability of the project and its social outcomes, but they have also complicated the project planning and implementation processes.

5.5.7 Contractor and Vendor Performance Issues

Contractor performance has been a major contributor to the outcome of water projects implementations, culminating in several water projects being seriously impacted by inadequate contractor capacity, poor contractor workmanship and even failure to meet the contract obligations (Abdullahi et al., 2023). Three issues underground water quality competent local contractors willing and able to participate in the water infrastructure projects are few due to the lack of supply created by a small number of willing and able local contractors; cost and quality of projects—due to the nature of the competent contractors with water infrastructure experience, the supply constraint can affect cost and quality of the water infrastructure projects and the tendency to select on the basis of lowest bid prices, which has led to inappropriate contractors with low capacity or having inadequate capacity.

International contractors, although found to possess better technical competence and financial abilities, have been found to falter in local situations and conditions as well as rules and regulations, and this has affected their relations with the local populace (Amadi et al., 2024). The joint venture between foreign and local contractors has been less successful with successful ventures being a result of well-established partnership arrangements and division of responsibilities.

There have been failures regarding contract management, supervision by the implementing agencies that have minimal technical capacity to assess contractor performance, facilitate quality standards and manage variations in the contract (Igbinosa, et al., 2024a,b). This has been a major factor in terms of acceptable poor performance of tasks, or through problems being overlooked until they become immense challenges to the project performance. One of the long-standing issues that has been causing challenges in the implementation of projects is payment delays to contractors, where most of them have ended up experiencing cash flow problems because of the slow method of the government making payments (Eboh et al., 2023a,b). This has caused labour holdups, contractor accounts and project overruns that raise total costs of the project and project delivery delays.

Development of local contractor capacity has not received sufficient attention in most projects and there has been a lack of technology transfer, training or mentoring aspects that may enhance the local supply chain in subsequent projects (Amadi et al., 2024). This promotes reliance on international contractors and constrains the advancement of sustainable local capability in implementing water infrastructures.

5.5.8 Monitoring and Evaluation Weaknesses

There have been weak monitoring and evaluation systems on the implementation of water projects, hindering the ability to track implementation progress, identify implementation issues early, and learn (Danladi et al., 2024a,b). A common problem among most of the discretely appointed implementing agencies is that they do not have appropriate project management information systems and depend on informal project reporting mechanisms, which cannot be helpful to effectively manage the projects.

In most cases, the performance indicators and targets have not been well defined in water projects, and this has proved to be unsound in the evaluation of whether the project has achieved its goals (Okafor et al., 2024). The focus on input and output indicators, as opposed to outcome and impact measures has constrained knowledge on the effectiveness and sustainability of projects. Follow up of project implementation milestones, budget use and quality parameters have been very poor in most cases. Project evaluation has not been done on an independent basis and rather most of the evaluation work has been done by the implementing agencies themselves (UNICEF Nigeria, 2024). This has minimised the objectivity and credibility of the project reporting and restricted the chances of external review and suggestions for improvement.

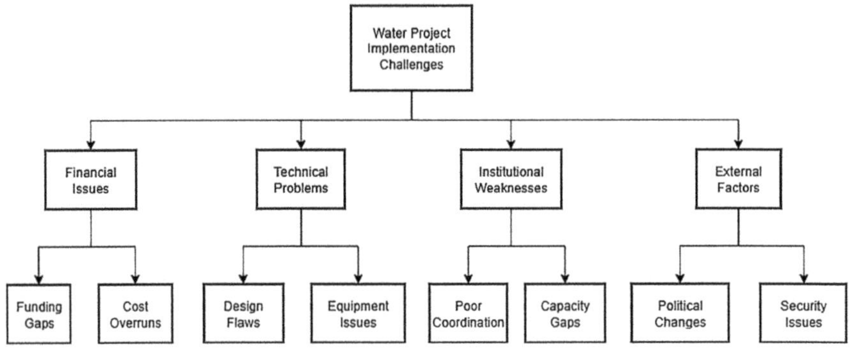

Fig. 5.3 Systemic challenges in water project implementation in Nigeria highlighting key issues such as financing gaps, technical limitations, governance inefficiencies, contractor performance problems, and weak monitoring and evaluation systems that collectively hinder project success

There has been inadequate documentation and knowledge management practice in most projects, whereby there is minimal systemisation of the knowledge on lessons learned, best practices and implementation challenges (Okafor et al., 2024). This has hampered the chances of refining the project design and implementation methods of projects through past experiences and has led to recurrence of similar issues in various projects.

The post-construction evaluation and impact analysis of water projects have been infrequent, thus hindering the knowledge of the long-term outcomes of projects and their sustainability (Suleiman et al., 2024a,b). This has posed a challenge to determine whether projects are generating expected amenities to beneficiary communities and whether application and implementation methodologies are apt in local requirements. These recurring issues highlight the systemic weaknesses in Nigeria's water project implementation process, as illustrated in Fig. 5.3.

5.6 Future Outlook, Emerging Trends and Recommendations

The nature of water supply in Nigeria is steadily developing under the influence of changes in the demographic situation, climatic variability, innovations, and policies (Ibrahim et al., 2024a,b). Due to urbanisation and population growth, there is a tendency to increase consumption of water services, especially in cities with rapid increase and in peri-urban regions (Nwachukwu et al., 2024). Climate change effects, probability of rainfall patterns, as well as rise in events of extreme weather may influence the occurrence of water resources and the resilience of water infrastructure.

The ongoing process, which lays stress on performance enhancement, the involvement of the private sector, and results-based financing, stretches across the sphere of activity of the utilities and their approach to providing customers with the service (Ajayi et al., 2024). Regional cooperation projects can create opportunities for sharing water resources and technical knowledge between state borders, and international cooperation will help to develop infrastructure and build capacities (World Bank, 2023).

Regarding unsolvable sector issues, emerging technologies including smart water systems, innovative water treatment methods, and renewable energy applications are offering hope (Adesina et al., 2024; Ezenwa et al., 2023a,b). However, adoption will need to be supported by appropriate funding mechanisms, the growth of technical capabilities, and supportive policy frameworks that will not only support innovative processes but also ensure equity and the fair delivery of the most basic services. Community participation and social accountability processes are also being recognised as an important factor of sustainable water service delivery, especially in the rural setting where community ownership and capacity to manage services is critical to long—term success (Hassan et al., 2023a,b).

The future of water project implementation in Nigeria is worth considering and would most likely relate to the persistence of urbanisation trends, the effects of climate change, advancement in technologies, and changes in governmental systems (Yakubu et al., 2023). Addressing persistent implementation challenges will require sustained, long-term efforts across several critical areas. These include reinforcing institutional frameworks, enhancing human and technical capacity, establishing robust and adaptable financing models, and promoting more transparent and effective governance structures. Each of these elements plays a vital role in ensuring that water and sanitation initiatives are not only launched but also maintained and scaled successfully over time.

Institutional reform must be aimed at ensuring documentation of the project management capacity, coordination mechanisms and frameworks that ensure accountability in terms of implementation to achieve set objectives (Ogundimu et al., 2023). This can involve the establishment of special project implementation units, standardisation of procedures and control systems and incorporation of performance-based management strategies.

The financing strategies should focus on diversification of sources of financing, establishment of long-term revenue streams, and development of risk-sharing and appropriate words which can solicit the support of the private sector (Jega et al., 2024). This could involve the setting up of water infrastructure funds, the development of blended finance funds and the ability to implement results-based financing.

Embracement of technology and innovation ought to be promoted by a pilot project, showcase programs and capacity-building consultations that develop local intelligence in contemporary project management and implementation methodologies (Nigerian Society of Engineers, 2024). This incorporates the creation of project observation with digital presence, quality control frameworks, and communication with the stakeholders.

The building of local capabilities in the implementation of all functions of the projects, ranging from design and engineering to construction and operation, should be given priority to minimise external resources importation and generate long-term implementation skills in Nigeria (Amadi et al., 2024). This involves concerted actions among and between educational institutions, professional bodies and industry associations to develop skills and knowledge to effectively develop the water infrastructure.

5.7 Conclusion

Nigeria's ongoing water and sanitation crisis stands as one of its most pressing developmental hurdles, with consequences that reach far beyond the basic provision of clean water and hygienic facilities. This chapter's in-depth analysis uncovers a complex network of interrelated challenges, ranging from deteriorating infrastructure and governance breakdowns to public health disparities and implementation setbacks, that have collectively hindered progress toward universal access to safe water and adequate sanitation.

Despite the country's rich water resources and decades of government investment, millions of Nigerians still face water scarcity. Abundance without access disproportionately affects rural areas and vulnerable populations, who continue to bear the brunt of poor service delivery. The effects of these deficits are both immediate and far-reaching. They perpetuate cycles of illness, poverty, and inequality, undermining national development goals. Waterborne diseases remain a leading cause of mortality and reduced quality of life, especially among children under five. The economic toll includes not only direct healthcare costs but also lost productivity, diminished educational outcomes, and entrenched social disparities. Women and girls face additional challenges, such as time-consuming water collection and limited access to menstrual hygiene resources factors that restrict their full participation in economic and social life.

Governance and institutional shortcomings further complicate the crisis. Fragmented responsibilities across government tiers have led to poor coordination, weakening policy alignment and execution. Political instability, inadequate funding structures, and lack of accountability have resulted in abandoned projects, budget overruns, and unsustainable infrastructure investments. The analysis also reveals systemic flaws in procurement processes, contractor performance, and monitoring systemsissues that have prevented policy goals from translating into tangible improvements.

Yet, there are signs of progress and innovation that offer hope. Community-driven initiatives have shown that local ownership and participatory models can yield lasting results. Technological advancements, such as solar-powered water systems and mobile payment platforms, are opening new avenues for service delivery in hard-to-reach areas. Growing awareness of climate resilience and the integration of environmental safeguards into project planning mark important steps toward sustainable infrastructure development. Reform efforts at both state and federal levels—focused on building institutional capacity and adopting performance-based financing—suggest promising directions for sectoral improvement.

References

Abdulkadir, N. (2018). Prevalence of waterborne diseases in relation to age and gender in Nakoloke Sub County Mbale District Uganda. https://doi.org/10.5281/ZENODO.1162962

Abdullahi, A., Garba, M., Usman, H., & Salisu, B. (2023). Contractor performance in Nigerian water infrastructure projects: Assessment of capacity and quality outcomes. *Construction Management Review, 41*(6), 178–195.

Adamu, H., Ibrahim, Y., Mohammed, A., & Hassan, S. (2023). Groundwater quality assessment in northern Nigeria: Implications for rural water supply sustainability. *Journal of Water Resources Management, 37*(8), 2145–2162.

Adebayo, K., Ogundimu, T., Williams, P., & Thompson, R. (2024). Public-private partnerships in water infrastructure: Nigerian experiences and lessons learned. *Infrastructure Partnership Journal, 18*(3), 89–106.

Adebayo, O., Okonkwo, C., Nwafor, P., & Ugochukwu, E. (2024). Urban water utility performance in West Africa: Comparative analysis of Nigerian cities. *Water Policy, 26*(4), 489–507.

Adeleke, B., Thompson, K., & Roberts, M. (2023). Community-based water management in Nigeria: Lessons from successful rural schemes. *International Journal of Water Development, 39*(2), 78–95.

Adeleke, O., Nwosu, C., Okoro, J., & Eze, M. (2023). Environmental compliance in water project implementation: Challenges and solutions in Nigeria. *Environmental Engineering Quarterly, 35*(7), 234–251.

Adesina, B., Olayemi, S., Roberts, K., & Davies, L. (2024). Community engagement in water infrastructure projects: Best practices from Nigerian case studies. *Community Development Review, 29*(4), 145–162.

Adesina, F., Olayemi, T., Williams, J., & Davies, L. (2024). Solar-powered water systems in sub-Saharan Africa: Nigerian case studies. *Renewable Energy and Water Systems, 12*(3), 201–218.

Adesogan, S. O., & Sasanya, B. F. (2025). Climate smart water conservation strategy for sustainable development in Nigeria: A systematic review. In R. U. Onyeneke, C. C. Emenekwe, & C. U. Nwajiuba (Eds.), *Energy transition, climate action and sustainable agriculture* (pp. 151–186). Springer Nature Switzerland. https://doi.org/10.1007/978-3-031-83165-2_9

Agboola, S., Egunjobi, L., Olatubara, C., & Alabi, M. (2023b). Lagos water sector reforms: Achievements and continuing challenges. *Urban Studies Quarterly, 45*(6), 112–129.

Agboola, T., Egunjobi, A., Olatubara, M., & Alabi, R. (2023a). Urban water infrastructure financing in Nigeria: Challenges and innovative mechanisms. *Urban Finance Studies, 22*(8), 267–284.

Agunwamba, J., Ugwu, H., Omole, D., & Ogbozor, P. (2024). Water quality challenges in Nigerian river systems: Sources and mitigation strategies. *Environmental Engineering Science, 41*(7), 445–462.

Ahmed, I., Bello, S., Sulaiman, A., & Abdullahi, K. (2024). Regional variations in water project implementation: Comparative analysis across Nigerian geopolitical zones. *Regional Infrastructure Studies, 16*(2), 67–84.

Ajayi, F., Ogundipe, P., Adeyemi, T., & Ogundimu, S. (2023). Cost escalation in water infrastructure projects: Nigerian patterns and mitigation strategies. *Project Cost Management, 31*(5), 178–195.

Ajayi, P., Ogundipe, S., Adeyemi, R., & Ogundimu, F. (2024). Private sector participation in urban water supply: Nigerian experiences. *Public-Private Partnership Review, 8*(2), 23–41.

Akinwale, A. A., & Udochukwu, I. S. (2025). The role of environmental policy in Nigeria's climate resilience: Evaluating the impact of political will and policy implementation, 11(1). https://doi.org/10.48028/iiprds/ijirsssmt.v11.i1.10

Akpabio, E. M. (2012). Water meanings, sanitation practices and hygiene behaviours in the cultural mirror: A perspective from Nigeria. *Journal of Water, Sanitation and Hygiene for Development, 2*(3), 168 181. https://doi.org/10.2166/washdev.2012.073

Akpan, V., Essien, U., Obot, I., & Udoh, E. (2023). Niger Delta water resources: Pollution impacts and community responses. *Environmental Management Journal, 52*(9), 178–196.

Alabi, O. S., Akintayo, I., Odeyemi, J. S., Oloche, J. J., Babalola, C. M., Nwimo, C., Popoola, O., Mogeni, O. D., Marks, F., & Okeke, I. N. (2024). Suboptimal bacteriological quality of household water in municipal Ibadan, Nigeria. *The American Journal of Tropical Medicine and Hygiene, 110*(2), 346–355. https://doi.org/10.4269/ajtmh.23-0134

Aliyu, B., Garba, M., Usman, D., & Tijjani, H. (2024). Fluoride contamination in northern Nigerian groundwater: Health implications and treatment options. *Water Quality Research, 15*(5), 234–251.

Amadi, C., Nwosu, J., Okeke, B., & Eze, P. (2023). Sachet water industry in Nigeria: Quality control and regulatory challenges. *Food and Water Safety, 29*(3), 145–163.

Amadi, N., Nwosu, P., Okeke, C., & Eze, O. (2024). Technology transfer in water infrastructure projects: Nigerian experiences with international contractors. *Technology Transfer Studies, 12*(4), 89–106.

Assembly, N. (2023). *Report of the committee on water resources: Assessment of federal water infrastructure projects.* Federal Republic of Nigeria.

Asuquo, M. I., Muhammad, A., & Asuquo, C. O.-A. (2025). Assessing community-based WASH interventions to reduce waterborne diseases in Akwa-Ibom state, Nigeria, *12*(3) https://tijer.org/tijer/papers/TIJER2503094.pdf

Balogun, S., & Islam, Z. (2025). Assessment of surface water quality of the Niger river in Lokoja town, Nigeria: An integration of water quality index (WQI), multivariate statistical analysis, and geographic information systems. *Nile Journal of Engineering and Applied Science, 2*(2), 1. https://doi.org/10.5455/njeas.194779

Bashir, A., Mohammed, Y., Abdulkarim, H., & Salisu, K. (2023). Security challenges in water project implementation: Northern Nigeria experiences. *Security and Development, 25*(8), 178–195.

Bashir, N., Mohammed, S., Abdulkarim, A., & Salisu, I. (2024). Small town water systems in northern Nigeria: Sustainability and management challenges. *Small Town Water Review, 11*(1), 45–62.

Bureau of Public Procurement. (2024). *Annual report on water infrastructure procurement: Challenges and improvements.* Federal Republic of Nigeria.

Chukwu, C., Okafor, B., Anyanwu, G., & Okoro, E. (2023a). Land acquisition challenges in water infrastructure development: Legal and social perspectives. *Land Rights Review, 30*(5), 145–162.

Chukwu, E., Okafor, N., Anyanwu, I., & Okoro, C. (2023b). South-East Nigeria water resources: Seasonal variations and supply implications. *Regional Water Studies, 28*(6), 134–152.

DAI. (2021). *Nigeria—Water supply and sanitation sector reform programme phase III*. WSSSR-PIII.

Danladi, A., Garba, I., Shehu, Y., & Musa, K. (2024a). Borehole sustainability in rural Nigeria: Technical and social factors. *Rural Water Systems, 16*(8), 178–195.

Danladi, S., Garba, K., Shehu, M., & Musa, A. (2024b). Monitoring and evaluation systems for water projects: Strengthening implementation oversight in Nigeria. *Project Monitoring Journal, 17*(3), 67–84.

Eboh, M., Ugwu, C., Nkwachukwu, O., & Chima, G. (2023a). Water vendor services in Nigerian cities: Pricing and accessibility patterns. *Urban Water Services, 22*(4), 23–40.

Eboh, P., Ugwu, N., Nkwachukwu, M., & Chima, S. (2023b). Payment systems and contractor cash flow in public water projects: Nigerian utility experiences. *Public Contract Management, 26*(6), 201–218.

Ezenwa, G., Madukwe, A., Okwu, H., & Alozie, T. (2023a). Political continuity and water project sustainability: Nigerian governance challenges. *Governance and Infrastructure, 15*(4), 123–140.

Ezenwa, P., Madukwe, D., Okwu, G., & Alozie, S. (2023b). Mobile payment systems for water utilities: Nigerian urban experiences. *Digital Water Management, 7*(2), 67–84.

Federal Ministry of Budget and National Planning. (2023). *Medium term expenditure framework: Water infrastructure investment analysis.* Federal Republic of Nigeria.

Federal Ministry of Water Resources. (2023). *National water resources master plan: Implementation progress report.* Federal Republic of Nigeria.

FMWR/NBS/UNICEF. (2022). *Water, sanitation and hygiene: national outcome routine mapping (WASHNORM) 2021: a report of findings*, p. 423. https://washdata.org/reports/nigeria-2021-washnorm-ii-report
Garba, H., Umar, B., Saleh, A., & Yakubu, M. (2024a). Hand-dug well water quality in rural northern Nigeria: Seasonal variations and health implications. *Rural Water Quality, 19*(5), 201–218.
Garba, M., Umar, S., Saleh, H., & Yakubu, B. (2024b). Equipment procurement and maintenance in water projects: Standardization and supply chain management. *Equipment Management Studies, 33*(7), 178–195.
Hassan, K., Abdullahi, A., Lawal, M., & Bello, I. (2023a). Institutional capacity constraints in water project implementation: State government analysis. *Public Administration Review, 42*(8), 234–251.
Hassan, M., Abdullahi, K., Lawal, O., & Bello, S. (2023b). Water committee effectiveness in rural Nigeria: Factors influencing sustainability. *Community Water Management, 14*(6), 145–162.
Ibrahim, S., Ahmed, H., Mohammed, U., & Suleiman, Y. (2024a). Environmental impact assessment in water projects: Compliance and implementation challenges. *Environmental Assessment Journal, 28*(5), 145–162.
Ibrahim, Y., Ahmed, A., Mohammed, B., & Suleiman, H. (2024b). Climate change impacts on water resources in northern Nigeria: Adaptation strategies. *Climate and Water Resources, 31*(4), 289–306.
Idowu, O., Martins, V., Oyebanji, F., & Adebisi, G. (2023a). Lagos state water sector transformation: Public-private partnership outcomes. *Water Partnership Review, 26*(8), 178–195.
Idowu, T., Martins, A., Oyebanji, R., & Adebisi, M. (2023b). Lagos state water infrastructure reforms: Implementation strategies and outcomes. *State Reform Studies, 19*(6), 267–284.
Igbinosa, E., Aighewi, O., Igbinosa, I., & Okuofu, A. (2024a). Construction supervision and quality assurance in water projects: Nigerian best practices and challenges. *Construction Supervision Review, 36*(3), 89–106.
Igbinosa, I., Aighewi, I., Igbinosa, E., & Okuofu, C. (2024b). Water treatment plant performance in Nigerian cities: Operational challenges and solutions. *Water Treatment Technology, 42*(3), 123–140.
Infrastructure Concession Regulatory Commission (2023). Infrastructure concession regulatory commission a report on the activities of the annual report and accounts. Retrieved from https://www.icrc.gov.ng/wp-content/uploads/2023/07/ICRC-AGM-2022-rep.pdf.
ISWSC. (2023). The role of strategic community engagement in sustainable WASH initiatives.
Jega, M., Kebbi, S., Bunza, A., & Argungu, H. (2024). Budget allocation and execution patterns in water infrastructure: Federal and state analysis. *Budget Studies Quarterly, 31*(1), 45–62.
Kaine, A., Briggs, L., Cookey, G., & Seiyabo, I. (2024). Port Harcourt water supply challenges: Infrastructure and service delivery analysis. *Urban Infrastructure Review, 29*(1), 56–73.
Kaine, L., Briggs, A., Cookey, T., & Seiyabo, O. (2023). Port Harcourt water supply expansion: Implementation challenges and lessons learned. *Urban Water Development, 24*(4), 123–140.
Lagos State Ministry of Works and Infrastructure. (2024). *Annual infrastructure development report: Water project implementation status*. Lagos State Government.
Lagos Water Corporation. (2023). *Annual performance report 2023: Service delivery and infrastructure development*. Lagos State Government.
Liu, K., Kinouchi, T., Tan, R., Heng, S., Chhuon, K., & Zhao, W. (2024). Unraveling urban hydro-environmental response to climate change and MCDA-based area prioritization in a data-scarce developing city. *Science of the Total Environment, 948*, 174389. https://doi.org/10.1016/j.scitotenv.2024.174389
Maina, A., Ado, S., Aliyu, M., & Garba, H. (2023). Foreign exchange constraints and water project implementation: Impact assessment and mitigation strategies. *International Finance Review, 39*(7), 178–195.
News Agency of Nigeria. (2025). Behavioural change, community action key to sanitation goals.
Nigerian Hydrological Services Agency. (2023). *Annual hydrological bulletin: Surface and groundwater resources assessment*. Federal Ministry of Water Resources.

Nigerian Society of Engineers. (2024). *Professional practice guidelines for water infrastructure projects*. Nigerian Society of Engineers.

Nwachukwu, M., Feng, H., Achilike, K., & Okpe, A. (2024). Rural-urban migration and water service demand in Nigeria: Implications for infrastructure planning. *Population and Resources, 38*(5), 267–284.

Ogbonna, E., Igwe, C., Ayogu, B., & Onwuka, A. (2024). Enugu state water infrastructure development: Institutional reforms and implementation improvements. *State Development Studies, 20*(2), 67–84.

Ogbonna, J., Igwe, E., Ayogu, N., & Onwuka, S. (2023). Enugu state water utility reform: Performance improvements and challenges. *State Water Utilities, 17*(4), 123–140.

Ogundimu, A., Adeosun, T., Olukanni, S., & Salami, R. (2023). Project management capacity in Nigerian water agencies: Assessment and development strategies. *Project Management Studies, 35*(6), 145–162.

Ogundimu, T., Adeosun, B., Olukanni, D., & Salami, A. (2024). Non-revenue water reduction strategies in Nigerian urban utilities. *Utility Management Review, 25*(6), 178–195.

Okafor, C., Ugwu, H., Okeke, J., & Eze, C. (2023). Household water treatment practices in southeastern Nigeria: Adoption and effectiveness patterns. *Household Water Studies, 12*(8), 201–218.

Okafor, J., Ugwu, A., Okeke, P., & Eze, N. (2024). Documentation and knowledge management in water projects: Improving institutional learning. *Knowledge Management Review, 18*(4), 89–106.

Okesanya, O. J., Eshun, G., Ukoaka, B. M., Manirambona, E., Olabode, O. N., Adesola, R. O., Okon, I. I., Jamil, S., Singh, A., Lucero-Prisno, D. E., Ali, H. M., & Chowdhury, A. B. M. A. (2024). Water, sanitation, and hygiene (WASH) practices in Africa: Exploring the effects on public health and sustainable development plans. *Tropical Medicine and Health, 52*(1), 68. https://doi.org/10.1186/s41182-024-00614-3

Okorafor, O. O., Chikwue, M. I., Dike, B. U., & Ujah., E.U. (2024). Challenges and solutions to water and sanitation in Nigeria: A review. *International Journal of Applied Science and Engineering Review., 05*(05), 01–13. https://doi.org/10.52267/IJASER.2024.5501

Okpasuo, O. J., Okafor, F. C., Aguzie, I., Ikele, C., & Anunobi, J. (2019). Spatiotemporal trend of waterborne disease in Enugu urban, Nigeria: A retrospective study. *International Journal of Tropical Disease & Health, 1*, –13. https://doi.org/10.9734/ijtdh/2019/v38i330188

Oladoja, N., Adelabu, I., Ahmad, A., & Ololade, I. (2024). Water quality monitoring in Nigerian river systems: Current status and improvement needs. *Environmental Monitoring, 46*(3), 145–162.

Olajide, T., Agunwamba, S., Ademiluyi, A., & Nwachukwu, P. (2024). Coordination mechanisms for interstate water projects: Institutional arrangements and effectiveness. *Interstate Cooperation Studies, 13*(3), 123–140.

Oyerinde, A. O., & Jacobs, H. E. (2022). The complex nature of household water supply: An evidence-based assessment of urban water access in Southwest Nigeria. *Journal of Water, Sanitation and Hygiene for Development, 12*(3), 237–247. https://doi.org/10.2166/washdev.2022.176

Pooi, C. K., & Ng, H. Y. (2018). Review of low-cost point-of-use water treatment systems for developing communities. *npj Clean Water, 1*(1), 11. https://doi.org/10.1038/s41545-018-0011-0

River Basin Development Authorities. (2023). *Annual performance report: Multi-state water infrastructure projects*. Federal Ministry of Water Resources.

Shehu, B., & Nazim, F. (2022). Clean water and sanitation for all: Study on SDGs 6.1 and 6.2 targets with state policies and interventions in Nigeria. In *The 9th international conference on sustainable development. 20 June 2022 MDPI* (p. 71). https://doi.org/10.3390/environsciproc2022015071

Suleiman, H., Ibrahim, A., Mohammed, S., & Garba, U. (2024a). Rural water project sustainability: Implementation factors and community ownership. *Rural Development Quarterly, 26*(1), 45–62.

Suleiman, L., Ibrahim, H., Mohammed, A., & Garba, B. (2024b). Traditional water sources in northern Nigeria: Cultural significance and modern adaptation. *Cultural Water Studies, 8*(2), 67–84.

Tijani, A., Okunlola, T., Omitogun, S., & Talabi, R. (2023a). Feasibility study quality and project outcomes: Analysis of Nigerian water infrastructure investments. *Feasibility Studies Review, 22*(7), 178–195.

Tijani, M., Okunlola, O., Omitogun, A., & Talabi, A. (2023b). Groundwater development potential in southwestern Nigeria: Hydrogeological assessment. *Groundwater Development, 31*(5), 178–195.

Udoh, E., Etok, A., Akpan, M., & Essien, I. (2024a). Coastal aquifer saltwater intrusion in southern Nigeria: Monitoring and management strategies. *Coastal Water Management, 27*(4), 123–140.

Udoh, I., Etok, S., Akpan, U., & Essien, M. (2024b). Site security and material theft in water projects: Risk assessment and mitigation strategies. *Project Security Management, 16*(5), 145–162.

UN OCHA. (2024). NIGERIA: Floods (Situation Report No. 3, p. 17). UN OCHA. https://www.unocha.org/attachments/0243ce22-9a00-4c2b-b91a-47a9702b2e26/Nigeria%20Floods%20Sitrep%203%20FINAL.pdf

UNICEF. (2022). Nigeria WASH sector finance advocacy strategy and plan.

UNICEF. (2023). Water, sanitation and hygiene UNICEF Nigeria.

UNICEF Nigeria. (2023). *Water, sanitation and hygiene annual report: Progress and challenges.* United Nations Children's Fund.

UNICEF Nigeria. (2024). *Water infrastructure implementation assessment: Challenges and opportunities report.* United Nations Children's Fund.

Water Resources Development Consortium. (2023). *Nigerian water infrastructure implementation review: Patterns, challenges and recommendations..*

Water Supply and Sanitation Collaborative Council. (2024). *Nigeria country water supply assessment: Rural and urban service delivery..*

World Bank. (2023). *Nigeria water supply, sanitation and hygiene program: Implementation status report*. World Bank Group.

World Bank. (2024). *Nigeria water supply, sanitation and hygiene program: Implementation completion report.* World Bank Group.

World Health Organization. (2023). *WHO global water, sanitation and hygiene: Annual report 2022* (p. 60). World Health Organization. https://www.who.int/publications/i/item/9789240076297

World Health Organization. (n.d.). *Targets of Sustainable Development Goal 3.* https://www.who.int/europe/about-us/our-work/sustainable-development-goals/targets-of-sustainable-development-goal-3

Yakubu, A., Medugu, S., Yusuf, H., & Abubakar, M. (2023). Climate resilience in water infrastructure design: Adaptation strategies for Nigerian conditions. *Climate Adaptation Studies, 32*(6), 201–218.

Chapter 6
Challenges in Urban Water Management

Ibrahim Muhammad Magami and Ja'afar Malam Ahmed

Abstract Urban water management is a critical component of sustainable development, integrating water supply, sanitation, stormwater control, and wastewater treatment to meet the needs of growing urban populations. This review examines the multifaceted challenges confronting urban water systems, particularly in developing regions such as Nigeria, with a focus on socio-economic, environmental, infrastructural, and governance policies. Rapid urbanization, population growth, and inadequate infrastructure have contributed a lot into these challenges. Additional pressures from climate change, weak governance, and financial constraints further undermine the resilience and efficiency of urban initiatives. The review highlighted inadequate treatment capacity, poor distribution networks and overdependence on informal water vendors prevail. Innovative solutions such as Integrated Water Resources Management (IWRM), Sustainable Urban Drainage Systems (SUDS), green infrastructure and smart water technologies are explored as pathways toward resilience and sustainability. Addressing urban water challenges requires a holistic governance framework integrating community participation, capacity building, policy reform, and financial innovation. Sustainable outcomes depend on combining technical solutions with institutional accountability, climate adaptation, social inclusion to ensure equitable and reliable water access for present and future urban generations.

Keywords Climate change · Governance · Sustainability · Urbanization · Water management · Governance · Climate change

I. M. Magami (✉)
Department of Biology, Usmanu Danfodiyo University, Sokoto, Nigeria
e-mail: ibrahim.magami@udusok.edu.ng

J. M. Ahmed
Department of Zoology, Usmanu Danfodiyo University, Sokoto, Nigeria

M. I. Adeoba et al. (eds.), *Water Resources Management in Nigeria*, World Water Resources 10, https://doi.org/10.1007/978-3-032-22813-0_6

6.1 Introduction

6.1.1 Urban Water Management and its Relevance to Sustainable Development

Urban water management is the integrated process of utilizing and managing water resources to meet the needs of urban populations, after planning, developing and distributing these natural resources at the same time minimizing environmental impacts (Siddique, 2022). It encompasses full water cycle, including water supply, wastewater collection, treatment and stormwater drainage systems that are essential in maintaining urban functionality (Butler et al., 2024). Modern urban water management goes beyond engineering solutions to involve socio-economic planning, environmental regulation, and risk reduction strategies required for sustainable urbanization in both developed and developing regions (Varma et al., 2024). The rapid expansion of cities worldwide has intensified pressure on water systems, leading to growing concern about water scarcity, pollution, and service delivery inequalities (Borah, 2025).

Nieuwenhuis et al. (2021) defined urban water management as a coordinated system integrating supply, wastewater, and stormwater services, while Siddique (2022) emphasized sustainability within growing urban populations.

The primary objective of urban water management is to ensure reliable access to safe drinking water, adequate sanitation services, and effective stormwater control to prevent flooding and protect urban infrastructure (Moura et al., 2026). Sustainable water management practices promote water-use efficiency, demand management strategies, and water conservation technologies such as rainwater harvesting and wastewater reuse (Siddique, 2022). In addition, nature-based solutions, such as artificial wetlands and green infrastructure, are increasingly integrated into urban systems to improve resilience and enhance ecosystem services (Oral, 2020). These approaches support circular water economies by reclaiming wastewater as a resource rather than treating it as waste, contributing to urban water security and climate adaptation strategies (Mannina, 2022).

Urban water management also plays a vital role in achieving global development agendas, particularly Sustainable Development Goal 6, which ensure availability and sustainable management of water and sanitation for all (Arora & Mishra, 2022). Access to safe water and improved sanitation helps to reduce the spread of waterborne diseases, enhance public health, and promote dignified living, especially in low-income urban communities (Dunga, 2026). Furthermore, efficient water management supports economic development by sustaining industries, agriculture, and urban services, while also safeguarding aquatic ecosystems and biodiversity (UNEP, 2021a,b). Therefore, effective urban water management is fundamental for sustainable urban development as it integrates social equity, economic growth, and environmental sustainability into city planning and governance frameworks (Kosoe & Ogwu, 2025).

6.2 Importance of Water in Urban Environments

Water plays a vital role in sustaining urban life by supporting essential domestic needs such as drinking, cooking, sanitation, and hygiene, which are fundamental to public health and human survival (Reader, 2022) In addition to direct domestic use, water is indispensable to industrial production processes, commercial services, energy generation, and transportation, making it a key driver of economic growth and urban development (Tella et al., 2025). Water also supports environmental sustainability in cities by maintaining urban ecosystems such as wetlands, rivers, and green spaces that improve biodiversity, air quality, and microclimatic regulation (Jandaghian & Colombo, 2024). Furthermore, urban agriculture and peri-urban farming rely on water resources to enhance food security and local livelihoods, particularly in rapidly growing cities in developing nations (Dagar et al., 2023). The availability, quality, and efficient management of water services are therefore crucial to improving quality of life, reducing poverty, and ensuring equitable urban development (UN-Water, 2022).

6.2.1 *Global and Regional Context*

Urban water challenges have become a major global concern due to increasing water demand, deteriorating infrastructure, climate variability, and growing economic inequality (World Bank, 2020). More than 2.2 billion people globally still lack access to safely managed drinking water services, with the majority living in low—and middle-income urban areas (WHO, 2022). Cities in sub-Saharan Africa and South Asia experience the most acute water crises due to population pressure, weak institutions, limited investment, and dependency on unreliable surface and groundwater sources (Nyika & Dinka, 2023). In Nigeria, urban water supply coverage remains below 30% despite decades of reform, and many urban residents rely on private boreholes, tankers, and sachet water vendors for daily water needs (Oyerinde & Jacobs, 2022). Northern Nigeria faces additional challenges due to low rainfall, prolonged dry seasons, advancing desertification, and overdependence on seasonal rivers and shallow aquifers (Ayejoto et al., 2023). Sokoto Metropolis in particular is characterized by inadequate water distribution networks, insufficient treatment facilities, and rural-urban migration that continue to exert pressure on already stressed water systems (Worgu & Ikuru, 2026).

6.2.2 *Purpose of the Chapter*

The purpose of this chapter is to critically assess the major challenges affecting urban water management, with a focus on how socio-economic, environmental, infrastructural, and governance factors contribute to water insecurity in growing cities. The review synthesizes evidence from global literature and contextualizes findings within the African and Nigerian urban water management experience (Adem-Esmail & Suleiman, 2020). Special attention is given to case studies from Nigerian cities including Lagos, Kano, Kaduna, and Sokoto to illustrate local realities and policy gaps (Muoghalu, 2025). The review also identifies innovative urban water management strategies such as Integrated Water Resources Management (IWRM), Sustainable Urban Drainage Systems (SUDS), smart water technologies, and nature-based solutions as emerging pathways toward sustainability (Delgado et al., 2024). Finally, it provides recommendations for strengthening urban water governance, financing mechanisms, infrastructure resilience, and public engagement to ensure sustainable water management in rapidly growing urban environments (Khan et al., 2024).

6.3 Overview of Urban Water Management Systems

Urban water management systems are integrated frameworks designed to plan, supply, distribute, treat, conserve, and monitor water resources within urban environments to meet domestic, industrial, and environmental needs sustainably (Siddique, 2022). These systems encompass both centralized and decentralized infrastructures that manage freshwater delivery, wastewater disposal, stormwater control, and water reuse initiatives (Younos, 2022). A well-designed urban water management system seeks to balance water demand and availability while minimizing environmental degradation and resource depletion (Nieuwenhuis et al., 2021). However, in many cities in developing nations, fragmented and poorly coordinated water systems have led to service deficits, inefficiencies, and widespread water insecurity (Robins, 2019). The complexity of urban water systems has increased under the influence of climate change, growing urbanization, and socio-economic inequality, which collectively strain water resources and infrastructure (World Bank, 2020). Moving forward, sustainable urban water management must incorporate resilience planning, technological innovation, and multi-stakeholder participation to ensure equitable and reliable service delivery (Nyika & Dinka, 2023).

6.3.1 Components of Urban Water Systems: Water Supply, Wastewater Treatment, Stormwater Drainage, Reuse, and Conservation

Urban water systems consist of interlinked components that collectively manage the urban water cycle. The first component is water supply, which includes extraction from surface and groundwater sources, treatment to meet drinking water standards, and distribution through piped networks (Reader, 2022). The second component is wastewater management, which involves collection, conveyance, treatment, and safe disposal or reuse of domestic and industrial effluents to protect public health and the environment (Kumari et al., 2023). Stormwater drainage systems form the third component and are essential for managing rainfall runoff to prevent urban flooding and waterlogging, particularly in rapidly urbanizing areas (Kong, 2021). The fourth component is water reuse and recycling, where treated wastewater is reused for irrigation, industrial cooling, toilet flushing, or groundwater recharge, helping to reduce demand on freshwater systems (Silva, 2023). Finally, water conservation and demand management strategies include metering, leakage reduction, water-saving technologies, and public awareness campaigns that promote efficient water use (Moglia et al., 2018). These components must function in synergy to support sustainable, resilient, and inclusive urban water systems, particularly in regions facing water scarcity and infrastructure challenges such as sub-Saharan Africa and South Asia (Khan et al., 2024).

6.3.2 Historical Evolution and Modern Approaches

Historically, urban water management focused on centralized supply networks designed primarily to meet domestic consumption needs, with little attention to sustainability or environmental impacts (Varma et al., 2024). During the mid-twentieth century, increasing industrialization led cities to expand water infrastructure, but the approach remained supply-oriented, often neglecting wastewater treatment and ecosystem protection (Biswas & Tortajada, 2019). The late twentieth century saw growing recognition of the need for integrated water planning, resulting in frameworks like Integrated Water Resources Management (IWRM), which promotes coordinated water development across sectors to enhance sustainability and equity (Global Water Partnership, 2017). Recently, modern water management approaches have emerged that emphasize resilience, decentralization, and the circular economy (Delgado et al., 2024). These include Sustainable Urban Drainage Systems (SUDS), nature-based solutions, and smart water technologies that use digital tools and sensors to monitor leakage, consumption, and water quality in real time (Pachouri et al., 2025). In developing countries like Nigeria, however, water management is still evolving from traditional supply-focused models to more holistic strategies that integrate climate adaptation and community participation (Oyerinde & Jacobs, 2022).

6.3.3 Institutional and Governance Frameworks

Institutional and governance frameworks play a decisive role in determining the effectiveness of urban water management. These frameworks define the roles, policies, regulations, and coordination mechanisms among ministries, utilities, private firms, and communities involved in water services (Tella et al., 2025). Weak governance, institutional overlaps, and lack of accountability have been identified as major barriers to water security in developing nations (Julio et al., 2021). In many African countries, including Nigeria, water governance is fragmented among several agencies with limited coordination, resulting in policy duplication and implementation gaps (Ukpai, 2022). In contrast, effective governance systems in developed countries incorporate regulatory oversight, stakeholder participation, financial transparency, and performance monitoring (Kaushik et al., 2026). Emerging governance reforms advocate for decentralized decision-making, public–private partnerships, and collaborative planning with local communities to enhance service delivery and sustainability (UN-Water, 2022). Moreover, governance frameworks must integrate climate resilience, equity considerations, and environmental protection to address the multifaceted challenges of urban water systems (Kaushik et al., 2026).

6.4 Major Challenges in Urban Water Management

Urban water management challenges are shaped not only by physical and infrastructural limitations but also by socio-economic and behavioral factors that significantly influence water availability, quality, and sustainability in both developed and developing countries (Santos et al., 2023). Recent studies emphasize that technical solutions alone are insufficient without addressing community awareness, institutional governance, and public behavior towards water resources (Ali & Kamraju, 2024). Scholars argue that water security cannot be achieved without strengthening public participation and behavioral change through education, policy reinforcement, and stakeholder collaboration (Kaushik et al., 2026). In Africa, particularly in countries like Nigeria, public engagement and awareness remain weak despite government and donor-funded interventions aimed at improving urban water services (Ukpai, 2022). These behavioral and institutional gaps form a critical barrier to water sustainability and justify the inclusion of social dimensions in modern integrated water management frameworks (Madias & Szymkowiak, 2022).

6.4.1 Population Growth and Rapid Urbanization

Urbanization and population expansion are among the most pressing challenges for urban water management, especially in developing regions where infrastructure development lags behind demographic change (World Bank, 2021). Rapid migration from rural areas to urban centers increases demand for potable water and sanitation services, often outstripping municipal supply capacity (UN-Habitat, 2020). This demographic pressure results in over-extraction of surface and groundwater resources, leading to aquifer depletion and declining water tables (Khair et al., 2019). In megacities such as Lagos, Mumbai, and Dhaka, accelerated growth has intensified informal settlements that lack water infrastructure, exacerbating water insecurity (Satterthwaite, 2016; Agbo et al., 2025). Urban sprawl further complicates water distribution planning, as cities expand horizontally into peri-urban zones where infrastructure networks are weak or absent (Mallick, 2024). Consequently, urban water utilities become overstressed, system efficiency declines, and service inequity widens, especially among low-income populations (Hoekstra et al., 2018).

Population-induced pressure on water services is also linked to an increase in per capita water consumption due to rising living standards and lifestyle changes (Moglia et al., 2018). In many rapidly growing African and Asian cities, informal water vendors have become the primary suppliers, often delivering water at prices far higher than municipal tariffs, creating socio-economic disparities (Adams, 2019). Additionally, urban population growth contributes to unplanned connections, illegal tapping, and increased non-revenue water losses that burden water authorities financially (Osman et al., 2024). Meeting the water needs of expanding populations requires planned infrastructure investment, integrated land-use management, and demand management strategies, yet many cities lack the institutional capacity or financial resources to implement such reforms (Kosoe & Ogwu, 2025). Urbanization, therefore, represents a structural challenge that requires holistic policy and technological responses if urban water sustainability is to be achieved (World Bank, 2021).

6.4.1.1 Aging and Inadequate Infrastructure

Aging water infrastructure poses a critical barrier to effective water management in both developed and developing countries (Perera et al., 2021). Many cities rely on aging water distribution networks that are over 50 years old, leading to frequent pipe bursts, high leakage rates, and intermittent supply (Kosoe & Ogwu, 2025). In Sub-Saharan Africa, inadequate infrastructure results in an estimated 35–50% non-revenue water loss, reflecting system inefficiencies and weak maintenance regimes. In older European cities such as London and Rome, infrastructure deterioration contributes to similar leakage problems, demonstrating that the issue transcends economic levels (Al-Mansoori, 2025). Poor infrastructure planning and lack of redundancy systems also increase vulnerability to supply disruptions during droughts or extreme rainfall events (Khan et al., 2024). Furthermore, many wastewater treatment facilities operate beyond design capacity, discharging untreated effluents into water bodies and threatening public health (WHO and UNICEF, 2019). Without renewal, aging systems continue to compromise service reliability, water safety, and environmental sustainability (Fred et al., 2024).

In many low-income nations, infrastructure inadequacy is rooted in years of underinvestment and neglect (World Bank, 2021). Financial constraints limit regular maintenance activities, resulting in reactive rather than preventive management approaches (Mallick, 2024). In Nigeria and other West African countries, only 40% of urban residents have access to safely managed drinking water due to poor infrastructure coverage (Ukpai, 2022). Dilapidated pipelines are also susceptible to contamination through cross-connections with sewer lines, increasing incidences of waterborne diseases such as cholera and typhoid (Renwick, 2019). Furthermore, infrastructure expansion is commonly donor-driven rather than planned within long-term government strategies, reducing sustainability (UN-Habitat, 2020). Addressing aging and inadequate infrastructure requires targeted investment, life-cycle asset management, and the adoption of smart maintenance technologies to improve service efficiency and longevity (Kosoe & Ogwu, 2025; Reader, 2022).

6.4.1.2 Water Pollution and Quality Degradation

Water pollution remains a pervasive challenge in urban water management due to the increasing discharge of untreated wastewater, industrial effluents, agricultural runoff, and domestic waste into water bodies (UNEP, 2021a,b). In most developing countries, more than 80% of wastewater is released into rivers, lakes, and wetlands without adequate treatment, exacerbating environmental degradation and public health risks (UN-Water, 2022). Rapid urbanization increases impervious surfaces, which accelerates stormwater runoff carrying hydrocarbons, heavy metals, plastics, and pathogens into surface water systems (Wang et al., 2025). Industrial activities, especially in urban industrial zones, contribute significantly to toxic contamination of water bodies through pollutants such as lead, cadmium, mercury, and persistent organic pollutants. In cities such as New Delhi, Johannesburg, and Cairo, the presence of industrial and municipal pollutants has resulted in declining water quality indices and the loss of aquatic biodiversity (Yadav et al., 2020). Additionally, eutrophication caused by nutrient loading from detergents and fertilizers leads to algal blooms, oxygen depletion, and fish kills, further compounding ecological imbalances in urban waterways (Paerl & Otten, 2016; UNEP, 2021a,b). Groundwater, often assumed to be safe, is increasingly threatened by leachate from poorly managed solid waste dumpsites and latrines in informal settlements (Fida et al., 2024).

Poor sanitation infrastructure contributes significantly to microbiological contamination of urban water sources, increasing the spread of waterborne diseases such as cholera, typhoid fever, hepatitis A, and diarrheal infections (WHO & UNICEF, 2019). In Sub-Saharan Africa and South Asia, inadequate sewerage networks force millions to rely on open defecation or poorly constructed latrines, which contaminate shallow aquifers used for drinking water (Nyika & Dinka, 2023). Water quality degradation also undermines the economic potential of urban water resources, increasing treatment costs and reducing water availability for industrial and domestic use (Akhtar et al., 2021; Rusprayunita et al., 2025). Furthermore, weak enforcement of environmental regulations, corruption, and lack of industrial compliance worsen pollution trends in many countries (Khanam et al., 2023). The failure to monitor water quality regularly due to inadequate laboratory facilities and technical expertise limits early detection of contamination risks (Ateia et al., 2024). Addressing water pollution requires stricter effluent regulations, investment in wastewater treatment plants, enforcement of polluter-pays principles, and community-based pollution monitoring systems (UNEP, 2021a,b; Kulkarni et al., 2026).

6.4.1.3 Climate Change and Extreme Weather Events

Climate change is altering the intensity, frequency, and seasonality of precipitation and temperature extremes, producing direct and indirect stresses on urban water supplies, wastewater systems, and stormwater infrastructure (IPCC, 2021). Intensified heavy rainfall events increase the likelihood of urban flooding by overwhelming drainage networks and causing rapid runoff from impervious surfaces, while prolonged dry spells and droughts reduce river flows and groundwater recharge that many cities depend on (Pizzorni et al., 2024). Sea-level rise and increased coastal storm surges threaten coastal urban water infrastructure, causing saline intrusion into shallow aquifers and contamination of freshwater supplies that serve millions of coastal residents (Habel et al., 2024). Droughts driven by changing climate patterns reduce raw water availability and force utilities to draw on lower-quality or more costly sources (e.g., long-distance transfers, deep aquifers, or desalination), thereby raising operational costs and exacerbating supply inequities (World Bank, 2020). Compound and cascading events such as a heatwave that increases demand for water followed by a storm that damages treatment plants create systemic stresses that traditional single-hazard planning does not adequately address (IPCC, 2021). Moreover, urban characteristics such as high imperviousness, subsidence from groundwater over-extraction, and encroachment on floodplains amplify climate impacts, leading to greater exposure and vulnerability in rapidly growing cities (Kudapa, 2025). In West African contexts, altered rainfall seasonality and increased inter-annual variability have already affected surface water reliability and increased uncertainty for cities like Kano and Sokoto that depend on seasonal rivers and shallow wells (Ayejoto et al., 2023). Floods frequently damage water treatment and distribution infrastructure, contaminate source waters with sewage and hazardous materials, and provoke service outages that in turn precipitate public-health crises such as diarrhoeal disease outbreaks (Acharya & Silori, 2025). Heat extremes similarly raise daily water demand (for cooling, hydration, and hygiene) while increasing evaporation losses from reservoirs and reducing effective supply, particularly where storage is limited (Moglia et al., 2018). The direct physical impacts of extreme events also translate into higher repair, rehabilitation, and insurance costs for utilities and municipal governments, constraining already thin budgets in low-income cities (Kosoe & Ogwu, 2025). Finally, while climate-informed adaptation options exist such as green infrastructure, SUDS, diversified source portfolios, climate-resilient plant design, and flexible operational protocols their adoption remains uneven due to financing gaps, limited technical capacity, and weak integration of climate risk into urban planning (Delgado et al., 2024). Early-warning systems, climate-smart water allocation, and investment in natural buffers can materially reduce risk, but scaling these measures demands cross-sectoral governance, robust data systems, and sustained financing that many cities have yet to mobilize (UN-Habitat, 2020).

6.4.1.4 Governance, Policy, and Institutional Weaknesses

Weak governance structures and fragmented institutional arrangements are among the fundamental barriers to effective urban water management globally. In many developing countries, responsibilities for water supply, sanitation, and wastewater treatment are distributed across multiple agencies with overlapping mandates and poor coordination, resulting in inefficiency and duplication of efforts (Mason et al., 2020). Policy incoherence between water, urban development, land use, and environmental regulations weakens holistic planning and hinders the implementation of integrated water management strategies (Kant & Wrat, 2025). Moreover, weak enforcement mechanisms mean that environmental laws regulating effluent discharge, groundwater abstraction, and pollution control are routinely violated without consequence (Nkiaka et al., 2020). In Sub-Saharan Africa and South Asia, institutional challenges often stem from political interference in public utilities, leading to mismanagement, inconsistent tariff policies, and lack of autonomy in service delivery (Gregory & Sovacool, 2019). This undermines accountability and perpetuates operational inefficiencies within urban water utilities. Additionally, corruption in contract awarding, billing systems, and procurement processes diverts financial resources away from maintenance and infrastructure investment (Matimolane, 2026). The absence of transparent regulatory oversight also contributes to service inequity, where peri-urban and low-income communities are underserved or excluded entirely from formal water access (Satterthwaite, 2016).

Governance deficiencies also extend to inadequate stakeholder participation in water decision-making processes, particularly the exclusion of civil society organizations, women, and local communities from water governance frameworks (Naiga et al., 2024). The lack of bottom-up participation reduces public trust and limits the sustainability of policy interventions (Mehta et al., 2019). Furthermore, governance systems in many low—and middle-income countries lack reliable water data and monitoring systems necessary for evidence-based planning (Khan et al., 2024). Weak institutional capacity also hampers the implementation of Integrated Water Resources Management (IWRM) approaches, which require coordination among sectors and long-term resource planning (Gallegos et al., 2026). Political instability in some countries disrupts water sector reforms and reduces investment confidence for international funding partners (World Bank, 2021). Privatization of water services, attempted in some cities as a governance solution, has had mixed results, sometimes improved efficiency but often resulting in tariff hikes that marginalize poor households (Twum & Abubakari, 2020). Strengthening governance requires reforms that emphasize regulatory accountability, institutional coordination, transparency in financial management, and inclusive decision-making structures that prioritize water equity and sustainability (Reader, 2022; Kosoe & Ogwu, 2025).

6.4.1.5 Economic and Financial Constraints

Limited public and private financing is a pervasive constraint for urban water infrastructure, where the scale of investment needed for water supply, sanitation, and flood protection often far exceeds available domestic budgets and development finance (Jiang, 2023). Governments in many low—and middle-income countries face competing fiscal priorities and constrained fiscal space, which reduces their ability to fund long-term capital works, rehabilitate aging assets, or subsidize services for the poor (World Bank, 2024). Even where financing is available, the project preparation gap weak pipeline of bankable, well-designed projects limits the ability to absorb commercial capital and blended finance at scale (Delmon, 2024). High levels of non-revenue water (NRW) and operational inefficiencies further undermine utilities' financial viability because lost or unbilled water reduces revenues needed for maintenance and investment (Santos, 2024). In sub-Saharan contexts, NRW ratios frequently exceed 30–40% in many cities, substantially eroding the revenue base and deterring lenders and investors who require predictable cash flows (GWL case data; regional assessments).

Tariff design and affordability dilemmas intensify financial constraints because full cost-recovery tariffs can be politically infeasible and socially regressive in low-income urban neighbourhoods, yet underpriced services lead to chronic underfunding and deferred maintenance (World Bank, 2021; "Troubled Tariffs"). Subsidy targeting, social tariffs, and cross-subsidies are possible policy responses, but they require robust billing systems, clear targeting mechanisms, and institutional transparency that many utilities lack (Mohamed et al., 2026). Moreover, donor dependency and short-term project funding can produce piecemeal infrastructure that is not aligned with long-term asset-management plans, reducing sustainability after project close-out (Delmon, 2024). The combined effect of these financial barriers is that many cities postpone critical upgrades, rely on low-quality stop-gap solutions (private tankers, informal vendors), or accrue mounting deferred maintenance liabilities that ultimately raise the lifetime cost of services (World Bank, 2024).

6.4.1.6 Public Awareness and Behavioural Issues

Public awareness, social norms, and household behaviour play an important but often overlooked role in urban water sustainability because consumption patterns, hygiene practices, and willingness to pay directly affect demand management and service outcomes (Ahmed et al., 2022). Studies across diverse contexts show that improved knowledge about household water footprints, risks of contamination, and the benefits of conservation increases adoption of lower-consumption behaviours and support for utility reforms when outreach is evidence-based and culturally adapted (Ahmed et al., 2022; Vanhille, 2018). In many urban low-income settlements, however, lack of trust in utilities, limited access to information, and competing survival priorities mean that behaviour change campaigns have limited reach unless paired with tangible service improvements and community engagement (Thakur, 2026).

Community participation and local stewardship have repeatedly been shown to improve the sustainability of water and sanitation interventions by enhancing ownership, enabling local monitoring, and supporting operations and maintenance activities (Thakur, 2026). Evidence-driven behaviour-change interventions that identify and prioritise specific protective behaviours for example, household-level safe storage, proper disposal of waste, and reduced pollutant discharge can significantly reduce pollution loads and protect drinking sources when scaled with local leadership (Khanam et al., 2023). Effective public engagement also strengthens demand for accountability, improves tariff acceptance when revenues are transparently managed, and mobilizes local finance and in-kind contributions for minor infrastructure maintenance. Conversely, weak outreach, poor communication of risks, and exclusion of marginalized groups from planning perpetuate behaviours that heighten contamination and wasteful use; therefore, integrated communication strategies, school-based education, and inclusion of women and youth in governance are essential complements to technical interventions (Khanam et al., 2023).

6.5 Regional Perspectives

6.5.1 Examples from Developed and Developing Nations

Comparative case studies from developed and developing countries illustrate both universal challenges and context-specific responses in urban water management. In developed settings, cities have demonstrated successful deployment of water reuse and demand-management strategies that combine technological innovation with strong regulatory oversight; for example, municipal, private partnerships and industrial reuse projects have been implemented in North America and Europe to reduce freshwater abstractions and close local water loops (EPA case studies on reuse). In contrast, developing megacities face more acute gaps in service coverage and greater reliance on informal water markets, private boreholes, and tanker deliveries, which create inequities and increase per-unit costs for poor households. Lessons from both contexts highlight the value of diversified source portfolios (including managed aquifer recharge, treated reuse, and rainwater harvesting), robust regulatory frameworks, and community engagement to ensure both reliability and equity. Successful interventions combine technical measures (e.g., reuse, leakage control), institutional reforms (clear mandates, accountability), and financing innovations (blended finance, targeted subsidies) to create resilient urban water systems that can adapt to population growth and climate stressors.

6.5.2 Examples from Nigerian Cities

Urban water management challenges and interventions in Nigerian cities reveal recurring themes of infrastructure deficit, pollution, informal supply dependence, and nascent innovation (Rivera-Williams, 2025). Lagos, as Nigeria's largest megacity, exemplifies acute water stress where formal piped coverage is limited in peri-urban areas and household reliance on boreholes and packaged water is widespread (Tayo, 2023). Recent analyses discuss managed aquifer recharge and other supply-diversification options as potential mitigation strategies (Alvarez, 2025). Kano and Kaduna illustrated common service delivery dynamics in northern Nigerian cities, with intermittent piped supplies, high dependence on private vendors, and variable water quality. Several municipal and state plans identify investment in distribution networks, expansion of wastewater treatment and regulation of informal vendors as priority actions, but implementation has been constrained by financing gaps and institutional coordination problems (Aparcana, 2017). Empirical studies of peri-urban settlements in Lagos and Kano show that informal markets and private provision often fill the gap left by utilities, but at much higher cost and with less sanitary assurance, underscoring the need for formalizing and regulating alternative providers while expanding public infrastructure (Baye, 2025).

6.5.3 Examples from North-Western Nigeria

North-Western Nigeria (including states such as Kano, Kaduna, Sokoto, and Zamfara) faces distinctive hydro-climatic and socio-economic constraints that shape water management outcomes: seasonal rainfall patterns, reliance on seasonal rivers and shallow aquifers, and pressures from population growth and agricultural demands (regional assessments and Kano studies). Research focused on northern communities' documents frequent water quality issues in both surface and groundwater sources, driven by inadequate sanitation, waste disposal practices, and hydrological vulnerability during dry seasons (peri-urban and rural WASH assessments). Water vending and private boreholes are particularly prevalent in this region, often reflecting municipal supply shortfalls; such informal provision can alleviate short-term scarcity but also entrenches inequalities and weakens incentives for utilities to expand service coverage. Regional planning documents emphasize the need for integrated approaches combining groundwater protection, seasonal storage, rainwater harvesting, and community-based monitoring to build resilience in the face of climatic variability and growing urban demand.

6.5.4 Examples from Sokoto Metropolis

Sokoto Metropolis typifies many medium-sized West African cities where limited piped distribution, microbial contamination risks, and dependence on shallow wells and boreholes persist. Local studies assessing water quality in schools, households, and public points in Sokoto report faecal contamination and spatial disparities in water service availability, highlighting acute public-health risks associated with inadequate treatment, leakages, and proximity of sanitation facilities to shallow abstraction points (Sokoto water quality studies). Municipal reports and academic assessments recommend investments in network rehabilitation, expansion of controlled public standpipes, and targeted community education campaigns to reduce contamination, alongside improved monitoring and enforcement of sanitation standards. Given Sokoto's climatic constraints (low rainfall, seasonal river flows), measures such as safe storage, point-of-use treatment, improved sanitation containment, and small-scale rainwater harvesting are commonly proposed as pragmatic, locally appropriate interventions to augment supply and reduce health risks.

6.5.5 Highlight Innovative Urban Water Management Approaches

Innovative approaches such as Sustainable Urban Drainage Systems (SUDS) and Integrated Water Resources Management (IWRM) have gained traction as frameworks that shift practice from sectoral, supply-centric responses toward multifunctional, resilience-oriented strategies. Nature-based solutions manage stormwater through permeable pavements, bioswales, green roofs, and urban wetlands, reducing flood risk while providing co-benefits for biodiversity and urban cooling; recent reviews demonstrate how SUDS decision-support systems and nature-based design improve downstream outcomes when adapted to local hydrology and land-use realities (Alves et al., 2024). IWRM promotes cross-sectoral coordination, stakeholder participation, and basin-scale planning to reconcile competing water uses and support long-term resource sustainability, though practical IWRM adoption often faces institutional and capacity barriers in many African settings (recent IWRM implementation studies). Case evidence suggests that the combination of SUDS at local scales with IWRM at catchment scales can reduce urban flood exposure, improve recharge, and create more flexible water portfolios but success depends on governance reforms, financing mechanisms, and technical capacity to design, monitor, and maintain these systems.

6.5.6 *Comparative Lessons and Transferability*

Comparative case evidence indicated that while developed cities often benefit from stronger regulation, economies of scale, and financial instruments to implement advanced technical solutions such as large-scale reuse, advanced metering (Vogel, 2018). Many innovations can be adapted to resource-constrained contexts when attention is given to affordability, institutional fit, and community engagement (Barquet et al., 2025). Transferable lessons include: diversify water sources (reuse, rainwater harvesting, managed aquifer recharge), prioritize non-structural measures (demand management, tariffs targeting, public education), adopt modular and low-maintenance technologies (green infrastructure, decentralized treatment), and strengthen regulatory and planning institutions to coordinate across sectors. Importantly, adaptation requires contextualization what works in a temperate European city will need modification to function under West African climatic, economic, and governance conditions and thus pilot projects, local capacity building, and iterative scaling strategies should precede full roll-out. The comparative evidence also underscores that technical fixes alone are insufficient: social acceptance, transparent financing, and accountable institutions are equally determinative of sustainable outcomes (Nauman et al., 2024).

6.6 Emerging Trends and Innovative Solutions

6.6.1 *Green Infrastructure and Nature-Based Solutions*

Green infrastructure and nature-based solutions (NbS) are increasingly promoted as cost-effective, multifunctional approaches that address stormwater management, urban heat, biodiversity loss, and water quality simultaneously, making them highly relevant for cities confronting climate risks and constrained finances (Oral, 2020; Mabrouk, 2025). Designed interventions such as constructed wetlands, bioswales, permeable pavements, green roofs, and urban wetlands attenuate peak runoff, enhance local infiltration and groundwater recharge, and provide co-benefits for recreation and urban cooling that conventional gray infrastructure does not deliver (Oral, 2020; Chapman, 2023). Systematic reviews show that NBS perform best when they are integrated into broader catchment and urban planning, adapted to local hydrology and soils, and paired with long-term maintenance plans and community stewardship to avoid performance degradation over time (Oral, 2020; Mabrouk, 2025). Despite strong interest, barriers to widescale adoption persist notably land-use competition, unclear maintenance responsibilities, limited technical design guidance for arid contexts, and financing models that favour capital-intensive gray works over smaller, distributed NbS investments (Chapman, 2023; Badini, 2025). For water-stressed West African cities like those in North-West Nigeria, modular NbS (e.g., small retention basins, vegetated swales, and school-yard rain gardens) can be tailored to seasonal rainfall patterns and combined with recharge measures to improve both stormwater management and dry-season supply resilience when implemented alongside governance reforms and local capacity building (Oral, 2020; Mabrouk, 2025).

6.6.2 Smart Water Technologies and Digital Monitoring

Smart water technologies including Advanced Metering Infrastructure (AMI), Internet of Things (IoT) sensors, automated pressure management, real-time quality sensors, and data analytics are transforming how utilities detect leaks, manage demand, and prioritize repairs, thereby improving efficiency and reducing non-revenue water (Palermo et al., 2022; Obunga et al., 2025). Evidence from recent reviews demonstrates that remote metering and continuous monitoring enable near-real-time identification of anomalies in flow and pressure, support predictive maintenance, and improve customer billing accuracy, which together enhance utility cash flows and service reliability (Newing, 2023; Palermo et al., 2022). However, technology adoption in low-income cities faces obstacles such as high upfront capital costs, limited digital infrastructure, data governance and privacy concerns, and shortages of trained personnel to operate and interpret complex datasets (Palermo et al., 2022; Obunga et al., 2025). Cost-effective pathways for adoption include phased deployment (pilot → scale), public–private partnerships for equipment and data services, capacity building within utilities, and leveraging low-cost sensor networks for priority uses like district-metered areas and leak detection rather than full rollout immediately (Obunga et al., 2025; Newing, 2023). For contexts like Sokoto and other Nigerian cities, smart tools paired with pragmatic institutional reforms (transparent billing, performance-based contracts) can substantially reduce operational losses and improve service equity if financing and technical assistance are aligned to local capacity.

6.6.3 Water Recycling, Reuse, and Desalination

Water recycling and reuse are central elements of circular urban water strategies that reduce freshwater withdrawals and provide alternative sources for non-potable uses such as irrigation, industrial cooling, and street washing (Mannina, 2022; Florides, 2024). Reviews indicate that decentralized and centralized reuse schemes — when backed by appropriate treatment trains, fit-for-purpose standards, and risk-based managemen can be implemented safely and cost-effectively, particularly where freshwater is scarce or where wastewater treatment capacity exists but effluent is underutilized (Mannina, 2022; Ramm, 2023). Desalination provides a drought-resilient supply option for coastal cities but remains energy-intensive and expensive unless paired with renewable energy and energy-recovery technologies; therefore, desalination is generally more appropriate as part of a diversified portfolio and for higher-income urban settings or critical coastal facilities (Florides, 2024). Social acceptance, regulatory clarity, and robust monitoring are essential for public trust in reuse projects; the "fit-for-purpose" regulatory frameworks and outreach campaigns that accompany reuse pilots in several countries are instructive for cities considering scaled reuse (Mannina, 2022; Florides, 2024). In the Nigerian context, lower-cost reuse options such as treated wastewater for urban agriculture or industrial parks and greywater reuse at building scale offer near-term benefits and lower technical demands than potable reuse or large desalination plants. Pilot reuse projects, clear public health safeguards, and incentives for reclaiming treated effluent can catalyze broader uptake while safeguarding human and environmental health.

6.6.4 Integrated Planning and Community Engagement

Integrated urban water planning that combines basin-scale IWRM perspectives with neighborhood-level interventions and participatory governance has emerged as a critical enabler of resilient and equitable urban water systems (Saikia et al., 2022; Tella et al., 2025). Frameworks such as the City Water Resilience Framework and other governance-based planning tools explicitly link hazard assessment, infrastructure planning, stakeholder engagement, and financing to produce actionable city water resilience plans (Saikia et al., 2022). Community engagement and co-design approaches are particularly important for ensuring that local knowledge, priorities, and maintenance capacity inform technical choices for example, community-managed wetlands or locally maintained rainwater harvesting systems tend to be more sustainable when residents are co-invested in their design and upkeep (Perera et al., 2021; Mugari et al., 2025). Evidence shows that integrating social and technical elements combining low-cost infrastructure with education, women's participation, and local accountability mechanisms improves outcomes for health, equity, and system longevity (Tella et al., 2025; Perera et al., 2021). Implementing integrated planning in practice requires improved data systems, cross-agency coordination mechanisms, long-term financing instruments (including blended finance and performance-based grants), and legal frameworks that clarify roles and responsibilities across municipal, state, and basin institutions (Tella et al., 2025; Saikia et al., 2022). For Sokoto and similar cities, adopting pragmatic integrated planning focused on a prioritized set of scalable interventions (leak reduction, targeted reuse, neighbourhood NBS, and community education) offers a realistic pathway to build immediate resilience while preparing for longer-term investments.

6.7 Policy Implications and Future Prospects

6.7.1 Need for Sustainable Water Governance Frameworks

Sustainable urban water governance requires clear institutional mandates, transparent regulatory systems, and mechanisms that align incentives across municipal, basin, and national levels to manage competing water uses and risks effectively (Tella et al., 2025; UN-Habitat, 2020). Strengthening governance means unbundling overlapping responsibilities, establishing independent regulatory oversight for tariffs and service quality, and embedding social equity into policy design so that poorer and peri-urban households gain access to affordable services (World Bank, 2021; UN-Water, 2022). Governance reforms should also prioritize data transparency and performance monitoring including routine disclosure of non-revenue water, service coverage, and financial metrics to attract blended finance and private investment by reducing perceived risk and improving creditworthiness (Kosoe & Ogwu, 2025). Participatory governance that includes civil society, women's groups, and local communities enhances policy legitimacy and the long-term sustainability of interventions by aligning technical solutions with social needs and local knowledge (Mehta et al., 2019; Naiga et al., 2024). Further, integrating climate risk into governance and planning through legally mandated climate proofing of water assets, contingency budgeting, and climate-smart procurement guidelines helps ensure that investments are resilient to future extremes and do not create stranded assets (IPCC, 2021; World Water Development Report, 2020).

6.7.2 Investment in Infrastructure and Capacity Building

Meeting urban water needs requires substantial capital investment not only in new infrastructure but in asset renewal, digital monitoring, and wastewater treatment that enable circular water use (World Bank, 2020; Delmon, 2024). Public budgets alone are unlikely to suffice; therefore, financing strategies should combine public finance, performance-based concessional loans, municipal bonds (where markets allow), and blended finance facilities designed to de-risk private participation while protecting affordability for low-income users (Jiang, 2023; World Bank, 2024). At the same time, investment must be matched with human capital development training utility staff in asset management, digital tools, and customer engagement; strengthening local planning agencies; and supporting universities and professional bodies to provide ongoing technical support (Palermo et al., 2022; Mosleh et al., 2021). Capacity building also includes improving regulatory agencies' ability to design evidence-based subsidies and social tariffs, to monitor environmental compliance, and to deploy early-warning and contingency systems during extreme events (UNESCO/WWAP, 2020; Reader, 2022). Ultimately, strategic and sequenced investments that prioritize high-impact, low-cost interventions (e.g., district metering, leakage control, localized treatment and reuse) can deliver rapid service improvements and create fiscal space for longer-term projects.

6.7.3 Cross-Sectoral Integration (Urban Planning, Sanitation, Environment)

Urban water outcomes are strongly shaped by decisions in land use, transportation, housing, and waste management; thus, policy coherence across sectors is essential (UN-Habitat, 2020). Integrating water-sensitive urban design into planning protecting floodplains, preserving recharge areas, and enabling permeable surfaces reduces risk and enhances natural buffering capacity; this requires coordinated planning instruments, joint budgets, and cross-agency monitoring (Oral, 2020; Saikia et al., 2022). Sanitation must be treated as an integral component of urban water management rather than an afterthought; investments in safe containment, conveyance, and treatment of excreta drastically reduce source contamination and improve reuse opportunities (WHO/UNICEF JMP, 2019). Environmental safeguards including stricter effluent limits, enforcement capacity, and restoration of urban waterways are equally critical to maintain downstream ecosystem services and to secure reliable raw water sources for cities (UNEP, 2021a,b). Cross-sectoral integration further benefits from national framework legislation that clarifies institutional roles across water, environment, and urban development portfolios, enabling coherent budgeting and joint accountability.

6.8 Research and Innovation Priorities

Research agendas should prioritize both applied and translational studies that address city-scale decision needs. First, operational research on low-cost, context-appropriate nature-based solutions for arid and semi-arid urban contexts (e.g., small retention basins, vegetated swales tailored to seasonal rainfall) will fill a key evidence gap for cities in Northwest Nigeria and similar regions (Oral, 2020; Dezfooli et al., 2023). Second, pilot studies and implementation research on targeted smart-metering, district-metered areas, and low-cost sensor networks can demonstrate the utility and cost-effectiveness of digital monitoring in reducing non-revenue water and improving billing (Palermo et al., 2022). Third, socio-technical research that explores governance modalities for formalizing and regulating informal water providers (tankers, vendors, borehole operators) can help design mechanisms that preserve affordability while improving quality and reliability (Thakur, 2026). Fourth, cost benefit and life-cycle analyses of decentralized reuse, greywater systems, and small-scale desalination (where applicable) will clarify which reuse options are viable under different urban and economic conditions (Mannina, 2022; Florides, 2024). Fifth, research on blended finance instruments, project preparation facilities, and municipal credit enhancement practices would support the development of bankable pipelines that attract long-term capital for urban water projects (Delmon, 2024; World Bank, 2024). Lastly, climate science integration into local water planning downscaling climate projections, modeling compound hazards, and stress-testing urban water systems remains a cross-cutting research priority to design robust, adaptive infrastructure and policies (IPCC, 2021).

6.9 Public Awareness and Behavioral Issues

Public awareness plays a fundamental role in determining water consumption patterns and conservation behavior in urban communities across the world (Singha et al., 2022). Studies have shown that even in regions with improved water supply systems, inefficient water use persists due to poor awareness of water scarcity and the environmental consequences of mismanagement (Tzanakakis et al., 2020). Behavioral attitudes such as negligence, excessive water use, and unwillingness to adopt water-saving practices are common in both developed and developing countries (Fielding et al., 2019). Inadequate community engagement often results in minimal ownership of water management processes, leading to weak support for conservation initiatives (Adom & Simatele, 2022). Researchers argue that sustainable urban water management requires proactive public sensitization campaigns and behavioral interventions that encourage shared responsibility (Tripathi et al., 2026).

6.9.1 Water Wastage, Poor Sanitation Practices, and Inadequate Public Participation

Water wastage in urban areas is often attributed to careless household behaviors, illegal water connections, and leaking storage facilities (Hossain et al., 2023). Poor sanitation practices such as improper disposal of wastewater and solid waste significantly degrade urban surface and groundwater resources (Abanyie et al., 2022). In many African cities, including those in Nigeria, open defecation and inadequate sewage disposal continue to contaminate local water bodies due to poor public hygiene awareness (Oluseyi & Nweke, 2020). Inadequate public participation in water governance results from community exclusion in decision-making processes and limited access to environmental information (Antwi et al., 2025). Studies have found that community involvement in planning and monitoring water projects significantly improves sustainability and reduces resource misuse (Nkiaka et al., 2020). Effective policy reforms must integrate public engagement strategies for behavior change to reduce urban water inefficiencies (Chuenchum et al., 2024).

6.10 Regional Perspectives

Urban water management challenges vary significantly between regions due to socio-economic, environmental, and governance differences (Santos et al., 2023). Developed countries typically contend with aging water infrastructure and climate-adaptation concerns, while developing regions struggle with inadequate service delivery and weak institutions (Younos, 2022). Water scarcity is now a global issue affecting cities across all continents regardless of development level (World Bank, 2021). Urbanization, population growth, and climate variability are major contributors to urban water stress worldwide (Borah, 2025). Understanding these regional variations is important for designing context-specific water management strategies (Santos et al., 2023).

6.10.1 Examples from Developed and Developing Nations

Developed nations such as the United Kingdom, Germany, and Japan face urban water challenges associated with aging distribution networks and leakage losses exceeding 20% in some cities (Baldino & Pujol, 2018). Climate change has increased flood and drought risks even in highly developed urban systems such as those in Australia and the United States (Keenan-Jones et al., 2025). Despite advanced technology, countries like Spain and Italy still experience seasonal water shortages due to overdependence on surface water (De-Luca et al., 2024). In contrast, developing countries such as India, Kenya, and Bangladesh face chronic water supply interruptions and poor sanitation services (Ahona et al., 2025). Informal settlements in cities like Nairobi and Mumbai lack access to safe drinking water, relying instead on water vendors at high economic cost (Sarkar, 2020). These disparities reflect differences in water governance, financial investment, and institutional capacity (Ali & Kamraju, 2024).

6.10.2 Examples from Nigerian Cities

Urban water supply in Nigeria suffers from inadequate infrastructure and high levels of non-revenue water due to leakages and illegal connections (Ali, 2024). Cities such as Lagos, Kano, Ibadan, and Port Harcourt rely heavily on groundwater due to unreliable public water supply (Ukpai, 2022). Water demand in Lagos exceeds 700 million liters per day, yet the Lagos Water Corporation supplies less than 40% of the demand (Oyebanji, 2022). Kano metropolis faces water shortages due to siltation and reduced capacity of the Challawa and Watari dams (Isyaku, 2026). In Kaduna, inadequate treatment capacity at Malali Water Works contributes to microbial contamination in household storage systems (Abdulazeez et al., 2025). Poor electricity supply further hampers water pumping and distribution across Nigerian cities (Ihuoma & Nwankwoala, 2021).

6.10.3 Examples from Northwestern Nigeria

Northwestern Nigeria experiences unique water supply challenges due to semi-arid climatic conditions and low surface water availability (Ayanlade et al., 2022). States such as Kebbi, Zamfara, Sokoto, and Katsina rely primarily on seasonal rivers and shallow aquifers (Mahi & Murana, 2024). High poverty levels in the region contribute to low household investment in safe water storage and sanitation facilities (Swe et al., 2021). In Gusau (Zamfara State), urban water scarcity is worsened by mining activities that contaminate groundwater with heavy metals (Ibrahim & Umar, 2024). Frequent breakdown of water supply infrastructure due to poor maintenance is common in Kebbi and Katsina (Muhammad et al., 2024). The gap between urban water demand and supply continues to widen due to rapid population growth and limited government investment (Liu et al., 2024).

6.10.4 Examples from Sokoto Metropolis

Sokoto Metropolis relies heavily on the Sokoto-Rima River Basin and groundwater abstraction for urban water supply (Shuaibu et al., 2025). The Sokoto Water Board provides piped water to only about 38–45% of residents due to limited treatment capacity and pipeline distribution coverage (Gidadawa et al., 2024). Communities such as Arkilla, Gwiwa Low Cost, Runjin Sambo, and Mana experience frequent water shortages caused by low pumping pressure and aging pipelines (Hussein et al., 2023). Many households in Sokoto rely on boreholes, wells, and private vendors, paying up to five times the tariff of public water (Gidadawa et al., 2024). Intermittent power supply and rising energy costs disrupt daily water pumping operations at the Sokoto Water Board (Oyebanjo et al., 2025). Water quality challenges also persist in Sokoto urban areas due to poor household storage hygiene and shallow well contamination (Adamu, 2026).

6.10.5 Innovative Urban Water Management Approaches (SUDS, IWRM, Etc.)

Sustainable Urban Drainage Systems (SUDS) are widely adopted in developed countries to control urban flooding and improve stormwater quality (Pachouri et al., 2025).

Table 6.1 Comparison of urban water challenges across regions

Region	Key water challenges	Service coverage (%)	Major causes
Developed nations	Aging networks, leakages, climate risks	90–99	Infrastructure age, rising demand
Developing nations	Intermittent supply, contamination	40–70	Weak governance, low investment
Nigerian cities	Water scarcity, overdependence on groundwater	35–50	Poor maintenance, illegal connections
Northwestern Nigeria	Low surface water, poverty effects	< 40	Semi-arid climate, weak utilities
Sokoto Metropolis	Shortage, high vendor dependence	38–45	Low pumping capacity, power failure

Source Flörke et al. (2018), Mahi and Murana (2024)

Integrated Water Resources Management (IWRM) promotes balanced water allocation between sectors and is now implemented in over 70 countries (Alaerts, 2019). Water-sensitive urban design (WSUD) integrates water reuse, green spaces, and stormwater harvesting into city planning (Meng, 2022). In developing countries, constructed wetlands and rainwater harvesting are emerging as low-cost solutions for water security (Zabidi et al., 2020). Nigeria adopted IWRM principles in its National Water Resources Policy, but weak enforcement limits effective implementation (Nkiaka et al., 2020). Adoption of innovative water technologies remains slow in Northwestern Nigeria due to funding and technical capacity constraints (Ayanlade et al., 2022) (Table 6.1).

6.11 Emerging Trends and Innovative Solutions

Emerging innovations in urban water management are transforming how cities respond to water scarcity, flooding, and pollution challenges worldwide (Sahu & Debsarma et al., 2023). Modern approaches now integrate technology, sustainability, and community participation to improve water resilience (Ali & Kamraju, 2024). These innovations include nature-based solutions, digital monitoring systems, and decentralized water treatment options (Adeoba et al., 2025). Developing countries are increasingly adopting hybrid water management systems that combine traditional and modern practices to enhance water security (Oluseyi & Nweke, 2020). Sustainable solutions are critical for achieving United Nations Sustainable Development Goal 6 on clean water and sanitation (Khan et al., 2024).

6.11.1 Green Infrastructure and Nature-Based Solutions

Green infrastructure uses natural systems such as wetlands, vegetated swales, and green rooftops to enhance water recharge and flood control (Chandratreya, 2024). Nature-based solutions improve water quality by filtering sediments and absorbing pollutants before water reaches surface bodies (Robotham et al., 2023). They help reduce flood risks in urban areas by enhancing infiltration and reducing stormwater runoff (Ren et al., 2020). Cities like Copenhagen and Melbourne have successfully adopted green stormwater infrastructure to address urban flooding (Webber et al., 2020). Nature-based approaches are cost-effective and provide co-benefits such as biodiversity conservation and aesthetic improvements (Dubois et al., 2025). Adoption in African cities remains low due to limited technical knowledge and institutional support (Muwafu et al., 2024).

6.11.2 Smart Water Technologies and Digital Monitoring

Smart water technologies enable real-time monitoring of water distribution networks using sensors, IoT systems, and satellite tools (Okoli & Kabaso, 2024). These technologies help detect leakages, reduce water losses, and optimize system operations (Olodu et al., 2025). Digital metering improves billing transparency and customer engagement in urban water utilities (Monks, 2019). Smart grids are being used in developed countries to manage pressure flow and reduce energy costs in water distribution. In countries like India and South Africa, smart water applications are being tested in pilot projects to improve service coverage (Tiwari et al., 2025). However, adoption in Nigeria remains slow due to infrastructural and funding limitations (Ihuoma & Nwankwoala, 2021).

6.11.3 Water Recycling, Reuse, and Desalination

Water recycling and reuse are increasingly adopted to reduce freshwater demand in water-scarce regions (Wah, 2026). Treated wastewater is used for irrigation, industrial processes, and groundwater recharge in many developed countries (Alawneh et al., 2024). Desalination provides an alternative water supply source for coastal cities with high water stress (Williams, 2022). Countries like Saudi Arabia, Israel, and Spain have invested in large-scale desalination plants to meet urban water demand (Al-Addous et al., 2024). Although energy-intensive, advances in membrane technology have reduced desalination costs (Osman et al., 2024). In Nigeria, water reuse remains minimal due to low investment in treatment facilities.

6.11.4 Integrated Planning and Community Engagement

Integrated planning links water supply, sanitation, drainage, and environmental protection, promoting efficiency and sustainability (Santos et al., 2023). Community engagement improves water governance by involving stakeholders in decision-making and project monitoring (Nkiaka et al., 2020). Participatory water management has been linked to higher project sustainability in rural and urban communities (Sarkar, 2020). Collaborations between government agencies, NGOs, and community groups improve service delivery outcomes (Tripathi et al., 2026). In Nigeria, community-based water initiatives have helped reduce system failures due to vandalism and illegal connections (Antwi et al., 2025). However, urban governance must integrate bottom-up strategies to reinforce accountability and transparency (Ali & Kamraju, 2024).

6.12 Strategies for Sustainable Urban Water Management

Sustainable urban water management involves a combination of institutional reforms, technological innovation, community participation, and long-term planning to balance water demand with available resources (UN-Water, 2020). It emphasizes efficient use, equitable distribution, and ecological sustainability in managing increasingly stressed urban water systems (Marlow et al., 2018). The adoption of sustainable strategies is critical due to population growth, climate variability, and infrastructure decay affecting water security globally and in developing countries like Nigeria (Ezenwaji et al., 2019). These strategies also support climate resilience in water supply and sanitation, aligning with Sustainable Development Goal 6 on clean water and sanitation (World Bank, 2022). Integrating economic, environmental, and social dimensions in water governance increases long-term system performance and public welfare (Schmidt & Matthews, 2017).

6.12.1 Demand Management Strategies

Demand management strategies aim to reduce water consumption by influencing user behavior, improving efficiency, and optimizing allocation (Mitchell et al., 2019). Many urban centers experience water scarcity not solely due to physical shortages, but from poor water-use efficiency and wastage by domestic and commercial users (Hoekstra et al., 2018). Water pricing, if designed equitably, can reduce excessive consumption while supporting financial sustainability for utilities (Dziegielewski, 2020). Public awareness campaigns have been successful in reducing per capita water usage in countries like Australia and Spain where community-based conservation programs are widely practiced (García & Ballester, 2017). In Sub-Saharan

Africa, increasing water demand due to population growth makes demand-side strategies essential for urban water resilience (Mugagga & Nabaasa, 2016). In Nigerian cities like Lagos and Kano, public campaigns on water conservation are limited and public attitudes towards water use remain unsustainable, increasing water production pressures (Ayeni & Oloruntoba, 2020). Household water-saving technologies such as low-flow taps and dual-flush toilets are rarely used in Nigerian households due to low awareness and affordability constraints (Okonkwo & Ezenwaji, 2021). Schools and religious institutions can serve as platforms for promoting responsible water usage in urban communities (Adeniran et al., 2019). Water metering is another effective demand management tool, though it is poorly implemented in Nigeria where over 70% of urban households remain unmetered (Olawuni et al., 2023). Without effective demand management, urban water utilities will struggle to expand supply or maintain service quality (Umar & Goni, 2022).

6.12.2 Water-Loss Reduction and Leakage Control

Water-loss reduction focuses on minimizing non-revenue water (NRW) caused by leaks, illegal connections, meter inaccuracies, and poor network maintenance (Liemberger & Wyatt, 2019). High levels of NRW can exceed 40% in urban water systems in developing countries, reducing financial sustainability and limiting service expansion (Kingdom et al., 2020). Leakage control is essential in aging water distribution systems where pipe bursts and corrosion occur frequently (Mutikanga et al., 2017). Pressure management, active leak detection, and network zoning are recommended operational strategies for leakage reduction (Lambert & Hirner, 2018). Developed countries like Germany and Japan have successfully reduced NRW to below 10% using modern monitoring technologies and rapid repair programs (Farley, 2018). In contrast, many African cities including Nairobi, Accra, and Kampala still lose 35–50% of treated water due to poor maintenance and outdated water infrastructure (Adah & Abok, 2016). In Nigeria, average non-revenue water is estimated at over 45% due to illegal connections and pipe leakages in urban systems (Adegbola & Enitan, 2022). Utility corruption and weak regulatory enforcement also contribute to losses in the Nigerian water sector (Aliyu & Hassan, 2020). Smart leak-detection technologies using supervisory control and data acquisition (SCADA) systems are increasingly being used worldwide to prevent water loss (Kwak et al., 2019). Implementing district metered areas (DMAs) is another cost-effective approach adopted in South Africa to quickly isolate and repair leaks (Meyer, 2021). Reduction of water losses improves water security, reduces energy costs, and delays capital-intensive water supply expansion (Mutikanga et al., 2017). For developing countries, leak detection combined with community surveillance can significantly reduce unauthorized water connections (Olagunju & Omole, 2023).

6.12.3 Water Recycling and Reuse Strategies

Water recycling and reuse strategies aim to recover wastewater for non-potable or potable purposes, reducing pressure on freshwater sources (Angelakis & Snyder, 2018). Treated wastewater can be reused for agriculture, industrial processes, and urban landscaping, promoting circular water use (Qadir et al., 2020). Growing water scarcity in arid regions has accelerated the adoption of water reuse, particularly in the Middle East, Singapore, and California (Sharma et al., 2019). Advanced treatment technologies such as membrane filtration and reverse osmosis enable high-quality recycled water production (Rodriguez et al., 2020). In agriculture, treated wastewater provides nutrients that can improve crop productivity while conserving freshwater resources (Jiménez & Asano, 2015). However, inadequate wastewater treatment in developing countries raises concerns about pathogen and heavy metal contamination (Kannan et al., 2022). In Sub-Saharan Africa, only 15% of wastewater is treated before disposal, limiting safe reuse potential (UNEP, 2021a,b). Nigeria has minimal wastewater recycling due to lack of policy frameworks and investment in treatment plants (Iorhemen, 2023). Some pilot projects in Lagos and Abuja have demonstrated feasible reuse for irrigation in parks and golf courses (Nwankwoala & Mamah, 2019). Public perception remains a major barrier to wastewater reuse due to concerns over health risks (Ghebreyesus et al., 2020). Strong regulation, hygienic guidelines, and public education are essential to increase acceptance of water reuse (Faria & Naval, 2022). Water recycling also supports climate adaptation by reducing reliance on rainfall-dependent water sources (Boelee et al., 2019).

6.12.4 Public Awareness and Behavior Change Strategies

Public awareness and behavior change are essential components of sustainable water management because user attitudes directly influence water consumption patterns (Fielding et al., 2019). Awareness campaigns that educate households on water-saving practices significantly reduce domestic water wastage (Russell & Knoeri, 2020). Media platforms, schools, and community associations can be effective in promoting water conservation behaviors (Dolnicar et al., 2017). In many developing countries, public perception treats water as a free resource, leading to misuse and low willingness to pay for services (Obi et al., 2021). Behavioral interventions combined with metering systems lead to more responsible water use (Arrojo-Agudo & Martínez, 2022). In Nigeria, public awareness programs on water conservation remain weak and poorly coordinated among government agencies (Ochuko & Odiase, 2023). Faith-based institutions and local leaders can help promote responsible water use through culturally sensitive campaigns (Isa & Usman, 2020). Educational programs on water hygiene and conservation improve both water quality and consumption efficiency (Ezenwaji et al., 2019).

6.12.5 Infrastructure Investment and Maintenance

Urban water systems require long-term investment to upgrade aging infrastructure and expand service delivery to growing populations (World Bank, 2021). Inadequate funding is a major constraint to water supply in developing countries where utilities rely heavily on government subsidies (Adekola & Lamond, 2018). Regular maintenance prevents system collapse and reduces non-revenue water losses (Mutikanga et al., 2017). Investment in treatment plants, storage reservoirs, and distribution networks increases service reliability and quality (Olawuni et al., 2023). International financing institutions such as the African Development Bank and the World Bank support water infrastructure projects in Africa, but poor project management reduces sustainability (AfDB, 2020). Public-private partnerships (PPPs) offer alternative water infrastructure financing but require strong regulation to prevent inequitable service pricing (Akinola & Yahaya, 2019). In Nigeria, most urban water infrastructure was constructed over 30 years ago and has deteriorated due to poor maintenance (Odetola & Adedeji, 2022). Rehabilitation programs are urgently needed to modernize urban water supply systems and improve efficiency (Nwankwoala, 2021).

6.12.6 Policy and Institutional Reforms

Effective water management requires strong policies and institutional frameworks that support regulation, accountability, and sustainable practices (Biswas & Tortajada, 2019). Many developing countries face institutional fragmentation where multiple agencies share overlapping roles, leading to inefficiencies (Mugagga & Nabaasa, 2016). Policy reform must address governance challenges such as corruption, weak enforcement, and poor tariff regulation (Aliyu & Hassan, 2020). Integrated Water Resources Management (IWRM) provides a holistic policy framework for balancing water, environment, and development needs (Rahaman & Varis, 2020). However, implementation of IWRM in Africa has been slow due to political and economic barriers (van der Zaag, 2021). In Nigeria, the National Water Resources Bill has faced delays due to controversies regarding water ownership and governance (Adeoti et al., 2023). Institutional reforms must prioritize decentralization, giving states and local governments greater responsibility for water supply (Ishaku et al., 2021). Transparent regulatory agencies improve utility performance by enforcing service standards and tariff guidelines (Kariuki & Schwartz, 2015). Without strong policies and effective institutions, technical solutions to water scarcity cannot be sustained (Schmidt & Matthews, 2017).

6.12.7 Climate Change Adaptation Strategies

Climate change significantly affects urban water availability by altering rainfall patterns, increasing evaporation, and intensifying droughts and floods (IPCC, 2022). Urban water systems must incorporate adaptation strategies that enhance resilience to climate-induced water stress (Conway et al., 2019). Diversifying water sources through rainwater harvesting, stormwater capture, and aquifer recharge can build climate resilience in cities (Fletcher et al., 2018). Flood-resilient drainage infrastructure reduces the risk of urban flooding during extreme rainfall events (Amoako & Inkoom, 2018). Climate-smart water planning improves long-term supply reliability by integrating hydrological projections into water management decisions (World Bank, 2021). Developing countries face higher climate vulnerability due to low adaptive capacity and limited investment in resilient water infrastructure (Nkiaka et al., 2020). In Nigeria, climate change has reduced seasonal water availability in northern states while increasing flood risks in southern cities (Adelekan, 2020). Adaptation strategies must include drought monitoring systems, watershed protection, and sustainable groundwater extraction (Olanrewaju & Popoola, 2023). Mainstreaming climate change into water policies is essential for achieving sustainable urban water security (Satterthwaite et al., 2018).

6.12.8 Adoption of Smart Water Technologies

Smart water technologies enhance efficiency, monitoring, and decision-making in urban water supply systems (Kwak et al., 2019). Technologies such as Supervisory Control and Data Acquisition (SCADA), smart meters, and real-time sensors improve data accuracy and enable quick leak detection (Li et al., 2020). Internet of Things (IoT)-based water systems allow for remote control and monitoring of water distribution networks (Pérez-González et al., 2021). Smart metering enhances water-use accountability by providing households with real-time consumption data (Anusha & Prasad, 2022). Artificial intelligence (AI) models help optimize water demand forecasting and operational planning (Hutton et al., 2020). However, adoption in developing countries is slow due to high costs and limited technical capacity (Mutchek & Williams, 2019). In South Africa, digital water monitoring has been piloted to manage droughts more efficiently (Muller, 2018). Nigeria has begun introducing smart meters in Lagos and Abuja, but coverage remains below 10% (Olawuni et al., 2023). Smart technology integration requires investment in ICT infrastructure and staff capacity building (Adeniran & Nwankwo, 2021).

6.12.9 Community Participation and Stakeholder Engagement

Community participation strengthens urban water management by ensuring local needs and priorities are reflected in decision-making (Marks & Zadoroznyj, 2019). Water governance that excludes users often leads to project failure due to low public acceptance (Cleaver & Whaley, 2018). Engaging communities increases water conservation awareness and reduces vandalism of infrastructure (Abubakar, 2019). Participatory planning allows marginalized groups such as women and informal settlement residents to contribute local knowledge to water solutions (Allen et al., 2020). Multi-stakeholder platforms involving government agencies, NGOs, and traditional leaders promote inclusive water governance (Nkiaka et al., 2020). In Kenya and Uganda, community-managed water schemes have shown improved service sustainability (Hope et al., 2020). In Nigeria, limited community engagement has contributed to weak ownership of water projects and poor maintenance culture (Ishaku et al., 2021). Strengthening citizen participation enhances transparency and accountability in the water sector (Schmidt & Matthews, 2017). Public involvement is also critical in Integrated Water Resources Management (IWRM) approaches (Rahaman & Varis, 2020).

6.13 Conclusion

Major challenges to urban water management including population growth and urbanization pressures, aging and inadequate infrastructure, pollution and quality degradation, climate-driven extremes, governance weaknesses, financial constraints, and public-behavioural issues has highlighted practical, scalable solutions across technical, institutional and social domains. The evidence indicated that no single intervention is sufficient, because sustainable outcomes require integrated, cross-sectoral governance frameworks, prioritized, sequenced investments, community engagement and adaptive planning that incorporates climate risk and local realities. For cities in Nigeria, Northwest Nigeria, such as Sokoto Metropolis in particular, pragmatic, low-cost interventions (e.g., targeted leak reduction, point-of-use treatment, small-scale NBS and regulated formalization of informal suppliers), combined with capacity building and financing innovations, can deliver immediate benefits while laying the groundwork for longer-term transformation. Research must support this transition by producing contextually relevant evidence on NBS in dry climates, affordable digital monitoring, governance models for mixed supply systems and mechanisms to make urban water services both financially sustainable and socially equitable. Policymakers, donors, utilities and communities should therefore align around a shared agenda that values resilience, inclusion and environmental stewardship to secure urban water for present and future generations.

Recommendations

1. Governments should strengthen water governance and institutional coordination.
2. Massive investment is required to rehabilitate aging infrastructure, expand water distribution networks, and modernize treatment facilities.
3. Urban utilities should promote adoption of smart and climate-resilient technologies
4. Public participation is essential to ensure behavioural change toward water conservation and sanitation.
5. Adopting integrated water resources management (IWRM) and sustainable urban drainage systems (suds) can address multiple urban water challenges simultaneously.
6. Continuous research and innovation are needed to develop context-specific solutions suitable for arid and semi-arid urban environments like Sokoto.

References

Abanyie, S. K., Amuah, E. E. Y., Douti, N. B., Antwi, M. N., Fei-Baffoe, B., & Amadu, C. C. (2022). Sanitation and waste management practices and possible implications on groundwater quality in peri-urban areas, Doba and Nayagenia, northeastern Ghana. *Environmental Challenges, 8*, Article 100546.

Abdulazeez, A.M., Agada, J., Mohammed, S.A., Saidu, A., Amira, M.K., Idujagi, O.S., Adamu, S.U., Winner, L., Ransom, U.J., Joseph, I.B., Jamila, M.M. (2025). Determination of Bacteriological Parameters in Raw water and Treated Water of the Environment of Malali Water Works, Kaduna State, Nigeria.

Abubakar, I. R. (2019). Exploring the determinants of open defecation in Nigeria using demographic and health survey data. *Science of the Total Environment, 637–638*, 1455–1465.

Acharya, A., & Silori, R. (2025). A probable framework for flood-induced risk evaluation for water supply, sanitation and hygiene. In *Water Sustainability and Hydrological Extremes* (pp. 237–253). Elsevier.

Adah, P. D., & Abok, G. (2016). Challenges of urban water supply in Nigeria: The way forward. *Journal of Environment and Earth Science, 6*(4), 35–40.

Adamu, M. N. (2026). Groundwater Development for Potable Water Supply in Sokoto State, Nigeria: Hydrogeological Potential, Water Quality Assessment, and Sustainability Challenges. *Journal of Science Innovation and Technology Research.*

Adams, E. A., Sambu, D., & Smiley, S. L. (2019). Urban water supply in Sub-Saharan Africa: Historical and emerging policies and institutional arrangements. *International Journal of Water Resources Development, 35*(2), 240–263.

Adegbola, A., & Enitan, A. (2022). Urban water management challenges in sub-Saharan Africa: A review of Nigeria's water supply deficiencies. *Water Practice & Technology, 17*(3), 678–691.

Adekola, O., & Lamond, J. (2018). A media framing analysis of urban flooding in Nigeria: Current narratives and implications for policy. *Urban Water Journal, 15*(9), 983–991.

Adelekan, I. O. (2020). Urban climate resilience in Africa: Challenges and opportunities for managing climate risks in sub-Saharan cities. *Current Opinion in Environmental Sustainability, 43*, 1–7.

Adem-Esmail, B., & Suleiman, L. (2020). Analyzing evidence of sustainable urban water management systems: A review through the lenses of sociotechnical transitions. *Sustainability, 12*(11), 4481.

Adeniran, A. E., & Nwankwo, O. (2021). Adoption of smart metering technology in Nigeria: Challenges and prospects. *Utilities Policy, 73*, 101301.

Adeniran, A. E., Oyedele, O. A., & Ajayi, O. (2019). Water usage and pollution awareness among secondary school students in Nigeria. *African Journal of Environmental Science and Technology, 13*(2), 46–54.

Adeoba, M. I., Odjegba, E. E., & Pandelani, T. (2025). Nature-based solutions: Opportunities and challenges for water treatment. *Smart Nanomaterials for Environmental Applications*, 575–596.

Adeoti, O. S., Kandasamy, J., & Vigneswaran, S. (2023). Water infrastructure sustainability in Nigeria: A systematic review of challenges and sustainable solutions. *Water Policy, 25*(11), 1094–1111.

Adom, R. K., & Simatele, M. D. (2022). The role of stakeholder engagement in sustainable water resource management in South Africa. In *Natural resources forum* (Vol. 46, No. 4, pp. 410–427). Oxford, UK: Blackwell Publishing Ltd.

AfDB (African Development Bank). (2020). *Water sector development strategy for Africa.* AfDB Publications.

Agbo, C., Jeffrey, P., & Sule, M. N. (2025). Evaluation of failings in urban water supply and sanitation systems in Sub-Saharan africa: A systematic review to inform future planning. *Journal of Water, Sanitation and Hygiene for Development, 15*(2), 148–165.

Ahmed, T., Sipra, H., Zahir, M., Ahmad, A., & Ahmed, M. (2022). Consumer perception and behavior toward water supply, demand, water tariff, water quality, and willingness-to-pay: A cross sectional study. *Water Resources Management, 36*(4), 1339–1354.

Ahona, N., Anik, A. H., & Tareq, S. M. (2025). A systematic review of water, sanitation and hygiene of Bangladesh: Prospects and challenges to achieve sustainable development. *Water and Environment Journal, 39*(4), 402–418.

Akhtar, N., Syakir Ishak, M. I., Bhawani, S. A., & Umar, K. (2021). Various natural and anthropogenic factors responsible for water quality degradation: A review. *Water, 13*(19), 2660.

Akinola, S. R., & Yahaya, M. (2019). Public-private partnership and water provision in developing nations: Evidence from Nigeria. *African Journal of Public Policy and Administration, 12*(1), 67–83.

Al-Addous, M., Bdour, M., Rabaiah, S., Boubakri, A., Schweimanns, N., Barbana, N., & Wellmann, J. (2024). Innovations in solar-powered desalination: A comprehensive review of sustainable solutions for water scarcity in the Middle East and North Africa (MENA) region. *Water, 16*(13), 1877.

Alaerts, G. J. (2019). Financing for water—Water for financing: A global review of policy and practice. *Sustainability, 11*(3), 821.

Alawneh, A., Shawaqfah, M., Nusier, O., Odat, A. S. A., & Al-Fugara, A. K. (2024). Treated Wastewater for Sustainable Groundwater Recharge. In *International Conferences on Engineering and Applied Sciences Optimization* (pp. 62–73). Cham: Springer Nature Switzerland.

Ali, A. Y. (2024). Analysis of water losses and the Management of non-Revenue Water by plateau state water Board in Jos, Nigeria. *Journal of Applied Sciences & Environmental Management, 28*(7).

Ali, M. A., & Kamraju, M. (2024). The role of community participation in sustainable integrated water resources management: challenges, opportunities, and current perspectives. *Integrated Management of Water Resources in India: A Computational Approach: Optimizing for Sustainability and Planning*, 325–344.

Aliyu, A., & Hassan, A. (2020). Governance challenges in Nigeria's water sector. *International Journal of Water Resources Development, 36*(4), 587–603.

Allen, A., Griffin, L., & Johnson, C. (2020). Participatory approaches to urban water governance in Africa. *Environment and Urbanization, 32*(2), 67–84.

Al-Mansoori, S. (2025). The Growth of London from a Roman Outpost to a Global Metropolis. *Global Review of Humanities & Social Thoughts, 1*(01), 1–12.

Alvarez Wong, S. (2025). Enhancing Water Supply Resilience & Sustainability: Potential Solutions to Address Water Scarcity in Bogotá,.

Alves, R. A., Rudke, A. P., Martins, L. D., de Melo Souza, S. T., Bovo, A. B., Goroski Rambalducci, M. J., & Martins, J. A. (2024). Urban stormwater management from the perspective of nature-based solutions: A bibliometric review. *Journal of Ecohydraulics, 9*(2), 218–238.

Amoako, C., & Inkoom, D. K. (2018). The production of flood vulnerability in Accra, Ghana. *International Journal of Urban Sustainable Development, 10*(3), 1–22.

Angelakis, A. N., & Snyder, S. A. (2018). Wastewater treatment and reuse: Past, present, and future. *Water, 10*(4), 371.

Antwi, S. H., Stephens, C. G., Rolston, A., Getty, D., & Linnane, S. (2025). Public participation in environmental decision-making: A water sector perspective. *Environmental and Sustainability Indicators, 26*, 100656.

Anusha, M., & Prasad, R. (2022). Smart water management using IoT for water distribution monitoring. *Water Resources Management, 36*, 991–1005.

Aparcana, S. (2017). Approaches to formalization of the informal waste sector into municipal solid waste management systems in low-and middle-income countries: Review of barriers and success factors. *Waste Management, 61*, 593–607.

Arora, N. K., & Mishra, I. (2022). Sustainable development goal 6: Global water security. *Environmental Sustainability, 5*(3), 271–275.

Arrojo-Agudo, P., & Martínez, V. (2022). Water pricing and consumption behavior: Evidence from conservation programs. *Water Resources Management, 36*(5), 1821–1834.

Ateia, M., Wei, H., & Andreescu, S. (2024). Sensors for emerging water contaminants: Overcoming roadblocks to innovation. *Environmental Science & Technology, 58*(6), 2636–2651.

Ayanlade, A., Oladimeji, A. A., Okegbola, O. M., Eludoyin, A. O., Eslamian, S., Ayinde, A. F., & Perkins, P. E. (2022). Effect of climate change on water availability and quality: An assessment of socio-resilience in Nigeria. *Disaster Risk Reduction for Resilience: Disaster and Social Aspects* (pp. 245–262). Springer International Publishing.

Ayejoto, D. A., Egbueri, J. C., Agbasi, J. C., Omeka, M. E., Unigwe, C. O., Nwazelibe, V. E., et al. (2023). Influence of seasonal changes on the quality of water resources in southwestern Nigeria: a review. *Climate change impacts on Nigeria: environment and sustainable development*, 423–447.

Ayeni, F. E., & Oloruntoba, E. (2020). Urban water consumption behavior in Lagos, Nigeria. *Journal of Water, Sanitation and Hygiene for Development, 10*(3), 548–557.

Badini, S. (2025). *Essays on adaptation to environmental change* (Doctoral dissertation, Wageningen University and Research).

Baldino, N., & i Pujol, D. S. (2018). Characterizing the recent decline of water consumption in Italian cities. *Investigaciones Geográficas (España)*, (69), 9–21.

Barquet, K., Gunnarsson, M., & Engström, E. (2025). A bottom-up framework for product development in humanitarian, poor, and emerging markets. *Journal of Innovation and Entrepreneurship, 14*(1), 102.

Baye, F. (2025). Exploring urban infrastructure challenges in informal peri-urban Woldia: Barriers, implications, and informal strategies. *Frontiers in Sustainable Resource Management, 4*, 1555564.

Biswas, A. K., & Tortajada, C. (2019). Water governance in a changing world. *International Journal of Water Resources Development, 35*(2), 1–12.

Boelee, N. L., Geerts, S., & Venselaar, J. (2019). Urban water reuse for agriculture: A guide for policymakers. *IWMI Research Report, 178.*

Borah, G. (2025). Urban water stress: Climate change implications for water supply in cities. *Water Conservation Science and Engineering, 10*(1), 20.

Butler, D., Digman, C., Makropoulos, C., & Davies, J. W. (2024). *Urban drainage.* CRC Press.

Chandratreya, A. (2024). Sustainable water management through green infrastructure. *Indian Scientific Journal Of Research In Engineering And Management.*

Chapman, C. M. (2023). *Green and grey drainage infrastructure: costs and benefits of reducing surface water flood risk* (Doctoral dissertation, University of Oxford).

Chuenchum, P., Meneesrikum, C., Teerapanuchaikul, C., & Sriariyawat, A. (2024). Community participation and effective water management: A study on water user organizations (WUOs) in Thailand. *World Development Perspectives, 34*, Article 100589.

Cleaver, F., & Whaley, L. (2018). Understanding process, power, and meaning in adaptive governance: A critical institutional reading. *Ecology and Society, 23*(2), 49.

Conway, D., Dalin, C., Landman, W., & Osborn, T. (2019). Climate and southern Africa's water-energy-food nexus. *Nature Climate Change, 5*(9), 837–846.

Dagar, J. C., Gupta, S. R., & Sileshi, G. W. (2023). Urban and peri-urban agroforestry to sustain livelihood and food security in the face of global environmental change and epidemic threats. In *Agroforestry for sustainable intensification of agriculture in Asia and Africa* (pp. 89–118). Singapore: Springer Nature Singapore.

De-Luca, D. A., Cocca, D., Egidio, E., & Lasagna, M. (2024). Earthwater resources: a quantitative evaluation and future challenges. In *Water reuse and unconventional water resources: a multidisciplinary perspective* (pp. 13–38). Cham: Springer Nature Switzerland.

Delgado, A., Rodriguez, D. J., Amadei, C. A., & Makino, M. (2024). Water in circular economy and resilience (WICER) framework. *Utilities Policy, 87*, Article 101604.

Delmon, J. (2024). *Innovative funding and financing for infrastructure: Addressing scarcity of public resources*. Cambridge University Press.

Dezfooli, D., Bolson, J., Arabi, M., Sukop, M. C., Wiersema, I., & Millonig, S. (2023). A qualitative approach to understand transitions toward integrated urban water management. *Water, 15*(14), 2499. https://doi.org/10.3390/w15142499

Dolnicar, S., Hurlimann, A., & Grün, B. (2017). Water conservation behavior. *Water Research, 44*(2), 611–622.

Dubois, E., Loghmani-Khouzani, T., Ottaviani, S., Serrao, L., & Starkey, E. (2025, December). Are water-related nature-based solutions (NbS) assessed for their full multi-benefit potential? A review from an urban perspective. In *Sustainability Nexus Forum* (Vol. 33, No. 1, p. 22). Berlin/Heidelberg: Springer Berlin Heidelberg.

Dunga, K. E. (2026). Waterborne Diseases and Public Health: Risks and Interventions. In *Water Quality and Safety in the Global South: Challenges, Solutions and Future Directions* (pp. 87–112). Cham: Springer Nature Switzerland.

Dziegielewski, B. (2020). Water demand management and conservation planning. *Journal AWWA, 112*(4), 18–25.

Ezenwaji, E. E., Phil-Eze, P. O., & Otti, V. (2019). Household water demand and supply in urban Nigeria. *Water Supply, 19*(5), 1534–1544.

Faria, D. C., & Naval, L. P. (2022). Wastewater reuse: Perception and social acceptance. *Water and Environment Journal, 36*(3), 433–447.

Farley, M. (2018). Non-revenue water — International best practice. *Water Utility Journal, 21*(1), 1–12.

Fida, M., Li, P., Alam, S. K., Wang, Y., Nsabimana, A., & Shrestha, P. S. (2024). Review of groundwater nitrate pollution from municipal landfill leachates: Implications for environmental and human health and leachate treatment technologies. *Exposure and Health, 16*(5), 1225–1249.

Fielding, K. S., Dolnicar, S., & Schultz, T. (2019). Public acceptance of recycled water. *Wiley Interdisciplinary Reviews: Water, 6*(4), e1355.

Fletcher, T. D., Shuster, W., & Hunt, W. (2018). SUDS, green infrastructure, and water-sensitive design. *Urban Water Journal, 12*(7), 525–542.

Florides, G. A. (2024). Desalination in the urban water mix: Energy, economics and environmental considerations. *Desalination Reviews, 12*(2), 101–120.

Flörke, M., Schneider, C., & McDonald, R. I. (2018). Water competition between cities and agriculture driven by climate change and urban growth. *Nature Sustainability, 1*(1), 51–58.

Fred, O., Ukagwu, K. J., Abdulkarim, A., & Eze, V. H. U. (2024). Reliability and maintainability analysis of Solar Photovoltaic Systems in rural regions: A narrative review of challenges, strategies, and policy implications for sustainable electrification. *KIU Journal of Science, Engineering and Technology, 3*(2), 103–122.

Gallegos, A., Grigg, N. S., & Llano, W. (2026). Enablers and Obstacles in Integrated Water Resources Management (IWRM) Implementation and Their Contributions to Sustainable Territorial Development. *Land, 15*(2), 270.
García, X., & Ballester, J. (2017). Environmental education and domestic water consumption. *Environmental Management, 60*(4), 640–653.
Ghebreyesus, D., Gebremeskel, T., & Tareke, T. (2020). Public perception of wastewater reuse. *Environmental Challenges, 1*, 100004.
Gidadawa, Z. S., Gatawa, B. M., Umar, A., & Muhammad, A. (2024). Domestic water sources and factors influencing urban household attitudes on water storage facilities in sokoto metropolis, North Western Nigeria. *Sokoto Journal of Geographical Studies, 1*(1), 160–162.
Global Water Partnership. (2017). Integrated water resources management: Background and principles. https://www.gwp.org/
Gregory, J., & Sovacool, B. K. (2019). Rethinking the governance of energy poverty in sub-Saharan Africa: Reviewing three academic perspectives on electricity infrastructure investment. *Renewable and Sustainable Energy Reviews, 111*, 344–354.
Habel, S., Fletcher, C. H., Barbee, M. M., & Fornace, K. L. (2024). Hidden threat: The influence of sea-level rise on coastal groundwater and the convergence of impacts on municipal infrastructure. *Annual Review of Marine Science, 16*(1), 81–103.
Hoekstra, A. Y., Chapagain, A. K., Aldaya, M. M., & Mekonnen, M. M. (2018). Water footprint assessment. *Water, 10*(3), 296.
Hope, R., Foster, T., & Thomson, P. (2020). Reducing rural water failures in Africa. *World Development, 130*, 104928.
Hossain, I., Haque, A. M., & Ullah, S. A. (2023). Assessment of domestic water usage and wastage in urban Bangladesh: A study of Rajshahi City corporation. *The Journal of Indonesia Sustainable Development Planning, 4*(2), 109–121.
Hussein Farh, H. M., Ben Seghier, M. E. A., Taiwo, R., & Zayed, T. (2023). Analysis and ranking of corrosion causes for water pipelines: A critical review. *NPJ Clean Water, 6*(1), 65.
Hutton, C., Kapelan, Z., & Savic, D. (2020). AI-based optimization of urban water networks. *Journal of Hydroinformatics, 22*(5), 1210–1225.
Ibrahim, I., & Umar, H. A. (2024). Hydrochemical Characterization, Spatial Analysis and Quality Indexing of Groundwater in part of Gusau Town, Zamfara State Northwestern Nigeria. *African Journal of Earth and Environmental Science, 6*(2).
Iorhemen, O. (2023). Wastewater recycling in Nigeria: Opportunities and challenges. *Environmental Technology & Innovation, 29*, 102944.
IPCC. (2021). Climate change 2021: Impacts, adaptation and vulnerability. In P. R. Shukla et al. (Eds.), *Contribution of working group II to the sixth assessment report of the intergovernmental panel on climate change*. Cambridge University Press. https://www.ipcc.ch/report/ar6/wg2/
IPCC. (2022). *Climate change 2022: Impacts*. Cambridge University Press.
Isa, S., & Usman, H. (2020). Faith-based environmental education and water conservation in northern Nigeria. *Journal of Islamic Environmental Studies, 3*(2), 54–63.
Ishaku, H. T., Majid, M. R., & Johar, F. (2021). Public participation and water governance reform in Nigeria. *Habitat International, 65*, 43–51.
Isyaku, A. (2026). Sustainable management of the Tiga dam and Kadawa irrigation area: addressing socio-economic and environmental challenges in Kano state, Nigeria.
Jandaghian, Z., & Colombo, A. (2024). The role of water bodies in climate regulation: Insights from recent studies on urban heat island mitigation. *Buildings, 14*(9), 2945.
Jiang, Y. (2023). Financing water investment for global sustainable development: Challenges, innovation, and governance strategies. *Sustainable Development, 31*(2), 600–611.
Jiménez, B., & Asano, T. (2015). *Water reuse: An international survey*. IWA Publishing.
Julio, N., Figueroa, R., & Ponce Oliva, R. D. (2021). Water resources and governance approaches: Insights for achieving water security. *Water, 13*(21), 3063.
Kannan, A., George, D., & Rajesh, R. (2022). Pathogen risks in wastewater reuse. *Environmental Monitoring and Assessment, 194*(3), 201.

Kant, N., & Wrat, G. (2025). A review of integrated groundwater and surface water management for environmental sustainability. *Acta Geochimica, 44*(5), 1120–1141.

Kariuki, M., & Schwartz, J. (2015). *Small-scale private water providers.* World Bank Working Paper.

Kaushik, G., Kondekar, K., Aher, R., Dhage, G., Chel, A. L., Chaudhery, H., & Fatima, S. (2026). Harmonizing water governance: Exploring governance structures, stakeholder engagement, and policy in water management. In *Advances in Hydrology* (pp. 225–246). Elsevier.

Keenan-Jones, D. C., Serra-Llobet, A., He, H., & Kondolf, G. M. (2025). Urban development and long-term flood risk and resilience: Experiences over time and across cultures. Cases from Asia, North America, Europe and Australia. *Urban Studies, 62*(3), 469–486.

Khair, S. M., Mushtaq, S., Reardon-Smith, K., & Ostini, J. (2019). Diverse drivers of unsustainable groundwater extraction behaviour operate in an unregulated water scarce region. *Journal of Environmental Management, 236*, 340–350.

Khan, T., Samiullah, M., Rouf, I., Sultana, S., Rahman, S., Rahman, B., & Khanum, R. (2024). The nexus of water scarcity and climate change: Understanding interconnected challenges and formulating resilient strategies. *International Journal of Environmental Sciences, 7*(3), 57–68.

Khanam, Z., Sultana, F. M., & Mushtaq, F. (2023). Environmental pollution control measures and strategies: an overview of recent developments. *Geospatial analytics for environmental pollution modeling: Analysis, control and management*, 385–414.

Kingdom, B., Liemberger, R., & Marin, P. (2020). The challenge of reducing non-revenue water. *Water Loss Journal, 35*(1), 12–27.

Kong, F., Sun, S., & Lei, T. (2021). Understanding China's urban rainstorm waterlogging and its potential governance. *Water, 13*(7), 891.

Kosoe, E. A., & Ogwu, M. C. (2025). Sustainable urban planning and environmental management. In *Evaluating environmental processes and technologies* (pp. 463–485). Cham: Springer Nature Switzerland.

Kudapa, V. K. (2025). Restoration Strategies for Rivers and Wetlands Affected by Overextraction of Water. *Global Water Scarcity: Causes, Impacts, and Management Strategies*, 331–352.

Kulkarni, S., Whig, P., Soubache, I. D., & Banerjee, S. (2026). Water resources pollution management: issues and perspectives. In *Sustainable Solutions for Environmental Pollution* (pp. 1–20). Elsevier.

Kumari, S., Dwivedi, S., Khan, M. E. R., Nayanam, S., Dhasmana, A., & Malik, S. (2023). The challenges of wastewater and wastewater management. In *Advanced and Innovative Approaches of Environmental Biotechnology in Industrial Wastewater Treatment* (pp. 99–121). Singapore: Springer Nature Singapore.

Kwak, Y., Seo, D., Kim, H., & Lee, J. (2019). Smart water management using ICT. *Water, 11*(5), 1088.

Lambert, A., & Hirner, W. (2018). *Losses from water supply systems.* IWA Water Loss Task Force.

Li, R., Zhao, Y., & Chen, Y. (2020). Real-time monitoring of leakage in water distribution systems. *Water Resources Management, 34*(1), 455–469.

Liemberger, R., & Wyatt, A. (2019). Quantifying the global non-revenue water problem. *Water Supply, 19*(3), 831–837.

Liu, Z., Ying, J., He, C., Guan, D., Pan, X., Dai, Y., et al. (2024). Scarcity and quality risks for future global urban water supply. *Landscape Ecology, 39*(2), 10.

Mabrouk, M. (2025). Toward climate-resilient cities: A review of nature-based solutions for urban flood management. *Advance in Sustainability, 5*(1).

Madias, K., & Szymkowiak, A. (2022). Residential sustainable water usage and water management: Systematic review and future research. *Water, 14*(7), 1027.

Mahi, A., & Murana, K. M. K. (2024). SWAT Hydrological Model of Zamfara Watershed of Sokoto-Rima River Catchment, North West Nigeria. *Nigerian Journal of Technology, 43*(4), 818–828.

Mallick, S. K. (2024). Peri-urban interface: Challenges and opportunities in spatial planning and management. *Modern Cartography Series, 11*, 529–542.

Mannina, G. (2022). Water reuse and recycling: Opportunities for circular cities. *Journal of Environmental Management, 305*, 114350.

Marks, J., & Zadoroznyj, M. (2019). Social acceptance of water reuse. *Sociological Research Online, 24*(3), 514–532.

Marlow, D., Moglia, M., Cook, S., & Beale, D. (2018). Towards sustainable urban water management. *Water Science & Technology, 68*(1), 1–10.

Mason, N., Oyaya, C., & Boulenouar, J. (2020). Reforming urban sanitation under decentralization: Cross-country learning for Kenya and beyond. *Development Policy Review, 38*(1), 42–63.

Matimolane, S. (2026). Governance failure and corruption perpetuate water insecurity in democratic South Africa. *Discover Water*.

Mehta, L., Basa, S., Anand, S., & Bhardwaj, A. (2019). Water governance and inclusion: Policy pathways. *Global Environmental Change, 57*, 101935.

Meng, X. (2022). Understanding the effects of site-scale water-sensitive urban design (WSUD) in the urban water cycle: A review. *Blue-Green Systems, 4*(1), 45–57.

Meyer, R. (2021). District metered areas for leakage reduction. *Civil Engineering and Environmental Systems, 38*(2), 101–116.

Mitchell, V., Inman, M., & Kazmierczak, A. (2019). Water demand management in cities. *Urban Water Journal, 16*(3), 150–163.

Moglia, M., Cook, S., & Tapsuwan, S. (2018). Promoting water conservation: Where to from here? *Water, 10*(11), 1510.

Mohamed, M., Harun, N. H., Jaaffar, A. H., & Idris, N. D. M. (2026). Every drop counts: A systematic literature review on the role of tariffs in sustainable water supply services. *Utilities Policy, 98*, Article 102112.

Monks, I., Stewart, R. A., Sahin, O., & Keller, R. (2019). Revealing unreported benefits of digital water metering: Literature review and expert opinions. *Water, 11*(4), 838.

Mosleh, L., Kazemi, A., & Delavar, M. R. (2021). Role of models in the decision-making process for integrated urban water management. *Water, 13*(9), 1252.

Moura, N. D., Pellegrino, P. R. M., & Martins, J. R. S. (2016). Best management practices as an alternative for flood and urban storm water control in a changing climate. *Journal of Flood Risk Management, 9*(3), 243–254.

Mugagga, F., & Nabaasa, B. (2016). The centrality of water resources to sustainable development. *European Journal of Earth and Environment, 3*(2), 1–15.

Mugari, E., Nethengwe, N. S., & Gumbo, A. D. (2025). A co-design approach for stakeholder engagement and knowledge integration in flood risk management in Vhembe district, South Africa. *Frontiers in Climate, 7*, 1517837.

Muhammad, B. I., Olanipekun, O. I., Nabage, Y. A., & Bala, J. (2024). Flood risk assessment and its implications on critical infrastructure resilience in Nigeria. *African Journal of Environmental Sciences and Renewable Energy, 17*(1), 69–85.

Muller, M. (2018). Cape Town's drought: Lessons for urban water security. *Ambio, 47*(7), 781–783.

Muoghalu, L. (2025). Sustainable Urban Development and Climate Resilence In Nigerian Cities. In *FESCON Conference Proceedings* (Vol. 5, No. 1, pp. 1–41).

Mutchek, M., & Williams, E. (2019). Moving towards smart water systems. *Environmental Science & Technology, 53*(2), 661–667.

Mutikanga, H., Sharma, S., & Vairavamoorthy, K. (2017). Water loss management strategies. *Water International, 42*(6), 863–877.

Muwafu, S. P., Celliers, L., Scheffran, J., & Máñez Costa, M. (2024). Community governance performance of nature-based solutions for sustainable urban stormwater management in Sub-Saharan Africa. *Sustainability, 16*(19), 8328.

Naiga, R., Ananga, E. O., & Kakumba, U. (2024). Gendered participation in water governance: Implications for successful communitybased water management and women empowerment. *International Journal of Rural Management, 20*(2), 255–270.

Nauman, M., Naheed, R., & Khan, J. (2024). Navigating sustainable horizons: Exploring the dynamics of financial stability, green growth, renewable energy, technological innovation, financial inclusion, and soft infrastructure in shaping sustainable development. *Environmental Science and Pollution Research, 31*(20), 29939–29956.

Newing, A., & Palermo, A. (2023). Smart metering and leak detection: Lessons for utilities. *Utilities Policy, 78*, 101–110.

Nieuwenhuis, E., Cuppen, E., Langeveld, J., & de Bruijn, H. (2021). Towards the integrated management of urban water systems: Conceptualizing integration and its uncertainties. *Journal of Cleaner Production, 280*, 124977.

Nkiaka, E., Taylor, R., & Dougill, A. (2020). Water security in sub-Saharan Africa. *Water International, 45*(6), 1–19.

Nwankwoala, H. O. (2021). Water resources management in Nigeria. *Journal of African Environmental Studies, 4*(2), 18–27.

Nwankwoala, H. O., & Mamah, L. (2019). Wastewater reuse in Lagos. *African Journal of Environmental Science, 13*(6), 244–251.

Nyika, J., & Dinka, M. O. (2023). *Water challenges in rural and urban sub-Saharan Africa and their management*. Springer.

Obi, C. N., Umeudu, A. N., & Okoh, D. (2021). Community perception of water as a free good in Nigeria. *Journal of Water Policy, 23*(4), 812–824.

Obunga, P. O., Rwanga, S. S., Dinka, M. O., & Otieno, B. O. (2025). Review of the emerging technologies in the water sector with a focus on the deployment of Internet of Things solutions. *npj clean water*, *8*(1), 53.

Ochuko, E., & Odiase, O. (2023). Public awareness of water conservation in Nigeria. *African Journal of Sustainable Development, 13*(1), 67–76.

Odetola, C., & Adedeji, O. (2022). Urban water infrastructure decay in Nigeria. *International Journal of Water Resources, 7*(2), 33–42.

Okoli, N. J., & Kabaso, B. (2024). Building a smart water city: IoT smart water technologies, applications, and future directions. *Water, 16*(4), 557.

Okonkwo, E., & Ezenwaji, E. (2021). Water-saving behavior in Nigerian households. *Water Conservation and Science, 5*(3), 120–130.

Olagunju, O., & Omole, D. (2023). Reducing illegal water connections in Nigeria. *Water Practice and Technology, 18*(1), 65–73.

Olanrewaju, O., & Popoola, A. (2023). Mitigating drought impacts through watershed protection. *Environment and Ecology Research, 11*(4), 58–66.

Olawuni, P., Adeyemi, O., & Ojo, F. (2023). The challenge of water metering in Nigeria. *Journal of Water Resource Economics, 11*(1), 78–87.

Olodu, D. D., Inegbedion, F., & Ihenyen, O. I. (2025). Smart water management systems: Engineering innovations for water conservation and distribution. *Journal of Studies in Advanced Technologies, 3*(1), 13–31.

Oluseyi, T. O., & Nweke, J. C. (2020). Implications of Improper Sewage Management on Public Health: A Case Study of Kosofe Local Government Area, Lagos State. *African Journal of Housing and Sustainable Development, 1*(1).

Oral, H. V. (2020). A review of nature-based solutions for urban water management in European circular cities. Biosystems Engineering, 2(1), 112–134. Publishing.

Osman, A. I., Chen, Z., Elgarahy, A. M., Farghali, M., Mohamed, I. M., Priya, A. K., et al. (2024). Membrane technology for energy saving: Principles, techniques, applications, challenges, and prospects. *Advanced Energy and Sustainability Research, 5*(5), 2400011.

Oyerinde, A. O., & Jacobs, H. E. (2022). The complex nature of household water supply: An evidence-based assessment of urban water access in Southwest Nigeria. *Journal of Water, Sanitation and Hygiene for Development, 12*(3), 237–247.

Oyebanji, A. (2022). *Developing sustainable water supply management strategies in Lagos*. Walden University.

Oyebanjo, O. O., Ojekunle, Z. O., Oderinde, F. O., & Osinusi, K. B. (2025). Factors Influencing Household Water Consumption and Revenue Generation in Urban Water Supply System in Southwest Nigeria. *Ijagun Journal of Social and Management Sciences, 9*(1), 37–50.

Pachouri, V., Kothari, P., Kathuria, S., Gehlot, A., Singh, R., Thakur, A. K., et al. (2025). Revolutionizing urban water resilience: Innovative strategies and advancements in sustainable urban drainage systems (SuDS). *Desalination and Water Treatment*, 101407.

Paerl, H. W., & Otten, T. G. (2016). Harmful cyanobacterial blooms: Causes, consequences, and controls. *Microbial Ecology, 57*(3), 315–330. https://doi.org/10.1007/s00248-009-9612-8

Palermo, A., Di Iaconi, C., Nanni, A., Mancini, M., Clemente, R., & Rossi, G. (2022). Digital transformation for water utilities: Sensors, analytics and institution building. *Journal of Water Resources Planning and Management, 148*(6), 04022033.

Perera, D., Smakhtin, V., Williams, S., North, T., & Curry, A. (2021). Ageing water storage infrastructure: An emerging global risk. *UNU-INWEH Report Series, 11*, 25.

Pérez-González, D., Mirzaei, P., & Keane, M. (2021). IoT technologies for smart water distribution. *Smart Cities, 4*(3), 1205–1224.

Pizzorni, M., Innocenti, A., & Tollin, N. (2024). Droughts and floods in a changing climate and implications for multi-hazard urban planning: A review. *City and Environment Interactions, 24*, Article 100169.

Qadir, M., Drechsel, P., & Jiménez, B. (2020). Wastewater reuse in the circular economy. *Resources, Conservation & Recycling, 165*, 105218.

Rahaman, M. M., & Varis, O. (2020). Integrated water resources management. *Water Policy, 22*(3), 6–19.

Ramm, A. (2023). Regulatory frameworks for water reuse: Lessons from global pilots. *Water Policy, 25*(4), 679–695.

Reader, G. T. (2022). Access to drinking water, food security and adequate housing: Challenges for engineering, past, present and future. *Symposium on responsible engineering and living* (pp. 1–41). Springer International Publishing.

Ren, X., Hong, N., Li, L., Kang, J., & Li, J. (2020). Effect of infiltration rate changes in urban soils on stormwater runoff process. *Geoderma, 363*, Article 114158.

Renwick, D. V., Heinrich, A., Weisman, R., Arvanaghi, H., & Rotert, K. (2019). Potential public health impacts of deteriorating distribution system infrastructure. *Journal-American Water Works Association, 111*(2), 42.

Rivera-Williams, J. (2025). A Multiple Perspective Analysis on Urban Poverty in Developing Nations: A Case Study on Nigeria, Colonial Implications, and Modern-Day Outcomes.

Robins, S. (2019). 'Day Zero', hydraulic citizenship and the defence of the commons in Cape Town: A case study of the politics of water and its infrastructures (2017–2018). *Journal of Southern African Studies, 45*(1), 5–29.

Robotham, J., Old, G., Rameshwaran, P., Sear, D., Trill, E., Bishop, J., et al. (2023). Nature-based solutions enhance sediment and nutrient storage in an agricultural lowland catchment. *Earth Surface Processes and Landforms, 48*(2), 243–258.

Rodriguez, C., Gonzalez, R., & Viera, M. (2020). Potable water reuse using membranes. *Water Environment Research, 92*(10), 1635–1650.

Russell, S., & Knoeri, C. (2020). Influence of socio-technical factors on water-saving behavior. *Environmental Sociology, 6*(4), 327–339.

Rusprayunita, N. R. A., Puspitasari, S., Mahiroh, H., Setyani, E. R., Citraningtyas, V. I., Wahyuningsih, W. S., & Rauf, A. U. (2025). Water Pollution Reduction For Sustainable Urban Development. In *Sustainable Urban Environment And Waste Management: Theory And Practice* (pp. 1–21). Singapore: Springer Nature Singapore.

Sahu, P., & Debsarma, C. (2023). Urban water scarcity: A global challenge and impending solutions. *Impacts of urbanization on hydrological systems in India* (pp. 129–142). Springer International Publishing.

Saikia, P., Ghosh, A., Ravindranath, N. H., & Bandyopadhyay, D. (2022). City water resilience framework: Integrating planning, infrastructure and governance. *Cities, 119*, 103424.

Santos, E., Carvalho, M., & Martins, S. (2023). Sustainable water management: Understanding the socioeconomic and cultural dimensions. *Sustainability, 15*(17), 13074.

Santos, E. (2024). Beyond leakage: Non-revenue water loss and economic sustainability. *Urban Science, 8*(4), 194.

Sarkar, A. (2020). Informal water vendors and the urban poor: Evidence from a Nairobi slum. *Water International, 45*(5), 443–457.

Satterthwaite, D. (2016). Informal settlements and their water and sanitation challenges: Global overview. *Environment and Urbanization, 28*(2), 415–431. https://doi.org/10.1177/0956247816663558

Satterthwaite, D., Archer, D., & Colenbrander, S. (2018). *Adapting to climate change in urban areas*. International Institute for Environment and Development Working Paper.

Schmidt, J., & Matthews, T. (2017). Transformational water governance. *Water. Alternatives, 10*(1), 177–196.

Sharma, S., Burn, S., & O'Connor, P. (2019). Urban water reuse technologies. *Water, 11*(1), 180.

Shuaibu, A. M., Musa, K. O., & Okiyi, I. M. (2025). Groundwater recharge modelling using SWAT analysis for groundwater reserve quantification of Ka watershed catchment area part of Sokoto-Rima Basin, North West Nigeria. *African Scientific Reports*, 255–255.

Siddique, I. (2022). Sustainable water Management in Urban Environments. *Chemistry Research Journal, 7*(4), 95–101.

Silva, J. A. (2023). Wastewater treatment and reuse for sustainable water resources management: A systematic literature review. *Sustainability, 15*(14), 10940.

Singha, B., Eljamal, O., Karmaker, S. C., Maamoun, I., & Sugihara, Y. (2022). Water conservation behavior: Exploring the role of social, psychological, and behavioral determinants. *Journal of Environmental Management, 317*, Article 115484.

Swe, K. T., Rahman, M. M., Rahman, M. S., Teng, Y., Abe, S. K., Hashizume, M., & Shibuya, K. (2021). Impact of poverty reduction on access to water and sanitation in low-and lower-middle-income countries: Country-specific Bayesian projections to 2030. *Tropical Medicine & International Health, 26*(7), 760–774.

Tayo, Y. (2023). *Assessment of water supply in the informal settlement of Lagos mainland local government area.* Master's thesis, Iowa State University.

Tella, T. A., Festus, B., Olaoluwa, T. D., & Oladapo, A. S. (2025). Water and wastewater treatment in developed and developing countries: Present experience and future plans. In *Smart nanomaterials for environmental applications* (pp. 351–385). Elsevier.

Thakur, R. (2026). A Community-Based Behavioural Model For Water Conservation: Mitigating Climate Change Impacts In Lowcost Housing Communities In South Africa. *Journal for Sustainable Development Law & Policy, 17*(2).

Tiwari, S., Botre, B. A., Sridhar, S., & Santos, C. M. (2025). Smart water grids in India: a systematic review of purification, conservation, and emerging digital trends. *Proceedings of the Indian National Science Academy*, 1–21.

Tripathi, I. M., Kumar, S., Modi, A., Bhat, A. M., Bhagat, C., Mohapatra, P. K., & Kansal, M. L. (2026). Towards Water Sensitivity: A Critical Review of Urban Water Management Strategies. *Water Resources Management, 40*(3), 94.

Twum, K. O., & Abubakari, M. (2020). Drops in the city: The puzzle of water privatization and consumption deficiencies in urban Ghana. *Water Policy, 22*(3), 417–434.

Tzanakakis, V. A., Paranychianakis, N. V., & Angelakis, A. N. (2020). Water supply and water scarcity. *Water, 12*(9), 2347.

Tzanakakis, V. A., Capodaglio, A. G., & Angelakis, A. N. (2023). Insights into global water reuse opportunities. *Sustainability, 15*(17), 13007.

Ukpai, S. N. (2022). Water policy reform in the Nigeria water governance system: Assessment of water resources management based on OECD Principles on Water Governance. *Water Policy, 24*(10), 1704–1722.

Umar, A., & Goni, A. (2022). Water utility performance in northern Nigeria. *Journal of Environmental Management, 59*(2), 142–151.

UNEP. (2021a). *Global environment outlook: Healthy planet*. Healthy People.
UNEP. (2021b). *Wastewater management in Africa*. United Nations Environment Programme.
UNESCO World Water Assessment Programme (WWAP). (2020). *The United Nations world water development report 2020: Water and climate change*. UNESCO. https://www.unwater.org/publications/world-water-development-report-2020
UN-Habitat. (2020). World cities report 2020: The value of sustainable urbanization. UN-Habitat. https://unhabitat.org/sites/default/files/2020/10/wcr_2020_report.pdf
UN-Water. (2020). *The United Nations world water development report*. UNESCO.
UN-Water. (2022). Water and Urbanization. https://www.unwater.org/water-facts/water-and-urbanization
Vanhille, J. (2018). Behavioural change approaches for water conservation. *Journal of Environmental Psychology, 58*, 1–8.
Varma, V. C., Rathinam, R., Suresh, V., Naveen, S., Satishkumar, P., Abdulrahman, I. S., et al. (2024). Urban waste water management paradigm evolution: Decentralization, resource recovery, and water reclamation and reuse. *Environmental Quality Management, 33*(4), 523–540.
Vogel, S. K. (2018). *Freer markets, more rules: Regulatory reform in advanced industrial countries*. Cornell University Press.
Wah, J. N. K. (2026). Solving water scarcity: Sustainable strategies & technological solutions.
Wang, Y., Zhang, X., Liu, H., Gong, Y., Li, X., & Li, J. (2025). A Comprehensive Review of the Status and Removal of Emerging Contaminants in Urban Water Bodies Mediated by Stormwater Over the Past 30 Years. *Water, Air, & Soil Pollution, 236*(15), 1023.
Webber, J. L., Fletcher, T. D., Cunningham, L., Fu, G., Butler, D., & Burns, M. J. (2020). Is green infrastructure a viable strategy for managing urban surface water flooding? *Urban Water Journal, 17*(7), 598–608.
WHO & UNICEF Joint Monitoring Programme (JMP). (2019). Progress on household drinking water, sanitation and hygiene 2000–2017: Special focus on inequalities. WHO/UNICEF. https://washdata.org/report/jmp-2019-report
Williams, J. (2022). Desalination in the 21st century: A critical review of trends and debates. *Water Alternatives, 15*(2), 193–217.
Worgu, Q. O., & Ikuru, S. E. (2026). Rural-Urban Migration and Socio-structural Challenges for Sustainable Development in Nigeria. *Central Asian Journal of Social Sciences and History, 7*(1), 233–251.
World Bank. (2020). *High and dry: Climate change, water, and the economy*. The World Bank. https://www.worldbank.org/en/topic/water/publication/high-and-dry-climate-change-water-and-the-economy
World Bank. (2021). *Water security in Africa*. World Bank Publications.
World Bank. (2022). *Tracking SDG 6 progress*. World Bank.
World Bank. (2024). *Financing urban water infrastructure: approaches and instruments. (Policy report)*.
World Health Organization. (2022). *State of the world's drinking water: an urgent call to action to accelerate progress on ensuring safe drinking water for all*. World Health Organization.
Yadav, A., Gupta, R., Kumar, S., Singh, P., & Sharma, M. (2020). *Urban water pollution sources and impacts in south Asian*.
Younos, T., Lee, J., & Parece, T. E. (Eds.). (2022). *Resilient water management strategies in urban settings: innovations in decentralized water infrastructure systems*. Springer Nature.
van der Zaag, P. (2021). IWRM: A water governance approach. *Water International, 46*(3), 350–356.
Zabidi, H. A., Goh, H. W., Chang, C. K., Chan, N. W., & Zakaria, N. A. (2020). A review of roof and pond rainwater harvesting systems for water security: The design, performance and way forward. *Water, 12*(11), 3163.

Chapter 7
Adaptive Capacity in Practice: A Conceptual Framework for Sustainable Transformation of Water Shortages in Nigeria

Oluwaseun Fadeyi, Enoch Bessah, and Olusola Ololade

Abstract As climate change forces global drinking water provision towards stress levels, human populations, especially city dwellers, have become increasingly vulnerable to water shortages. This has prompted the need to adapt effectively. In recent times, adaptation pathways to water shortages have focused mainly on societal attributes, with little or no consideration on other external qualities that society can imbibe to survive. This leaves a significant gap in assessing what constitutes ideal adaptation to water shortages. We adapted critical indicators from existing models to develop a framework that helps to assess and improve adaptive capacity based on the concept of *role-playing* by individuals, households, communities, and institutions, who are key stakeholders in drinking water management and governance in Nigeria. Our framework contributes two unique ideas to the body of adaptive capacity literature. First, it reveals how adaptive capacity ideologies can be linked and jointly adapted to solve water-related problems in a resource-constrained context. Second, it provides a guideline as to how developing nations can source and leverage new knowledge by asking and making efforts to answer the right questions. This empowers them to exercise greater capacity to improve on inherent challenges of the water system. If properly executed, collaborative efforts can significantly mitigate the impacts of stressors on water provision, supply, access, and usage. This can have tremendous long-term benefits on the water system.

Keywords Adaptation · Adaptive capacity · Climate change · Developing countries · Framework · Water shortage

---

O. Fadeyi (✉) · O. Ololade
Centre for Environmental Management, Faculty of Natural and Agricultural Sciences, University of the Free State, Bloemfontein, South Africa
e-mail: phar2kind@gmail.com

E. Bessah
Department of Agricultural and Biosystems Engineering, Kwame Nkrumah University of Science and Technology, Kumasi, Ghana

M. I. Adeoba et al. (eds.), *Water Resources Management in Nigeria*, World Water Resources 10, https://doi.org/10.1007/978-3-032-22813-0_7

7.1 Introduction

Globally, urban residents face disproportionate challenges, with water shortages being the predominant issue. Across cities in sub-Saharan Africa, water shortages has resulted in consequences such as severe health risk linked to the consumption of contaminated water, hovering between diverse adaptive choices to access safe drinking water, and increasing adaptation-induced expenses (Dos Santos et al., 2017; Isukuru et al., 2024; Niasse & Varis, 2020). It is therefore critical and logical to enhance the capacity of cities to effectively cope and respond to water scarcity in the face of additional drinking water challenges (e.g., those imposed by climate change). Multiple frameworks have been developed to guide adaptive capacity in water management, yielding diverse theoretical and practical implications (Brooks & Adger, 2005; Eisenack & Stecker, 2012; Juhola & Kruse, 2015; Kelly & Adger, 2000). Nevertheless, research that localise these frameworks to suit real-time water-constrained societies is limited. This hinders the progress of strategic efforts that can drive adaptive water governance, particularly in developing countries.

Smit and Wandel (2006) report that adaptive capacity fundamentally focuses on the potentials of socio-ecological systems to thrive in the face of heightened climate change impacts. As a consequence, Engle (2011a, 2011b) maintains that existing frameworks have excessively emphasised the influential drivers of adaptive capacity such as technology, local institutions, infrastructural development, and knowledge. Freduah et al. (2019) nonetheless believes that the benefits accrued to adaptive capacity and components are context-driven. For example, concepts such as institutional quality/capacity and community resilience are linked to the ability of the institution/community to manage climate change impacts and develop innovative plans, which offer long-term success (Gupta et al., 2010). Wagner et al. (2014) notes that adaptation thrives on the effective management and prioritisation of resources, technology, local institutions, and knowledge. This is because investing scarce resources alongside new innovations (such as technology) into community adaptation only follows the path of effective knowledge mobilisation. By doing so, the relevance of adaptation is duly emphasised to the population with clear-cut responsibilities ascribed to adaptive capacity outcomes. Although local adaptive capacity (LAC) is referred to as such, Smit and Wandel (2006) stressed that it is the product of events beyond the local level, suggesting an interaction of diverse conditions that drive LAC.

Notwithstanding the extent of scholarship associated with the concept of adaptive capacity, a disconnect exists between the trajectories of existing frameworks and the core aspects of social-ecological systems. This is particularly so if one considers the drivers linked to factors that drive effective and sustainably coping with water shortages in cities. While Brooks and Adger (2005) have identified '*governance*' as the most critical indicator for measuring adaptive capacity regardless of resource or context, Engle (2011a, 2011b) believes that operational management of institutions is the ideal construct. Furthermore, Juhola and Kruse (2015) suggest that the nature of interaction between science and policy may serve as the best approach to assessing adaptive capacity. According to Nixon et al. (2022), social influence, which combines institutional and cross-sectoral collaborations, controls adaptive capacity, while Nkhata and Breen (2016) recommend that system performance indicators (knowledge, resources, infrastructure, and flexibility) offer a comprehensive assessment, albeit with technical and market-based aspects. According to Freduah et al. (2019), assessing socio-ecological systems through the lens of diverse climatic and non-climatic factors provide a broader perspective on numerous areas of focus.

A critical examination of existing frameworks reveals significant attention to 'what is in existence', i.e., the societal situation (e.g., resources, institutions, socio-economic makeup, and even societal challenges) upon which adaptation assessment frameworks are built. For example, Freduah et al.'s (2019) adaptive capacity framework focused on climate and non-climate stressors in small scale fishing systems, which can be grouped as inherent challenges to a socio-ecological unit. In the same vein, Wagner et al. (2014) focused on building adaptation against Hurricane Sandy. While these approaches are good at least in the context of what societies possess and can offer to protect themselves against climate change and other stressors, we argue that adaptation cannot be sustained when the focus is only on what exists with respect to how societies approach coping. This is because certain adaptive actions barely align with global best practices and can only offer short-term societal benefits. For example, developing countries in the global south mostly suffer from water shortages accompanied with many kinds of unhealthy adaptive practices developed by households and communities. Thus, building an adaptive framework only on the need to adapt and cope would likely lead to more complicated challenges. Relying only on what societies have and can utilise to adapt would imply encouraging the barest minimum. This may limit the rationale for progress (i.e., seeking to adapt rightly), especially in the context of environmental, health, and social best practices. As a result, we posit that that the current direction of adaptive capacity assessments poses a methodical gap which arises from how scholars view socio-ecological systems merely as what they are as against what they should be when the need arises to mobilise adaptive capacity against climate change. Based on this understanding, this study aims to develop forward-looking framework that not only helps cities assess their current standing in terms of quality and performance of existing water system, but also provides a road-map as to how they can harness additional resources and expertise to achieve improved performance of the water system.

7.2 Method

7.2.1 Indicator Sourcing and Synthesis

To develop the proposed framework, we conducted a systematic review of the literature on adaptive capacity and urban water resilience. Four databases; Scopus, Web of Science (WoS), Water Resources Abstracts (WRA), and ProQuest were searched using the keywords: ADAPATATION, ADAPTIVE CAPACITY, WATER, URBAN, and RESILIENCE. Synonyms of some of the keywords (e.g., COPING, CITY) were also used to enhance search results (Salvador-Oliván et al., 2021). Boolean operators were introduced to combine keywords and achieve robust search output (Alharbi & Stevenson, 2020). A set of inclusion/exclusion criteria was adopted as limiters to standardise search execution (Patino & Ferreira, 2018). For example, location (i.e., developing countries), language (English), Interest (frameworks or models), time-frame (2005–2025), and others relevant attributes (publisher's reputation and text completeness). The search produced a total of 409 documents, which were saved in Zotero, a bibliography data management tool. Using the inclusion/exclusion criteria, the documents were manually screened and scanned by the research team to arrive at eleven (11) documents useful for the new framework. The Preferred Reporting Items for Systematic Reviews and Meta-Analyses (PRISMA) in Fig. 7.1 shows the steps adopted to select the documents.

Following the suggestions by Ahmed et al. (2025) and Mitchell and Schmitz (2023), we adopted qualitative thematic identification and clustering to derive key indicators from the selected documents after carefully reading through over and over again. This was achieved using Atlas.ti (version 25.0.1) qualitative analysis software. Choice of the thematic identification and clustering approach is due to the narrative nature of the documents. The indicators were therefore selected based on links with the concepts of *climate change*, *population*, and *migration*, which are critical stressors associated with drinking water in cities. This step produced 18 indicators associated with four (4) frameworks/concepts. We embark on further reading about each of the indicators to ascertain their relevance to urban drinking water system, especially in developing countries.

7.2.2 Framework Development: Adoption and Analytical Approach

Following the identification of 18 indicators, we developed questions related to each to facilitate model testing. This step is useful for assessing a city's adaptive capacity level (i.e., whether a city can be assigned a prevailing, intermediate, or ideal/enhanced adaptive capacity status). The questions were designed in two forms: thirty-three (33) current capacity items, which assess what exists, and twenty-five (25) futuristic questions, which seek to assess ongoing plans and efforts towards achieving ideal capacity.

Designed to be used by city water managers and policymakers, the framework can be analysed using a binary rating scale (YES coded as 1 and NO/UNSURE coded as 0), with all questions scored using Microsoft Excel. The choice of this rating approach is to discourage cognitive overload and minimise ambiguous interpretation (Suárez-García et al., 2024) when using the framework. Since the goal of the new framework is to enable users to promptly understand the city's adaptive capacity level in response to water shortages, we posit that framework adoption must always

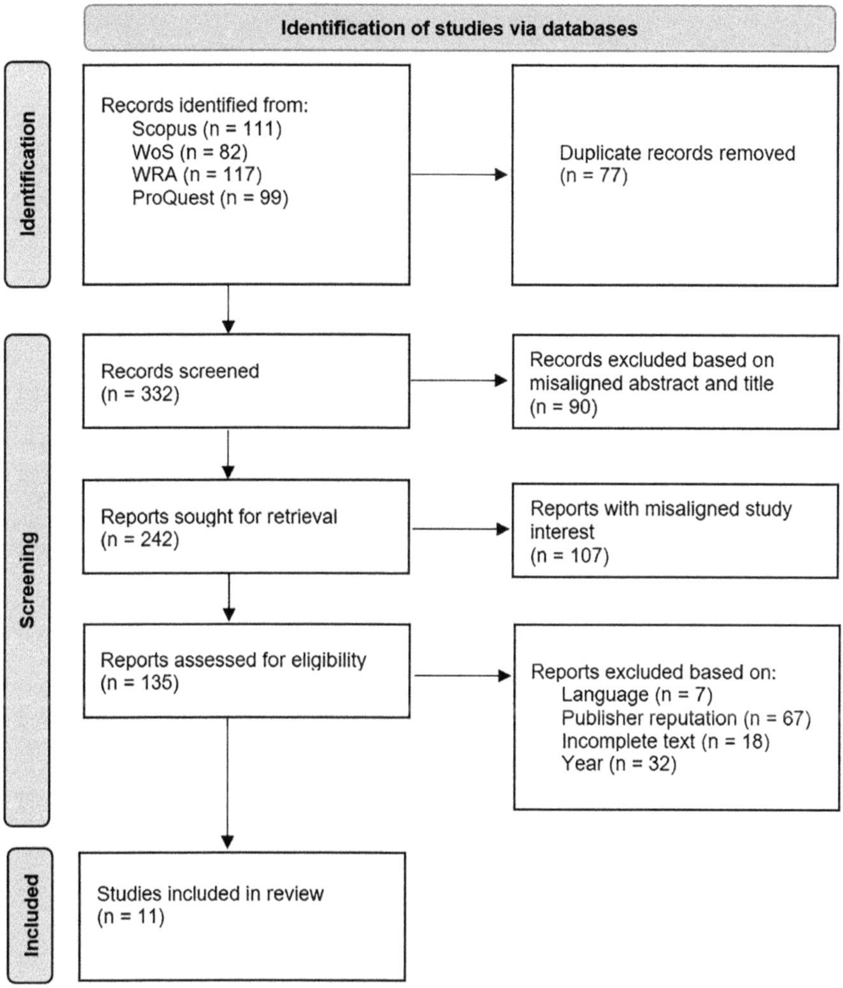

Fig. 7.1 Identification of studies for framework development (Page et al., 2021)

commence with the assessment of the current capacity of the water system before progressing to attributes that should be imbibed. Table 7.1 shows the rating threshold for each adaptive capacity level.

Table 7.1 Adaptive capacity scoring scale

Adaptive capacity level	Rating	Description	Characteristics
Prevailing	1–25	Prevailing adaptive capacity refers to a *business-as-usual* scenario within the water system marked with water shortages, and in which the government (through its institutions) is not doing enough with respect to water provision and supply. Individuals, households, and communities therefore adapt, albeit in a negative manner	Severe water shortages, increased spending on adaptation to water shortages, marked sanitary challenges, access to water sources with unverified water quality
Intermediate	26–45	Intermediate adaptive capacity is a phase in the drinking water system where the city actors are aware of the challenges and are willing to improve the water situation. Initial collaborative steps commence and actors all look to contribute and play a role	Gradual improvement to water provision and supply, culture of water-use and re-use is imbibed, sanitary challenges reduce gradually, strategic regulation of water sources, few regulatory bottlenecks
Ideal/enhanced	46–58	The enhanced adaptive capacity is a phase where collaboration by all city actors have been repeatedly exercised towards the goal of achieving optimal performance of the water system. The government plays its role effectively, notwithstanding conflicting priorities. Individuals, households, and community also seek to prioritise positive adaptation	The water system is fully functional, regulatory bottlenecks no longer exist, strategic and sustainable infrastructural development plans are underway

7.3 Results

Table 7.2 presents the four frameworks/concepts: national adaptive capacity (NAC), local adaptive capacity (LAC) frameworks, the city water resilience framework (CWRF), the concept of role-playing along with their corresponding indicators. The ideas from each of the identified frameworks were integrated to create a new reframed framework, which then offers a path to improving adaptive capacity to water shortages.

Table 7.2 Description of model indicators

Framework/concept (selected literature)	Indicator/ priority areas	Description	Additional reading
NAC (WRI, 2009)	*Assessment*	New decisions within the water system must be data-driven (e.g. water distribution or purification data). This must be prioritised by policy makers to ensure that decisions help to achieve global standards This is relevant since adaptation to water shortages requires long-term iterations and reviews	–
	Prioritisation	Prioritising areas most affected by water scarcity is crucial. Differential vulnerability and anticipated climate impacts can be used to determine this. When prioritisation is done correctly, stakeholders will be consulted and the process will be transparent. Additionally, it will guarantee that priorities are continuously adjusted as new water-related situations arise	Acharjee et al. (2020), Alamgir et al. (2023), Becher et al. (2024)
	Coordination	Coordination lowers the possibility of effort duplication or overlaps that could result in economies of scale in structural responses to water difficulties, since adaptation to water challenges require multi-stakeholder and multi-sectoral initiatives. The coordination process can be conducted directly between national departments, state-level water departments, and international or global partners (e.g., World Bank, World Water Institute); inter-sectorally between national departments and private sector companies; or horizontally among all stake holding national departments (e.g., FMWR, Federal Ministry of Environment, Federal Ministry of Infrastructure Development, etc.)	Clar (2019), Chai et al. (2025)

(continued)

Table 7.2 (continued)

Framework/concept (selected literature)	Indicator/ priority areas	Description	Additional reading
	Information management	Research and development for unique creation of knowledge, dissemination, and management is a guarantee for successful water policy formation and regulatory measures. These exercises could be helpful in promoting recently recommended adaptive habits. Information about climate factors, natural and human systems (such as settlement, biodiversity, and terrain), and current strategies for dealing with water scarcity would be important to these knowledge management initiatives. All stakeholders must be able to readily access and use the information included in a knowledge management platform once it has been established	Karki et al. (2011), Deslatte et al. (2024)
	Climate risk reduction	It is important to highlight particular and prevalent climate risks to water availability based on prioritisation and geographic location. A risk mitigation strategy must then be created when a risk evaluation process has been designed. It should be mentioned that risk reduction will probably differ depending on the region and may depend on the infrastructure that is available or lacking	Skinner et al. (2024)

(continued)

Table 7.2 (continued)

Framework/concept (selected literature)	Indicator/ priority areas	Description	Additional reading
LAC (Fuchs et al., 2018; Jones et al., 2010, 2017; Matewos, 2020)	*Asset-base*	The degree to which households and communities have access to water, functional water infrastructure, and operating funds (such as loans or savings) necessary for infrastructure expansion and upkeep must be assessed by policy actors. This stage must also aim to comprehend social capital, which includes collaborative community-level planning to mobilise strong adaptive actions for water management, and the skills and competences at the household and community levels that are essential for efficient water management	Jones et al. (2017)
	Knowledge and information	Understanding the extent to which publicly available climate and water management data are efficiently used to support local decision-making is vital within the NAC framework's information management component, as it can improve decisions regarding the dissemination of water management information	Iza et al. (2015)
	Innovation and flexibility	This dimension reflects the adoption of water-efficient innovations, local experimentation in water management, and flexible adaptation to unforeseen impacts on water availability	Cohen et al. (2016), Peat et al. (2017)
	Institutions and governance	This component evaluates local-level agreements, leadership frameworks, and decision-making procedures relevant to resolving disputes pertaining to water-related matters	Sultana and Thompson (2017), Gisevius et al. (2025)

(continued)

Table 7.2 (continued)

Framework/concept (selected literature)	Indicator/ priority areas	Description	Additional reading
	Access to external support	This applies to the community's and household's capacity to efficiently utilise all outside resources and assistance from national, international, or non-governmental organisations	Rosengren et al. (2023)
CWRF (Arup Global Water & Stockholm International Water Institute (SIWI), 2018; Saikia et al., 2022)	*Leadership and strategy*	High level governance structure and decision-making mechanism of the water system. This component describes the political willingness to develop and manage water resources and infrastructure in a way that makes the system crises-ready	–
	Planning and finance	This applies to how the water system manages revenue and expenditure, and the planning and execution of infrastructure upgrade. It also entails plans for unexpected shocks	Liang (2018)
	Infrastructure and ecosystem	Linked to planning and finance but diverges with respect to the quality of the infrastructure and its ability to maintain, deliver, and sustain service delivery in a way that protects the environment	Asghari et al. (2021), Sinha et al. (2023)
	Health and wellbeing	This aspect views the water system as a function of human health through experiences derived from water consumption. It assesses the extent to which the water system meets population need	Gaffan et al. (2023)
Role-playing (Fadeyi et al., 2025; Macheve et al., 2015; Olabode & Comte, 2024)	*Institution*	Mainly water institutions (e.g., state water boards) responsible to water provision and supply in cities. Within the framework of role-playing these entities are expected to perform optimally, and also ensure regulation of alternative sources such as water vendors and kiosks. Maintenance of water infrastructure is also a key role	George-Williams et al. (2024)

(continued)

Table 7.2 (continued)

Framework/concept (selected literature)	Indicator/ priority areas	Description	Additional reading
	Community	The highest level of societal co-operation with government. With regards to role-playing, communities are expected to maintain good sanitary condition in collaboration with head of household. The monitoring of water supply pipes is also a vital role communities can play	Bisung (2021)
	Household	Households must enforce agreements reached at community level. Practicing water re-use and other effective water management practices is crucial within the framework or role-playing by households	Moghayedi (2024)
	Individual	Individuals are expected to embrace positive adaptation. This means that only regulated water sources can be patronised during possible down-time of government water supply	Baddianaah et al. (2024)

Adapted from Fadeyi et al. (2025)

7.3.1 National Adaptive Capacity (NAC) Framework

The NAC framework is a function-based model that seeks to complement society's asset-focused details. It is premised on projected national institutional actions *(i.e.,* what national water organisations can do*) to build* urban water resilience (WRI, 2009). Given marked shortages in drinking water access across cities in many developing countries (Salehi, 2022), the integration of the NAC framework into the new model ensures that institutions are strongly embedded, aiming to achieve improved adaptive capacity in a manner that differs slightly from their traditional role-playing characteristics. While the proposed framework explains role playing as a concept that takes effect from intermediate adaptive capacity level up until the full maturity of the model, i.e., when adaptive capacity has been enhanced, the mainstreaming of NAC at prevailing adaptive capacity levels helps to assess where city-specific institutions stand in promoting global best practices with respect to national water guidelines and water governance structure. Hence, assessing the five (5) core functions that underpin the NAC framework: *vulnerability and impact assessment* (Borzì, 2023; Stathatou et al., 2016), *risks and resources prioritisation* (Gonen et al., 2010; Okeola & Raheem, 2016), *cross-sectoral coordination* (Lukat et al., 2023; Nieuwenhuis et al.,

2023), *information management* (Wybrands et al., 2021), and *proactive climate risk reduction* (Wang et al., 2019) in the context of water governance may be useful to analyse the alignment of a city's water attribute with national plans. In the case of Nigeria's multi-level approaches to water policies which allow state and national government authorities to exhibit some level of control, embedding NAC functions readily helps rank a city in terms of readiness to adopt enhanced adaptation, even amid prevailing challenges. This is achievable by seeking to link key components of the framework with existing city-specific process mostly derivable from regulations, reports, operational guidelines, or some undocumented/unstructured but practical water-related events. Overall, NAC in the context of this work provides a transformative mechanism to adaptation from a survival-driven mode to well-structured urban resilience practices.

7.3.2 Local Adaptive Capacity (LAC) Framework

Unlike the NAC, which draws on a much broader perspective, the LAC framework is concerned with the abilities of city components (individuals, households, communities, and institutions) to predict and prepare for climate-induced and other stressors affecting city water supply. The LAC framework therefore identifies aspects such as: asset base (Shen et al., 2025), collective action and social institutions (Bisung, 2021), knowledge and information (Hering et al., 2018; Wybrands et al., 2021), innovation (Diao & Ulanicki, 2025), and governance (Rosso, 2025), useful to manage the impacts of stressors, and explain the potential enactment of adaptation via day-to-day decision making linked to sourcing, storage, and usage of water.

The asset base aspect for example, is useful for assessing adaptive capacity with respect to finance or infrastructure to cope with water shortages (Le Gat et al., 2025; Pathirana et al., 2021). Should the city fall short in terms of existing resources, an assessment of plans and programs, as well as an estimated timeframe for the proposed implementation of such, may suffice, so that it represents a forward-looking and positive action. Simialrly, with collective action, adaptive capacity assessment can be viewed at community levels through community collaboration such as localised plans to enforce hygienic practices (Mittra & Sugam, 2020). Hence, the integration of LAC components within the new framework reflects a realistic trend that moves from national (NAC) to local (LAC) levels, drawing on the actions of local-level actors.

7.3.3 City Water Resilience Framework (CWRF)

The multi-dimensional nature of the CWRF makes it ideal for inclusion in the proposed framework. As with LAC, CWRF also tries to assess and measure vulnerability of urban water systems to shocks and stresses. It, however, differs from LAC due to its focus on cities, and the emphasis it places on assets alone being not enough to achieve resilience (Saikia et al., 2022). Hence, CWRF embeds strategic leadership (Meng, 2024), planning and finance (Babamiri et al., 2020), infrastructure and ecosystems (Adeniran et al., 2021), as well as health and well-being (Abanyie et al., 2023; Sheel et al., 2024), each representing a critical domain for a city to mobilise adaptive capacity and achieve urban water resilience.

As part of the proposed framework, the components of CWRF offer structural foundation for transitioning from prevailing to enhanced adaptive capacity. For example, the leadership aspect assesses the political commitment needed to build strong institutions and regulatory mechanisms for prioritising urban water resilience. On the other hand, the planning and finance aspect supports how financial resources are directed towards adaptation in a bid to become fully enhanced, while the infrastructure and ecosystems aspect ensures that physical inputs are in the shape in which urban water supply can be sustainably maintained. Finally, the health and well-being aspect of the CWRF reinforces the idea that linking adaptive capacity to water resilience strategies offers safety to population health. Summarily, for the new model, the adoption of the CWRF is based on the idea that the convergence of institutional mandates, financial mechanisms, ecological protection, and public health safety will promote individual-, household-, community-, and institutional-based adaptation. This will cause adaptation to generally evolve from short-term coping to a sustained, regulated, and safe adaptive pathway.

7.3.4 The Concept of Role-Playing

As part of the building blocks for the proposed framework, we adopt the definition of adaptive capacity in water management by Fadeyi et al. (2025): *the ability of consumers either at the individual, household, or community level to plan and implement water collection, storage, and use, in ways that ensure effective adaptation to institutionally-driven water shortages and the effects of climate change* (p. 16). This definition establishes the concept of role-playing, which is defined as a situation where all city actors (individuals, households, or communities) are aware of their unique responsibilities to improve the water systems and collaborate and take actions to achieve this goal.

Whilst role-playing focuses only on the actions within the community, household, and individual, our framework improves on this idea to suggest that for urban water resilience to be achieved, water institutions must also rise to the occasion and ensure the provision and supply of potable drinking water. This action, if well-established, readily provides a baseline for adaptation at the institutional level and necessitates corresponding, responsible consumption on the part of individuals, households, and communities. By implication, lower level city components must then cope (*role-play*) with shortages through experiential learning from past incidents and seek to implement effective water use (Fadeyi et al., 2025). The efforts of water institutions to provide and supply water notwithstanding inherent challenges, combined with those of individuals, households, and communities to manage the resource support the progress of the water system. Adaptive capacity can be therefore be said to minimise risks and promote the ability of the water system to potentially maintain resilience (Crozier et al., 2024; Li et al., 2016). Our proposed framework acknowledges that water scarce cities (a stressor to the water system), and the adaptation capacity of the city to manage the situation, are both intertwined and complex experiences within the water (socio-ecological) system (Kumar et al., 2020).

7.3.5 *Conceptual Framework*

Figure 7.2 is the new conceptual framework with key aspects such as prevailing adaptive capacity, capacity at ideal levels, and role playing by city components.

At the prevailing level of adaptive capacity, efforts at both upper (institutional) and lower (community) levels are often disjointed in many developing countries. The government is not performing optimally in terms of water provision, forcing individuals and households to cope in the best way they can. Hence, the concept of *role-playing* is absent at this stage. In contrast, enhanced adaptive capacity has the attributes of role-playing where the government through its institutions, makes significant efforts to provide potable drinking water (based on available resources and capability), and takes additional steps to regulate how people adapt by monitoring the activities of water vendors and water kiosk operators. This means frequent water quality assessment to maintain population health.

While the enhanced and prevailing capacities are mutually exclusive, achieving optimised role-playing on the part of government or community is a gradual process, and may be a function of how the city prioritises urban water resilience. As soon as relevant questions (Table 7.3) are asked and solutions are sought, the city enters a phase of *intermediate adaptive capacity*. This is the phase of awareness where city water managers clearly understand the water challenges and are continuously seeking solutions. It is also a stage where communities are more willing to co-operate with water institutions, and where the government regardless of its conflicting priorities is poised to solve prevailing water problems. It is noteworthy to emphasise that at intermediate levels, the challenges that has affected the water systems for so long may consistently resurface until role-playing is mastered. This process may sometimes be time-consuming as the city focuses on positive adaptation. A good example is government's regulation of private boreholes or surface wells, and regular water quality checks by water kiosks.

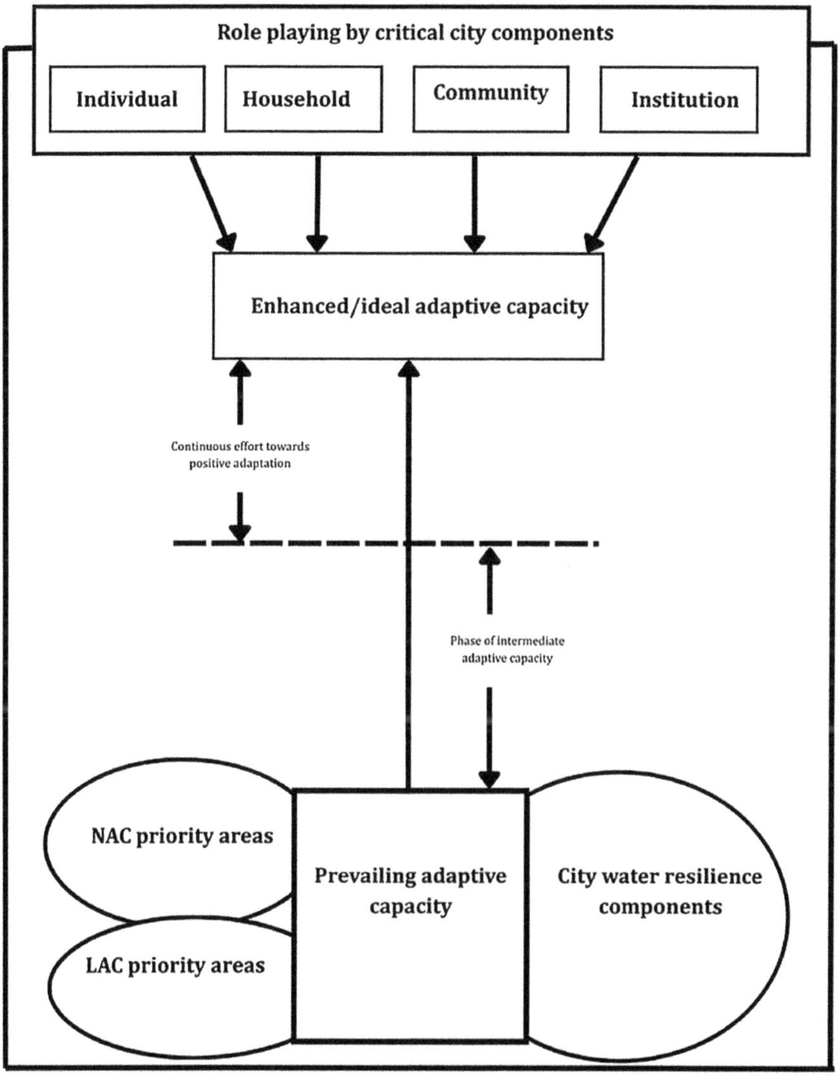

Fig. 7.2 A conceptual framework for assessing adaptive capacity to water shortage for a water-constrained city

7.4 Discussion

The need to develop tailored approaches to cope with the effects of climate change and other stressors on drinking water systems in developing countries has led to the development of many adaptive capacity frameworks. Nevertheless, given their unique socio-economic and leadership challenges (Azeez et al., 2023; Bose et al., 2024),

Table 7.3 Indicators for assessing adaptive capacity to water shortages

Model	Component	Capacity assessment question (what exists)	Best-practices aligned question (ideal scenario)
NAC	Assessment	(i) Does the waterboard service reach an appreciable number (e.g. 50%) of city residents?	(i) Will the city plan to implement frameworks or techniques that are future focused in order to effectively predict and evaluate water and climate risks? (ii) Are there strategic efforts to improve water board service coverage to capture all city residents?
	Prioritisation	(i) Is the city a hotspot for water shortage relative to others? (ii) Is climate change the main driver of water shortages in the city? (iii) Are there other inherent drivers of water shortages?	(i) Is the city preparing to put into practice proactive methods for prioritising risk and resources to prevent future scarcity? (ii) Are there strategies in place to regularly review risk and resource priorities to lessen the possibility of future scarcity?
	Coordination	(i) Does the adaptation process to water scarcity combine multi-stakeholder and multi-sectoral actions? (ii) Are there perceived overlaps in efforts?	(i) Will the city work with all sectors in a persistent, coordinated manner to deal with future water-related issues? (ii) Does the city plan to maintain cross-sector collaboration for water resilience over the long term?
	Information management	(i) Is there a city-sponsored drinking water information hub or research centre? (ii) Are there clear-cut information about the predominant coping practice(s)/source(s) to water shortages in the city?	(i) Is there an intention to generate and disseminate real time water and climate information for decision making?

(continued)

Table 7.3 (continued)

Model	Component	Capacity assessment question (what exists)	Best-practices aligned question (ideal scenario)
	Climate risk reduction	(i) Is climate change considered a risk to the city's water system? (ii) Is there any operational risk management and mitigation protocol?	(i) Will the city's planning include preventive risk reduction measures to deal with anticipated climate shocks early? (ii) Does the city plan to set up procedures for initiating risk reduction ahead of climate events
LAC	Asset-base	(i) Does the city possess a functional drinking water-related financial mechanism (e.g., for infrastructural re-development)? (ii) Are there social capital structures in place for effective city water management	(i) In times of severe water scarcity, does the city intend to assist all citizens in obtaining clean drinking water? (ii) Is there a plan in place to ensure that everyone in the community will have fair access to clean drinking water during future scarcity?
	Knowledge and information	(i) Does the city utilize research information hubs generated from city hubs to address and improve water challenges?	(i) Are there plans by the city to guide the residents on how to utilise water safely and efficiently?
	Innovation and flexibility	(i) Are there significant efforts to develop water efficient technologies/ innovations to prevent future water shortages?	(i) Does the city plan to promote innovation that helps communities move beyond short-term coping toward long-term sustainable adaptation?
	Institutions and governance	(i) Does the city possess adequately granular governance arrangements to address and resolve water-related conflicts?	(i) Are there any plans by the city to ensure trust and shared accountability among communities and institutions for future water and climate management?
	Access to external support	(i) Does the city benefit from marked external support with respect to water shortages?	(i) Does the city intend to make sure that technical and financial assistance is accessible before resilience limitations are reached?

(continued)

Table 7.3 (continued)

Model	Component	Capacity assessment question (what exists)	Best-practices aligned question (ideal scenario)
CWRF	Leadership and strategy	(i) Does the city's leadership structure prioritise drinking water resilience strategies? (ii) Are there well-established guidelines that inform decision-making on urban water resilience?	(i) Is the city planning to make water resilience a long-term urban priority beyond political cycles? (ii) Will the city put plan in place to guarantee that, despite shift in political leadership, water resilience will continue to be a top priority?
	Planning and finance	(i) Does the city frequently translate its water resilience plans into actions?	(i) Will the city implement sustainable regulations and funding to ensure long term resilience while protecting vulnerable populations?
	Infrastructure and ecosystem	(i) Does the city integrate natural and artificial drinking water infrastructure to enhance resilience against shocks?	(ii) Are there plans to integrate engineered and natural systems to ensure a reliable future water supply?
	Health and wellbeing	(i) Does the city carry out periodic water quality checks? (ii) Is the city's water quality in agreement with international drinking water standards? (iii) Does the city water conditions suggest the need for water treatment infrastructure upgrade? (iv) Is the city's water supply reliable and adequate to support population health over both short- and long-term horizons?	(i) Are there plans to ensure that resilience measures provide safe, equitable, and consistent access to water for all residents?
Role-playing	Individual	(i) Is there enough information to sensitise individuals about water shortages? (ii) Can city people be considered interested in water information?	(i) Are there plans to continuously raise the awareness level of individuals about water shortages?

(continued)

Table 7.3 (continued)

Model	Component	Capacity assessment question (what exists)	Best-practices aligned question (ideal scenario)
	Household	(i) Does the population possess household-level water management and competencies? (ii) Is water-reuse a common household practice across the city?	(i) Are households aware of climate change impacts on drinking water?
	Community	(i) Is there a good institution-community rapport within the city with respect to water management?	(i) Are there plans to sustainably execute community sanitation practices that may affect drinking water? (ii) Are there planned measures to protect community water supply infrastructure?
	Institution	(i) Is the city water board working optimally? (ii) Do waterboard employees possess the requisite skills to manage water infrastructure (iii) Are employees sufficient trained on the best way to react during severe water shortages? (iv) Are city water policies/ guidelines considered effective?	(i) Are there plans to update water policies based on current/future realities?

existing frameworks have yet to provide sustained solutions to the prevailing challenges surrounding water access in these climes. This has consistently led to severe impacts on population health (Shayo et al., 2023). Across sub-Saharan Africa, adaptation to water shortages is largely unstructured (Ampaire et al., 2017; Sithole et al., 2025). In many cities across West Africa, water vendors and water kiosk operators, whose services are often unregulated, have become significant actors within the water system. Furthermore, Isukuru et al. (2024) notes that the majority of the household in cities across Nigeria rely on own-sourced water, with government taking a backseat. This situation has persisted for many decades, prompting a search for tailored solutions.

Whilst the improvement of drinking water systems in many developing countries (e.g., Nigeria) is heavily reliant on government, the emergence of a role-playing approach, upon which the framework described in this work is founded, stresses a more holistic approach (Fadeyi et al., 2025). Through role-playing, the government is not the only custodian of solutions to city water challenges, as efforts by individuals, households, and community members are crucial for a functional water system. Hence, role-playing by one city component (e.g., households) reduces pressure on the other (e.g., government) and vice versa. For example, while governments play a regulatory role to deter unsafe hygienic and sanitary practices in cities across many developing countries, Hutton and Chase (2017) suggest that communities and households must ensure that latrines and open defecation sites are located far from water supply pipes to avoid faecal contamination when the pipes are damaged. As a result, these efforts are complementary, thereby ensuring the safety of the population.

Whilst communities and households in developing nations must continue to adapt to water shortages in the face of climate change, conflicting priorities faced by decision-makers when the need arises to invest in drinking water also require some sort of adaptive counter-measures that can help governments better prioritise prevailing water issues (Macchiaroli et al., 2023; Schwartz et al., 2025). This hierarchical form of adaptation must be benchmarked against the usefulness of potable water as a necessity for life, health, and well-being. The proposed framework therefore suggests a social responsibility approach to coping at both the top and bottom.

The transition from prevailing to enhanced adaptation in a water system requires significant efforts in conducting inside-out assessments and reassessments by policymakers, due to the need across the main building blocks: NAC, LAC, CWRF, and role-playing frameworks. Nevertheless, when either of these models are adopted singly, there is a likelihood that each may fall short of the solutions needed in a water-constrained setting. This is not to say that these frameworks have not provided solutions in the past (e.g. Freduah et al. (2019)). Nevertheless, the continuous changes in the conditions that must be met from 'what exists' to what the ideal water situation should be may hinder successful model adoption. For instance, currently, less than half of Lagos, Nigeria's most commercialised urban centre, is linked to the state water board (Balogun et al., 2017; Olabode & Comte, 2024). Ultimately, the adoption of the LAC framework, which deals with asset (e.g., infrastructure) improvement, may be useful in improving the situation. However, this may not be sustainable without community members engaging in role-playing to ensure local infrastructure maintenance.

The proposed framework highlights critical implications for adaptive capacity in low-resource settings where water shortages are intertwined with institutional weaknesses and community/household socio-economic standing (Fadeyi et al., 2025). By recognising areas where the water system is challenged as well as areas of strength, our framework creates a path for progress. This is achieved by not relying on past achievements but making efforts to incorporate newer features that strengthen all aspects of the water management system, from infrastructural development and maintenance to water use and reuse. Identifying the cumulative effects of multi-component efforts through role-playing ensures strategic progress that not only improves the capacity to adapt to water shortages but also provides an avenue to address other inherent issues facing the water system (Mazzucato, 2023).

Our framework advances existing knowledge by offering a practical approach to assessing adaptive capacity to water shortages. Bearing in mind that this challenge is multifaceted in many developing countries, it provides a holistic picture of the existing water shortages and coping scenarios. It also moves forward to suggest relevant actions that can help improve the situation to attain global best practices.

7.5 Limitations

The interpretation of the proposed adaptive capacity framework described herein is limited to settings of a developing country with water challenges, such as Nigeria, where there are marked socio-economic and leadership issues affecting the drinking water sector. Whilst the framework offers to bridge the gaps between current societal efforts (using what exists) and what societies must do (i.e., linked to global best practices) to improve the state of the drinking water system, successful adoption of the framework is a function of how well city actors master the act of *role-playing*. Since optimal framework performance would mean effective role-playing by all city components, the framework's adoption may be flawed due to excessive interdependence among operational units. This means that water sufficiency will remain far-fetched if one of four city components fails in their role-playing performance. Furthermore, the fact that the proposed framework relies on what already exists and what needs improvement may introduce some level of bias. Whilst responses to the capacity assessment questions will be based on what currently exists in terms of a city's water system, it may be difficult to provide false responses. In contrast, the ideal and/or projected situation may be repeatedly falsified to reflect plans that are not underway. This may cause repeated setbacks to the city's water system, as objective responses will ensure that city water managers and policymakers reflect and work together on challenging areas of the water system.

7.6 Conclusions

This study proposes a framework to address the issue of water shortages in cities in developing countries. It draws on the idea that existing frameworks largely focus on societal attributes, neglecting global best practices in the quest to proffer solutions on how best to adapt to water scarcity as a socio-ecological challenge. Drawing on the attributes of the LAC, NAC, and CWRF frameworks alongside the concept of role-playing, the new framework suggests that city water managers and policy implementers can rate the water system based on responses to strategic questions about its operations and use the responses to chart a path forward. Depending on the outcome of this evaluation, strategic roles are then taken up by individuals, households, communities, and institutions to push the water system to an enhanced level of adaptation, where water shortages soon become a thing of the past. It is recommended that water manager in nations facing similar water issues may adapt the proposed framework in a bid to solve their water problems.

References

Abanyie, S. K., Ampadu, B., Frimpong, N. A., & Yahans Amuah, E. E. (2023). Impact of improved water supply on livelihood and health: Emphasis on Doba and Nayagnia, Ghana. *Innovation and Green Development*, *2*(1), 100033. https://doi.org/10.1016/j.igd.2023.100033

Acharjee, T. K., Hellegers, P., Ludwig, F., van Halsema, G., Mojid, Md. A., & van Scheltinga, C. T. (2020). Prioritization of adaptation measures for improved agricultural water management in Northwest Bangladesh. *Climatic Change, 163*(1), 431–450. https://doi.org/10.1007/s10584-020-02852-w

Adeniran, A., Daniell, K. A., & Pittock, J. (2021). Water Infrastructure Development in Nigeria: Trend, Size, and Purpose. *Water, 13*(17), Article 17. https://doi.org/10.3390/w13172416

Ahmed, S. K., Mohammed, R. A., Nashwan, A. J., Ibrahim, R. H., Abdalla, A. Q., M. Ameen, B. M., & Khdhir, R. M. (2025). Using thematic analysis in qualitative research. *Journal of Medicine, Surgery, and Public Health, 6*, 100198. https://doi.org/10.1016/j.glmedi.2025.100198

Alamgir, M., Uddin, M. N., Hasan, M. M., Wang, X., Shiru, M. S., & Shahid, S. (2023). Prioritization of sectoral adaptation strategies and practices: A case study for Bangladesh. *Environmental Development, 45*, Article 100813. https://doi.org/10.1016/j.envdev.2023.100813

Alharbi, A., & Stevenson, M. (2020). Refining Boolean queries to identify relevant studies for systematic review updates. *Journal of the American Medical Informatics Association: JAMIA, 27*(11), 1658–1666. https://doi.org/10.1093/jamia/ocaa148

Ampaire, E. L., Jassogne, L., Providence, H., Acosta, M., Twyman, J., Winowiecki, L., & van Asten, P. (2017). Institutional challenges to climate change adaptation: A case study on policy action gaps in Uganda. *Environmental Science & Policy, 75*, 81–90. https://doi.org/10.1016/j.envsci.2017.05.013

Arup Global Water & Stockholm International Water Institute (SIWI). (2018). *City Water Resilience Approach.* ILRI. https://www.arup.com/globalassets/downloads/insights/c/city-water-resilience-approach/city-water-resilience-approach.pdf

Asghari, M., Abdi-Benemar, H., Maheri-Sis, N., Salamatdoust-Nobar, R., Salem, A. Z. M., Zamanloo, M., & Anele, U. Y. (2021). Effects of emulsified essential oils blend on performance, blood metabolites, oxidative status and intestinal microflora of suckling calves. *Animal Feed Science and Technology, 277*, Article 114954. https://doi.org/10.1016/j.anifeedsci.2021.114954

Azeez, O., Henderson-Mitchell, R. J., LaFevor, M. C., & Gregg, A. (2023). Socioeconomic predictors of access to improved water sources, sanitation facilities, and household water treatment in Nigeria. *Journal of Water, Sanitation and Hygiene for Development, 13*(11), 875–884. https://doi.org/10.2166/washdev.2023.169

Babamiri, A. S., Pishvaee, M. S., & Mirzamohammadi, S. (2020). The analysis of financially sustainable management strategies of urban water distribution network under increasing block tariff structure: A system dynamics approach. *Sustainable Cities and Society, 60*, Article 102193. https://doi.org/10.1016/j.scs.2020.102193

Baddianaah, I., Dongzagla, A., & Salifu, S. N. (2024). Navigating access to safe water by rural households in sub-Saharan Africa: Insights from north-western Ghana. *Sustainable Environment, 10*(1), 2303803. https://doi.org/10.1080/27658511.2024.2303803

Balogun, I. I., Sojobi, A. O., & Galkaye, E. (2017). Public water supply in Lagos State, Nigeria: Review of importance and challenges, status and concerns and pragmatic solutions. *Cogent Engineering, 4*(1), 1329776. https://doi.org/10.1080/23311916.2017.1329776

Becher, O., Smilovic, M., Verschuur, J., Pant, R., Tramberend, S., & Hall, J. (2024). The challenge of closing the climate adaptation gap for water supply utilities. *Communications Earth & Environment, 5*(1), 356. https://doi.org/10.1038/s43247-024-01272-3

Bisung, E. (2021). Community based collective action for safe water in rural sub-Saharan Africa. *Water Security, 13*, Article 100088. https://doi.org/10.1016/j.wasec.2021.100088

Borzì, I. (2023). Vulnerability assessment of water supply infrastructures through multiple indicator methodology. *Journal of Water and Climate Change, 14*(11), 3967–3984. https://doi.org/10.2166/wcc.2023.148

Bose, D., Bhattacharya, R., Kaur, T., Banerjee, R., Bhatia, T., Ray, A., Batra, B., Mondal, A., Ghosh, P., & Mondal, S. (2024). Overcoming water, sanitation, and hygiene challenges in critical regions of the global community. *Water-Energy Nexus, 7*, 277–296. https://doi.org/10.1016/j.wen.2024.11.003

Brooks, N., & Adger, W. N. (2005). Assessing and enhancing adaptive capacity. In B. Lim, E. Spanger-Siegfried, I. Burton, E. Malone, & S. Huq (Eds.), *Adaptation policy frameworks for climate change* (pp. 165–181). Cambridge University Press.

Chai, Y., Chen, X., Pan, S., & Zeng, Y. (2025). Integrative coordination and state-reinforced adaptive governance of small-scale irrigation systems in the Yangtze River basin, China. *Ecology and Society, 30*(2). https://doi.org/10.5751/ES-15900-300211

Clar, C. (2019). Coordinating climate change adaptation across levels of government: The gap between theory and practice of integrated adaptation strategy processes. *Journal of Environmental Planning and Management, 62*(12), 2166–2185. https://doi.org/10.1080/09640568.2018.1536604

Cohen, P. J., Lawless, S., Dyer, M., Morgan, M., Saeni, E., Teioli, H., & Kantor, P. (2016). Understanding adaptive capacity and capacity to innovate in social–ecological systems: Applying a gender lens. *Ambio, 45*(3), 309–321. https://doi.org/10.1007/s13280-016-0831-4

Crozier, A., Lence, B. J., & Weijs, S. V. (2024). Resilience framework for urban water supply systems planning. *Sustainable and Resilient Infrastructure, 9*(4), 386–406. https://doi.org/10.1080/23789689.2024.2340378

Deslatte, A., Adams, J. A., Cheema, F. S., Vicario, S. A., Barnes, J. L., & Koebele, E. A. (2024). How state-reinforced knowledge infrastructure influences adaptive urban water governance. *Ecology and Society, 29*(4). https://doi.org/10.5751/ES-15454-290428

Diao, K., & Ulanicki, B. (2025). Cutting-edge innovations in drinking water management. *AQUA—Water Infrastructure, Ecosystems and Society, 74*(2), iii–iv. https://doi.org/10.2166/aqua.2025.001

Dos Santos, S., Adams, E. A., Neville, G., Wada, Y., de Sherbinin, A., Mullin Bernhardt, E., & Adamo, S. B. (2017). Urban growth and water access in sub-Saharan Africa: Progress, challenges, and emerging research directions. *Science of the Total Environment, 607–608*, 497–508. https://doi.org/10.1016/j.scitotenv.2017.06.157

Eisenack, K., & Stecker, R. (2012). A framework for analyzing climate change adaptations as actions. *Mitigation and Adaptation Strategies for Global Change, 17*(3), 243–260.

Engle, N. L. (2011a). Adaptive capacity and its assessment. *Global Environmental Change, 21*(2), 647–656. https://doi.org/10.1016/j.gloenvcha.2011.01.019

Engle, N. L. (2011b). Adaptive capacity and its assessment. *Global Environmental Change. Part A, Human and Policy Dimensions, 21(2):647–656, 21*(2), Article PNNL-SA-76881. https://doi.org/10.1016/j.gloenvcha.2011.01.019

Fadeyi, O. O., Bessah, E., & Ololade, O. O. (2025). Accelerating the achievement of SDG 6 via adaptive capacity mainstreaming into water governance strategies: A case study of Nigeria. *Journal of Water, Sanitation and Hygiene for Development, 15*(9), 732–758. https://doi.org/10.2166/washdev.2025.038

Freduah, G., Fidelman, P., & Smith, T. F. (2019). A framework for assessing adaptive capacity to multiple climatic and non-climatic stressors in small-scale fisheries. *Environmental Science & Policy, 101*, 87–93. https://doi.org/10.1016/j.envsci.2019.07.016

Fuchs, L., Njuguna, L. W., Robinson, L. W., Zampaligre, N., Coulibaly, J. Y., Crane, T., Nebie, E., Nganga, T., Some, B., & Ontiri, E. (2018). *Characterizing local adaptive capacity at various scales: A framework for adaptation planning* (Briefing Paper No. 2). Local Governance and Adapting to Climate Change in Sub-Saharan Africa (LGACC). https://www.cifor-icraf.org/knowledge/publication/25595/

Gaffan, N., Kpozehouen, A., Degbey, C., Ahanhanzo, Y. G., & Paraïso, M. N. (2023). Effects of household access to water, sanitation, and hygiene services on under-five mortality in Sub-Saharan Africa. *Frontiers in Public Health, 11*. https://doi.org/10.3389/fpubh.2023.1136299

George-Williams, H. E. M., Hunt, D. V. L., & Rogers, C. D. F. (2024). Sustainable water infrastructure: visions and options for Sub-Saharan Africa. *Sustainability, 16*(4), 1592. https://doi.org/10.3390/su16041592

Gisevius, K., Niesters, L.-M., & Braun, B. (2025). The role of community leadership in building community adaptive capacity to coastal hazards—Insights from neighborhood networks in Semarang, Indonesia. *Environmental Science & Policy, 163*, Article 103963. https://doi.org/10.1016/j.envsci.2024.103963

Gonen, A., Zeitouni, N., Gonen, A., & Zeitouni, N. (2010). Risk management of water resources in a changing climate. In *Advances in risk management*. IntechOpen. https://doi.org/10.5772/intechopen.83886

Gupta, J., Termeer, C., Klostermann, J., Meijermink, S., van den Brink, M., Jong, P., Nooteboom, S., & Bergsma, E. (2010). The adaptive capacity wheel: A method to assess the inherent characteristics of institutions to enable the adaptive capacity of society. *Environmental Science & Policy, 13*(6), 459–471. https://doi.org/10.1016/j.envsci.2010.05.006

Hering, J. G., Nunnenmacher, L., & von Waldow, H. (2018). Perspectives from a water research institute on Knowledge Management for Sustainable Water Management. In *Handbook of knowledge management for sustainable water systems* (pp. 15–33). Wiley. https://doi.org/10.1002/9781119271659.ch1

Hutton, G., & Chase, C. (2017). Water supply, sanitation, and hygiene. In *Injury prevention and environmental health* (3rd ed). The International Bank for Reconstruction and Development/The World Bank. https://doi.org/10.1596/978-1-4648-0522-6_ch9

Isukuru, E. J., Opha, J. O., Isaiah, O. W., Orovwighose, B., & Emmanuel, S. S. (2024). Nigeria's water crisis: Abundant water, polluted reality. *Cleaner Water, 2*, Article 100026. https://doi.org/10.1016/j.clwat.2024.100026

Iza, A., Müller, A., & Nozza, V. (2015). Adaptive water governance: Lessons learned from implementing an ecosystem approach in Mesoamerica. In K.-L. Koh, I. Kelman, R. Kibugi, & R.-L. Eisma Osorio, *Adaptation to climate change* (pp. 57–81). World Scientific. https://doi.org/10.1142/9789814689748_0003

Jones, L., Ludi, E., & Levine, S. (2010). *Towards a characterisation of adaptive capacity: A framework for analysing adaptive capacity at the local level* (p. 8). Overseas Development Institute. https://media.odi.org/documents/6353.pdf

Jones, L., Ludi, E., Jeans, H., & Barihaihi, M. (2017). Revisiting the local adaptive capacity framework: Learning from the implementation of a research and programming framework in Africa. *Climate and Development, 11*(1), 3–13. https://doi.org/10.1080/17565529.2017.1374237

Juhola, S., & Kruse, S. (2015). A framework for analysing regional adaptive capacity assessments: Challenges for methodology and policy making. *Mitigation and Adaptation Strategies for Global Change, 20*(1), 99–120.

Karki, M. B., Shrestha, A. B., & Winiger, M. (2011). Enhancing knowledge management and adaptation capacity for integrated management of water resources in the Indus River Basin. *Mountain Research and Development, 31*(3), 242–251. https://doi.org/10.1659/MRD-JOURNAL-D-11-00017.1

Kelly, P. M., & Adger, W. N. (2000). Theory and practice in assessing vulnerability to climate change and facilitating adaptation. *Climatic Change, 47*(4), 325–352. https://doi.org/10.1023/A:1005627828199

Kumar, P., Avtar, R., Dasgupta, R., Johnson, B. A., Mukherjee, A., Ahsan, Md. N., Nguyen, D. C. H., Nguyen, H. Q., Shaw, R., & Mishra, B. K. (2020). Socio-hydrology: A key approach for adaptation to water scarcity and achieving human well-being in large riverine islands. *Progress in Disaster Science, 8*, Article 100134. https://doi.org/10.1016/j.pdisas.2020.100134

Le Gat, Y., Curt, C., Werey, C., Caillaud, K., Rulleau, B., & Taillandier, F. (2025). Water infrastructure asset management: State of the art and emerging research themes. *Structure and Infrastructure Engineering, 21*(4), 539–562. https://doi.org/10.1080/15732479.2023.2222030

Li, Y., Degener, J., Gaudreau, M., Li, Y., & Kappas, M. (2016). Adaptive capacity based water quality resilience transformation and policy implications in rapidly urbanizing landscapes. *Science of the Total Environment, 569–570*, 168–178. https://doi.org/10.1016/j.scitotenv.2016.06.110

Liang, X. (2018). Integrated economic and financial analysis of China's Sponge City program for water-resilient urban development. *Sustainability, 10*(3), 669. https://doi.org/10.3390/su10030669

Lukat, E., Lenschow, A., Dombrowsky, I., Meergans, F., Schütze, N., Stein, U., & Pahl-Wostl, C. (2023). Governance towards coordination for water resources management: The effect of governance modes. *Environmental Science & Policy, 141*, 50–60. https://doi.org/10.1016/j.envsci.2022.12.016

Macchiaroli, M., Dolores, L., & De Mare, G. (2023). Multicriteria decision making and water infrastructure: An application of the analytic hierarchy process for a sustainable ranking of investments. *Applied Sciences, 13*(14), 8284. https://doi.org/10.3390/app13148284

Macheve, B., Danilenko, A., Abdullah, R., Bove, A., & Moffitt, L. J. (2015). *State water agencies in Nigeria: A performance assessment* (pp. 17–30). The World Bank. https://doi.org/10.1596/978-1-4648-0657-5_ch2

Matewos, T. (2020). The state of local adaptive capacity to climate change in drought-prone districts of rural Sidama, southern Ethiopia. *Climate Risk Management, 27*, Article 100209. https://doi.org/10.1016/j.crm.2019.100209

Mazzucato, M. (2023). *Why public-private collaboration is key to achieving SDGs on water*. World Economic Forum. https://www.weforum.org/stories/2023/09/public-private-collaboration-key-achieving-sdgs-water/

Meng, M. (2024). Research on the impact of leadership on improving urban water efficiency and water conservation policies. *Natural Resource Modeling, 37*(4), Article e12410. https://doi.org/10.1111/nrm.12410

Mitchell, G. E., & Schmitz, H. P. (2023). Using model-based clustering to improve qualitative inquiry: computer-aided qualitative data analysis, latent class analysis, and interpretive transparency. *VOLUNTAS: International Journal of Voluntary and Nonprofit Organizations, 34*(1), 162–169. https://doi.org/10.1007/s11266-021-00409-8

Mittra, S., & Sugam, R. (2020, November 25). *Collective Action for water security and sustainability*. Council on Energy, Environment, and Water. https://www.ceew.in/publications/collective-action-water-security-and-sustainability

Moghayedi, A. (2024). Enhancing water recycling adoption in South African residential properties: A multi-dimensional analysis for sustainable water management. *Discover Sustainability, 5*(1), 285. https://doi.org/10.1007/s43621-024-00469-y

Niasse, M., & Varis, O. (2020). Quenching the thirst of rapidly growing and water-insecure cities in sub-Saharan Africa. *International Journal of Water Resources Development, 36*(2–3), 505–527. https://doi.org/10.1080/07900627.2019.1707073

Nieuwenhuis, E., de Bruijn, H., Cuppen, E., & Langeveld, J. (2023). Understanding cross-sectoral innovations for urban water management through the lens of organizational ambidexterity. *Urban Water Journal, 20*(9), 1183–1195. https://doi.org/10.1080/1573062X.2023.2251973

Nixon, R., Ma, Z., Khan, B., Birkenholtz, T., Lee, L., & Mian, I. A. (2022). Social influence shapes adaptive water governance: Empirical evidence from northwestern Pakistan. *Ecology and Society, 27*(3). https://doi.org/10.5751/ES-13546-270337

Nkhata, B., & Breen, C. (2016). Assessing and measuring adaptive capacity: The experiences of African countries in developing meaningful metrics for water management. *Current Opinion in Environmental Sustainability, 21*, 9–14. https://doi.org/10.1016/j.cosust.2016.10.008

Okeola, O. G., & Raheem, K. A. A. (2016). Prioritization of water resources management problems in North Central Nigeria using Rapid Impact Assessment Matrix (RIAM). *Journal of Water Resource and Protection, 8*(3), 345–357. https://doi.org/10.4236/jwarp.2016.83029

Olabode, O. F., & Comte, J.-C. (2024). Water scarcity in the fast-growing megacity of Lagos, Nigeria and opportunities for managed aquifer recharge. *Wires Water, 11*(5), Article e1733. https://doi.org/10.1002/wat2.1733

Page, M. J., McKenzie, J. E., Bossuyt, P. M., Boutron, I., Hoffmann, T. C., Mulrow, C. D., Shamseer, L., Tetzlaff, J. M., Akl, E. A., Brennan, S. E., Chou, R., Glanville, J., Grimshaw, J. M., Hróbjartsson, A., Lalu, M. M., Li, T., Loder, E. W., Mayo-Wilson, E., McDonald, S., … Moher, D. (2021). The PRISMA 2020 statement: An updated guideline for reporting systematic reviews. *BMJ, 372*, n71. https://doi.org/10.1136/bmj.n71

Pathirana, A., den Heijer, F., & Sayers, P. B. (2021). Water infrastructure asset management is evolving. *Infrastructures, 6*(6), 90. https://doi.org/10.3390/infrastructures6060090

Patino, C. M., & Ferreira, J. C. (2018). Inclusion and exclusion criteria in research studies: Definitions and why they matter. *Jornal Brasileiro De Pneumologia: Publicacao Oficial Da Sociedade Brasileira De Pneumologia E Tisilogia, 44*(2), 84. https://doi.org/10.1590/s1806-37562018000000088

Peat, M., Moon, K., Dyer, F., Johnson, W., & Nichols, S. J. (2017). Creating institutional flexibility for adaptive water management: Insights from two management agencies. *Journal of Environmental Management, 202*, 188–197. https://doi.org/10.1016/j.jenvman.2017.06.059

Rosengren, L. M., Schinko, T., Sendzimir, J., Mohammed, A.-R., Buwah, R., Vihinen, H., & Raymond, C. M. (2023). Interlinkages between leverage points for strengthening adaptive capacity to climate change. *Sustainability Science, 18*(5), 2199–2218. https://doi.org/10.1007/s11625-023-01327-y

Rosso, R. (2025). On seven principles of water governance. *Water, 17*(6), 896. https://doi.org/10.3390/w17060896

Saikia, P., Beane, G., Garriga, R. G., Avello, P., Ellis, L., Fisher, S., Leten, J., Ruiz-Apilánez, I., Shouler, M., Ward, R., & Jiménez, A. (2022). City water resilience framework: A governance based planning tool to enhance urban water resilience. *Sustainable Cities and Society, 77*, Article 103497. https://doi.org/10.1016/j.scs.2021.103497

Salehi, M. (2022). Global water shortage and potable water safety; Today's concern and tomorrow's crisis. *Environment International, 158*, Article 106936. https://doi.org/10.1016/j.envint.2021.106936

Salvador-Oliván, J. A., Marco-Cuenca, G., & Arquero-Avilés, R. (2021). Development of an efficient search filter to retrieve systematic reviews from PubMed. *Journal of the Medical Library Association : JMLA, 109*(4), 561–574. https://doi.org/10.5195/jmla.2021.1223

Schwartz, K., Gupta, J., Mazzucato, M., J. Bosch, H., & van Vliet, L. W. (2025). Prioritising the purpose: A Water System Justice framework for water utilities. *Water Policy, 27*(4), 540–560. https://doi.org/10.2166/wp.2025.284

Shayo, G. M., Elimbinzi, E., Shao, G. N., & Fabian, C. (2023). Severity of waterborne diseases in developing countries and the effectiveness of ceramic filters for improving water quality. *Bulletin of the National Research Centre, 47*(1), 113. https://doi.org/10.1186/s42269-023-01088-9

Sheel, V., Kotwal, A., Dumka, N., Sharma, V., Kumar, R., & Tyagi, V. (2024). Water as a social determinant of health: Bringing policies into action. *Journal of Global Health Reports, 8*, e2024003. https://doi.org/10.29392/001c.92160

Shen, J., Huang, X., Nie, Y., Li, S., & Sun, F. (2025). Evaluating asset-based management of water resources based on the water asset stocks and changes table. *Journal of Cleaner Production, 505*, Article 145473. https://doi.org/10.1016/j.jclepro.2025.145473

Sinha, S. K., Davis, C., Gardoni, P., Babbar-Sebens, M., Stuhr, M., Huston, D., Cauffman, S., Williams, W. D., Alanis, L. G., Anand, H., & Vishwakarma, A. (2023). Water sector infrastructure systems resilience: A social–ecological–technical system-of-systems and whole-life approach. *Cambridge Prisms: Water, 1*, Article e4. https://doi.org/10.1017/wat.2023.3

Sithole, M. Z., Morepje, M. T., & Mokoena, T. I. (2025). Unpacking water scarcity adaptation strategies for sustainable food production systems in Sub-Saharan Africa. *Sustainability, 17*(23), 10627. https://doi.org/10.3390/su172310627

Skinner, W., Bardsley, D. K., & Drew, G. (2024). Post-crisis risk management: Water, community, and adaptation in a South Australian irrigation district. *Ecology and Society*, *29*(1). https://doi.org/10.5751/ES-14789-290110

Smit, B., & Wandel, J. (2006). Adaptation, adaptive capacity and vulnerability. *Global Environmental Change, 16*(3), 282–292. https://doi.org/10.1016/j.gloenvcha.2006.03.008

Stathatou, P. -M, Kampragou, E., Grigoropoulou, H., Assimacopoulos, D., Karavitis, C., Porto, M. F. A., Gironás, J., Vanegas, M., & Reyna, S. (2016). Vulnerability of water systems: A comprehensive framework for its assessment and identification of adaptation strategies. *Desalination and Water Treatment, 57*(5), 2243–2255. https://doi.org/10.1080/19443994.2015.1012341

Suárez-García, A., Álvarez-Hernández, M., Arce, E., & Ribas, J. R. (2024). Exploring the efficacy of binary surveys versus Likert scales in assessing student perspectives using Bayesian analysis. *Applied Sciences, 14*(10), 4189. https://doi.org/10.3390/app14104189

Sultana, P., & Thompson, P. M. (2017). Adaptation or conflict? Responses to climate change in water management in Bangladesh. *Environmental Science & Policy, 78*, 149–156. https://doi.org/10.1016/j.envsci.2017.09.011

Wagner, M., Chhetri, N., & Sturm, M. (2014). Adaptive capacity in light of Hurricane Sandy: The need for policy engagement. *Applied Geography, 50*, 15–23. https://doi.org/10.1016/j.apgeog.2014.01.009

Wang, H., Asefa, T., Bracciano, D., Adams, A., & Wanakule, N. (2019). Proactive water shortage mitigation integrating system optimization and input uncertainty. *Journal of Hydrology, 571*, 711–722. https://doi.org/10.1016/j.jhydrol.2019.01.071

World Resources Institute. (2009). *The national adaptive capacity framework: Key institutional functions for a changing climate* (p. 12). WRI. http://pdf.wri.org/working_papers/NAC_framework_2009-12.pdf

Wybrands, M., Frohmann, F., Andree, M., & Gómez, J. M. (2021). WISdoM: An information system for water management. *Progress in IS*, 131–146. https://ideas.repec.org//h/spr/prochp/978-3-030-61969-5_10.html

Part III
Sustainable Practices and Community Engagement

Chapter 8
Water Conservation and Sustainable Practices

Sesan Abiodun Aransiola, Amos Musa Sumakada, Aishat Olorunkemi Salisu, Bolaji Oluwatosin Akanbi, Juliana Okwena Pondei, Olabisi Peter Abioye, and Naga Raju Maddela

Abstract Water is a fundamental resource on earth that sustains life, yet its availability is becoming scarce daily. Excessive consumption, increase in industry activities, pollution and climate change due to global increase in population are reasons for water scarcity. Water resources should be given top priority due to the importance of water to life. Planning the availability and future development of this essential and scarce resource can be achieved through water sustainability practices. This paper explores methods of water conservation, the role of microorganisms in recycling water, mitigating the impacts of pollution, climate change and promoting sustainable environmental practices. Water scarcity and deterioration of the environment are issues faced globally and can be resolved using microorganisms in wastewater treatment, wastewater recycling, and reuse. This paper also discussed the benefits, challenges, and offers predictions for the future on how microorganisms will be employed in water conservation efforts. The findings ultimately provided scalable solutions to water conservation and sustainable practices making water treatment, wastewater recycling and reuse economically viable.

S. A. Aransiola (✉)
Centre for Artificial Intelligence & Multidisciplinary Innovation Studies, Department of Auditing, College of Accounting Sciences, University of South Africa, Pretoria 0002, South Africa
e-mail: sesan.aransiola@uniabuja.edu.ng

S. A. Aransiola · A. M. Sumakada · A. O. Salisu · B. O. Akanbi · J. O. Pondei
Department of Microbiology, Faculty of Life Science, University of Abuja, PMB 117, Abuja, Nigeria

O. P. Abioye
Department of Microbiology, School of Life Sciences, Federal University of Technology, Minna, Nigeria

Department of Public Health, School of Life Sciences, Federal University of Technology, Minna, Nigeria

N. R. Maddela
Departamento de Ciencias Biológicas, Facultad la Ciencias de la Salud, Universidad Técnica de Manabí, Portoviejo, Manabí, Ecuador

M. I. Adeoba et al. (eds.), *Water Resources Management in Nigeria*, World Water Resources 10, https://doi.org/10.1007/978-3-032-22813-0_8

Keywords Water scarcity · Microbial water treatment · Bioaugumentation · Water conservation · Microbial consortia and electrocatalysis

8.1 Introduction

One of the most important but yet paradoxically short in supply commodities on Earth is clean water. As the population of cities and towns increases, climate change and industrial activity also increase, and the demand for potable water is far higher than the supply. It is true that about 70% of the planet's earth surface is covered with water; still less than 1% of it is clean for human consumption and domestic use. The increasing global scarcity and freshwater crisis is because of the lopsided discrepancy where only about 0.3% of the freshwater on Earth is easily accessible for direct human use, according to UNESCO (2022a, 2022b). World Bank (2021) opined that More than two billion people across the globe are currently experiencing severe water stress.

The increasing rate of scarcity of clean water has far-reaching consequences, including the deepening of social inequalities, environmental degradation, public health risks, and increased food insecurity. Access to safe drinking water in many developing countries, such as Nigeria, remains a major challenge. As a result, people are often rely on untreated water sources that are frequently contaminated with germs and pathogens such as coliform bacteria, protozoan parasites and significantly increasing the risk of waterborne diseases (Tabe et al., 2021).

For generations, people have relied on practical strategies like collecting rainwater reusing treated wastewater, and improving irrigation, to help manage water use wisely. These conventional methods have played a crucial role in reducing the burden on limited freshwater supplies. For instance, recycling and reusing treated wastewater not only helps conserve clean water but also offers a steady and safe source for city needs and farming (Sharma et al., 2023). Likewise, combining greywater reuse with rainwater harvesting has proven to be an effective way to reduce much pressure on the amount of wastewater we produce and tap water we use, especially in highly-populated urban areas where water demand is very high (Domínguez-Chávez et al., 2016). Microbes like algae, fungi, and bacteria help clean water by cycling the nutrients and breaking down waste. By putting these microorganisms to work through techniques like sustainable agriculture, bioaugmentation, and biological water treatment, we can use these natural systems to boost water quality and make it sustainable for the future.

In Nigeria, for example, drinking water often becomes contaminated during distribution, even when it's properly treated at the source. The cause of these problems lies in poor maintenance of the pipelines and old pipelines. These pipes are vulnerable to leaks, which give room for recontamination. Even though they have been treated before by corrosion, back-siphonage, and inconsistent pressure, this undermines both water quality and disinfection (WHO, 2022).

To protect and maintain good public health, maintaining a residual level of chlorine in drinking water is essential, as it helps ensure the microbiological integrity of distribution pipelines during water transit (WHO, 2017). These microbiological safeguards are particularly critical in low-resource settings where infrastructure may be vulnerable to contamination. This paper explores the intersection of microbiological science, sustainable water management and highlighting how innovative approaches such as microbial source tracking and decentralized disinfection technologies can complement traditional water conservation methods, especially in low- and middle-income countries (String et al., 2021). It also addresses shortcomings of institutions along with policy barriers and economic, that have hindered the widespread adoption of microbial water conservation strategies.

8.2 Globe Water Scarcity and the Need for Conservation

Water scarcity or crisis happens when the demand for clean water surpasses what is available for supply. Though water is about 70% of the planet's earth surface, still less than 1% of it is clean for domestic use and human consumption. As shown in Fig. 8.1, 97% of the water body is ocean while 3% is freshwater. This lopsided discrepancy is because of the increasing global freshwater crisis and scarcity. It is a daily growing issue globally and some of the factors contributing to this include rapid urbanization, climate change, poor management of resources, and a booming global population. Right now, about two billion people are living with some level of water scarcity, and it is a crisis that contributes to nearly 3.4 million deaths each year (UN FAO, 2017; Water, 2020). This shortage makes it harder for communities to access clean, safe water for drinking, farming, sanitation, and industry.

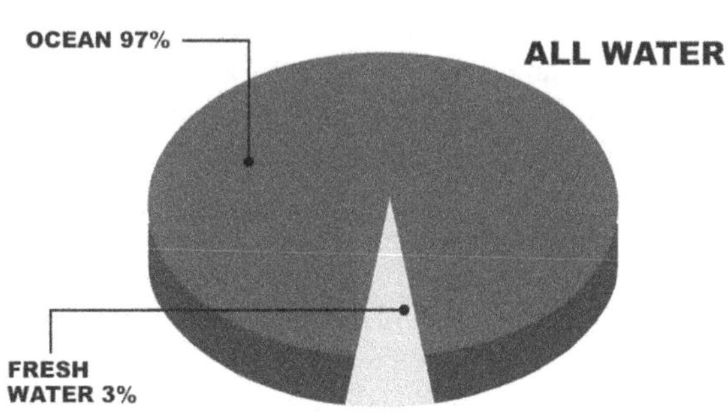

Fig. 8.1 Global water distribution

8.2.1 Causes of Water Scarcity

Environmental changes, economic pressures, and strain on freshwater resources influence global water scarcity. Factors like urbanization, industrialization, agricultural practices, and climate change affect water quality and microbial community structures. Urban runoff introduces contaminants, industrial effluents discharge heavy metals, agricultural runoff contributes nutrients, and climate-related shifts alter microbial ecologies. Addressing water scarcity requires integrating macro-level hydrological shifts and micro-level microbial dynamics (Trevathan-Tackett et al., 2020).

Urban growth in Nigeria has led to limited access to drinking water due to inadequate sanitary services and wastewater treatment. The crisis is exacerbated by small-capacity facilities, old pipes, and unplanned settlements, and by urban development-related solid surfaces (UN-Habitat, 2022).

Mining, textiles, energy production, pharmaceuticals, and food processing industries heavily rely on freshwater, causing pollution with heavy metals, hydrocarbons, and synthetic chemicals. Limited access to cleaner technologies and lax enforcement impact water quality and competition (WWAP, 2021).

Agriculture uses 70% of global freshwater withdrawals, but traditional methods like flood irrigation cause significant water losses (FAO, 2020). Water-intensive crops in unsuitable climates increase aquifer depletion and surface water scarcity. Fertilizer and pesticide runoff pollute freshwater ecosystems, reducing biodiversity.

Climate change and evaporation rates are causing global changes in the hydrological cycle, leading to longer droughts and floods. In semi-arid and desert regions, climate change threatens long-term water security and complicates resource planning. Water governance structures must be robust and flexible (IPCC, 2021).

Urbanization and industrial development worsen freshwater resource scarcity, causing pollution and over extraction. Climate change-induced droughts worsen insufficient irrigation infrastructure, constraining water availability. Integrated Water Resources Management strategies are essential to address these interdependencies, but fragmented implementation can undermine sustainability goals. Holistic frameworks align policies across multiple domains (UNESCO & UN-Water, 2023).

8.2.2 Water Scarcity in Nigeria

Across Nigeria, many water systems are struggling primarily due to years of inadequate maintenance. Cities such as Lagos and Port Harcourt, for instance, face increasing difficulty in supplying clean and reliable water through their municipal infrastructure. This challenge is not recent; rather, it stems from a long history of systemic neglect, mismanagement, and chronic underfunding that has weakened water delivery capacity and infrastructure resilience (Adeniran et al., 2022).

Many people have stopped relying on the public water supply altogether; rather, they are turning to buying water from vendors or private boreholes. These options can be expensive and do not always guarantee safe water. According to Chellaraj et al. (2018), more than 30% of Nigeria's upgraded water infrastructure is not functioning properly.

With Nigeria's population now topping 220 million and still growing rapidly, the pressure on this already broken system is only getting worse. Around the edges of these cities, informal settlements are sprouting up, many without proper drainage, clean water, or sanitation (Adelekan, 2020). According to Orebiyi et al. (2010), across the country, shallow aquifers, rivers, and lakes are increasingly polluted by industrial waste, household sewage, and runoff from farms. The widespread contamination and pollution Nigeria puts water security under serious threat not just from scarcity. In Kaduna State, for instance, textile factories discharge biochemical oxygen demand, untreated waste laden with dyes, heavy metals, and high levels of chemical directly into the Kaduna River, drastically reducing its water quality (Mohammed et al., 2015). Similarly, slaughterhouses near rivers contribute high-strength organic waste, which raises heavy metals, high concentration of suspended solids, organic pollutant and nutrients downstream (Kwadzah & Iorhemen, 2015).

Climate change and environmental issues are making things worse. In northern Nigeria, rainfall has become unpredictable, and long periods of drought have taken a serious toll on water supplies for farms, households, and livestock (Odjugo, 2010). One of the most striking examples is Lake Chad, which has lost over 90% of its size since the 1960s due to both climate change and unsustainable water use (FAO, 2021a, 2021b). Humanitarian groups frequently provide emergency water supplies to internally displaced people (IDPs) in camps, but these are erratic and insufficient (UNICEF, 2020). Even when water is physically present, it may not always be available or safe to use due to treatment facilities, damaged pipes, and boreholes (UN OCHA, 2021).

8.3 Importance of Water Conservation

The impact of drinking water contaminated with waterborne diseases is not just something about the reports or data; it is part of everyday life for many Nigerians, especially children, the elderly, and the immunocompromised. Contaminated sources of water often carry harmful organisms like *Salmonella*, *E. coli*, intestinal worms, and other bacteria and parasites. These can lead to sudden illnesses such as diarrhea or, worse, long-term damage to the digestive system. Across the globe, poor water quality is responsible for more than 80% of diseases in developing countries. According to UN Water (2020), diarrheal illnesses alone claim nearly 485,000 lives each year most of them children under the age of five. In Nigeria, this crisis is made worse by the lack of access to clean, treated water and major gaps in sanitation infrastructure. Many communities, especially in the outskirts of Abuja, depend on open streams, rivers, or hand-dug wells that are rarely treated. A study by Abdulsalam et al. (2022) revealed dangerously high levels of fecal contamination in these water sources far exceeding what the World Health Organization considers safe. Because of this, outbreaks of cholera, typhoid, and dysentery have become all too common, placing heavy financial and emotional burdens on families and overwhelming public health services. Children are the ones who suffer the most from the consequences. Their young immune systems are still developing, and they are more likely to be exposed to dirty water playing outside, fetching water, or drinking it out of necessity. In crowded slums and rural villages, where clean water is hardest to come by, diarrheal diseases continue to be one of the leading causes of death among children under five. A study by Igbinosa and Okoh (2012) confirmed a strong link between poor water quality and high rates of childhood infections across several Nigerian states. In addition, these illnesses do more than make kids sick, they can stunt growth, harm brain development, and hold them back from thriving in school and later in life. In some areas, pollution from unregulated development, industry, or mining has led to water mixed with heavy metals and nitrates. These can cause damage to the liver, kidneys, neurological issues, and even child developmental delays. Research by Yisa and Jimoh (2010) shows that groundwater in parts of southern Nigeria has been seriously affected by human activity, leading to unsafe levels of chemical exposure. Children and pregnant women face the greatest risks, as these toxins can interfere with organ development and brain function. Many years of underfunding in water infrastructure, weak enforcement of safety standards, and a lack of public awareness about hygiene are major factors behind Nigeria's water crisis and are deep-rooted problems: While national programs like the WASH (Water, Sanitation, and Hygiene) plan have been introduced, implementation has always been inconsistent, leaving many communities to consuming unsafe and contaminated water. Nigeria needs to invest in reliable, safe water systems, strengthen its systems for monitoring and regulation and support community-level treatment solutions like chlorination and filtration. Addressing this crisis will require more than policies; it demands real action. Every Nigerian deserves clean water, as it is a fundamental human right and not luxury.

8.3.1 Traditional Approaches in Water Conservation

8.3.1.1 Rainwater Harvesting (RWH)

Rainwater harvesting (RWH) is an ancient, conventional and sustainable technique for managing water resource and preserving water by collecting and storing rainwater for future use. It is now more relevant in the recent years. This method assist with both non-potable and potable water requirements, RWH massively reduces the impact of freshwater scarcity (Yannopoulos et al., 2019). This method is more beneficial to rural communities with little access to centralized and municipal water infrastructures and in cities with increasing population growth.

RWH is very vital for reducing the stress on municipal systems, which saves energy, treatment and costs related to water distribution, reducing to the bearest minimum improving groundwater recharge, and stormwater runoff (Rajagopalan et al., 2020). RWH is a simple method to climate-change water planning because of these environmental and financial advantages. Harvested rainwater storage acts as a very vital remedy in drought-prone areas, which guarantees the availability of water during dry season. RWH's adaptability and acceptability to a wide range of settings is further shown by its scalability from household-level systems into more extensive institutional undertakings.

8.3.1.2 Improved Irrigation Techniques

The biggest user of clean water resources is the agricultural sector and about 70% of usable, freshwater is used. (FAO, 2017). Surface irrigation or flood are traditional irrigation ways and are characterized with large water wastes from evaporation, percolation, and runoff. Better irrigation methods have been developed to maximize water utilization that can guarantee more sustainable and long-lasting agricultural production,

Modern irrigation technologies like the drip and sprinkler systems are a method that delivers water straight to the plants root zones which reduce water waste and increase water uptake. Drip irrigation has shown to be very effective with up to 95% water utilization efficiency (Howell, 2001). Conserving water and increase in crop output and be achieved through maintaining constant moisture levels. Furthermore, combining real-time soil moisture data, smart irrigation systems, automated controls and weather forecasts have become an advanced option.

8.3.1.3 Wastewater Recycling and Reuse

Reusing and recycling wastewater efficiently and effectively manage water resources, especially in rural areas with little or restricted water supplies and meagre income. According to Khan et al. (2020), this method involves treating wastewater to a very good quality appropriate for a range of non-potable applications, including groundwater recharge, agricultural irrigation, urban landscaping, and industrial processes. The contemporary treatment methods such as nutrient removal, membrane filtration, and ultraviolet (UV) disinfection, wastewater can be transmogrified from a pollutant into a good resource.

Making use of treated wastewater helps to save the environment by reducing the amount of toxic chemicals discharged into the ecosystems. Additionally, it advances the idea of a circular water economy, by increasing system resilience and continuously recycling and again reusing water (Maron & Shafran, 2019; UNESCO, 2023a, 2023b). However, the support from public, strong regulatory frameworks, ultraviolet (UV) disinfection, and safety guarantees, are necessary for the effective execution of wastewater reuse projects. That will confirm the illustration of the rising acceptance and viability of this approach, treated municipal wastewater is being explored for non-potable purposes in Abuja, Nigeria (Nnamani et al., 2021).

8.3.1.4 Traditional Water Storage Structures

Different cultures and Communities particularly in Africa have device native ways of water storage system that best suits their weather, climate and surroundings. These include constructions that have historically been essential to guaranteeing year-round water availability, particularly in arid and semi-arid areas, such as cisterns, stepwells, kunds, tanks, and baolis. In India and some other countries for instance, stepwells, also known as baolis, were used as community social hub and cultural events Centre in addition to storing rainwater. Rainwater collected during the rainy season is stored in the underground cisterns for use throughout the dry season in North Africa and the Middle East (UNESCO, 2018).

8.3.2 Microbial Strategies for Water Treatment and Conservation

8.3.2.1 Biological Treatment Processes

The eco-friendly and low operating costs of microbial treatment methods in water conservation and treatment are becoming more increasingly and widely used. Algae-based systems, Microbial fuel cells (MFCs), and artificial wetlands are notable examples of these. Constructed wetland–microbial fuel cell systems as shown in Fig. 8.2 are gaining attention on a daily basis for dual role of combined wastewater treatment and energy generation (Jacobs et al., 2024; Montoya Vallejo et al., 2025).Similar to this, algae-based remediation systems use microbial photoautotrophs to absorb nutrients from wastewater, such as phosphorus and nitrogen, while also generating biomass and oxygen that can be used to make biofuels or fertilizers. More so, electrogenic bacteria are used by microbial fuel cells to oxidize organic pollutants and move electrons to an anode, removing germs and contaminants by producing power at the same time as a result (Montoya Vallejo et al., 2025).

8.3.2.2 Biofilms and Anaerobic Digestion

Microbial colonies attached to the surface encased in an extracellular polymeric substance (EPS) matrix are microbial biofilms. The improved metabolic cooperation and robustness of these biofilms enable efficient pollutant sequestration and biotransformation. They are very effective at degrading stubborn organic pollutants, including pesticides, and pharmaceuticals, which immobilize heavy metals through

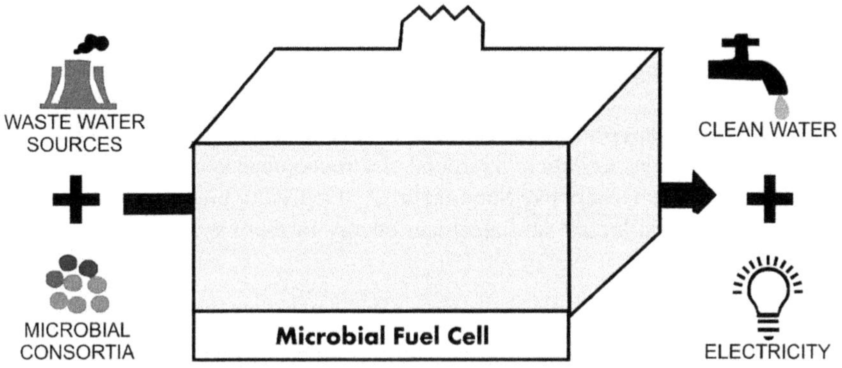

Fig. 8.2 Schematic diagram of a microbial fuel cell

microbial and biochar-mediated mechanisms (Manikandan et al., 2023), Technologies for wastewater treatment have advanced and thanks to their use in membrane bioreactors biofilm reactors, and trickling filters. In addition, complicated organic matter can be broken down in oxygen-limited environments by anaerobic digestion, which is usually carried out by groups of hydrolytic, methanogenic microbes, acetogenic, and acidogenic. This meets the objectives of renewable energy and waste minimization by producing methane-rich biogas and reducing the volume of sludge.

8.3.2.3 Bioaugmentation

This is the introduction of advantageous microorganisms to water systems in order to increase the quality of the water. By introducing particular strains of microorganism with high metabolic efficiency into contaminated environments, bioaugmentation elevates the capacity of the microorganisms to debase the contaminants. In cleaning up settings tainted with hydrocarbons, heavy metals, and chlorinated solvents, this process is being used. Genetically engineered or naturally adapted microbial strains have shown traits capable of debauching persistent environmental contaminants under site-specific conditions, (Singh et al., 2019).

8.3.2.4 Microbial Consortia and Electrocatalysis

A reliable approach offered by microbial consortia is suitable for degrading a wide range of complex pollutants and contaminants in wastewater because they are involved in enigmatic microbial interactions and metabolic complementarity. System stability and metabolic flexibility are made easier by synergistic relationship within microbial consortia, which is useful for the bioremediation of industrial effluents containing high levels of xenobiotic compounds (Joutey et al., 2013). Consortiums possess synergistic effects, where the metabolic output of one species acts as a substrate for another thereby decomplexifying breakdown processes. Microbial electrocatalysis makes use of specific bacteria with the capacity to move electrons to external electrodes, especially in microbial electrochemical systems (MESs) unlike microbial consortia. Genera like Shewanella and Geobacter play beneficial roles by changing organic pollutants into electrical energy in these systems (Logan et al., 2019).

8.3.2.5 Microbial Encapsulation and Immobilization Techniques

A developing microbial approach to water treatment and conservation is the encapsulation, trapping and immobility of beneficial microbes within polymeric matrices like silica gels, polyurethane foams, or alginate beads. Proliferation in the resilience of microbial communities to external stressors like changes or toxins in salinity are due to stabilization of this technique. Encapsulated or trapped microorganisms are economical for long-term operations due to the microbes being reused many times in the treatment cycles and keeping high metabolic activity. Operational stability and biodegradation of persistent organic pollutants such as polycyclic aromatic hydrocarbons and chlorinated compounds are ways in which effectiveness and efficiency are revealed (Kumar et al., 2016).

8.4 Sustainable Water Conservation Practices

Sustainable water conservation strategies ascertain the safeguarding and effective management of the planet's restricted freshwater resources. These strategies provide workable, long-term solutions to guarantee water availability for future generations, particularly as water scarcity worsens due to population growth; industrial expansion and climate change (Gleick & Palaniappan, 2023). The purpose of sustainable water conservation is not only to meet human needs but also to protect ecosystems and fosters environmental resilience through the combination of technology, natural processes, and behavioral change across domestic, agricultural, and industrial sectors. The global distribution of water use is depicted in Fig. 8.3 as this aids in identifying the sectors with most water usage. (UNEP, 2022; UNESCO, 2021).

8.4.1 Sustainable Water Conservation for Domestic Use

The demand for fresh water is reduced through efficiency, reuse, and behavioral modifications of sustainable water conservation approach in residential settings. A crucial

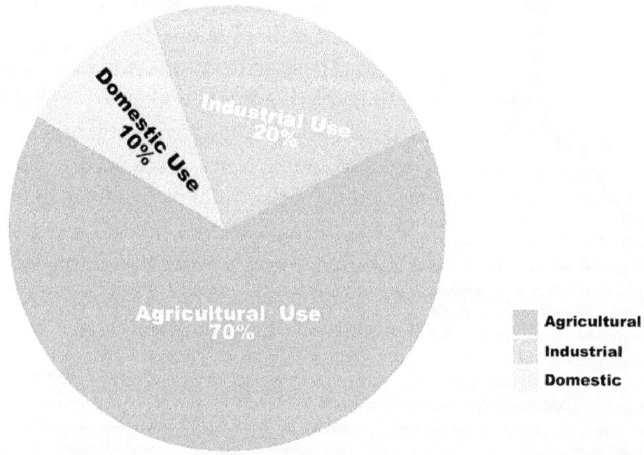

Fig. 8.3 Global distribution of water use in the domestic, agricultural and industrial sectors (Adopted from FAO, 2021a, 2021b)

strategy that can reduce home water usage by up to 50% is through the adoption of water-efficient fixtures, such as low-flow showerheads, faucets, and toilets (EPA, 2023). These technologies are feasible for long-term usage since they lower water use without compromising functionality. Greywater recycling systems are essential to sustainable household practices because they enable the filtration and reuse of lightly used water from sinks, showers, and laundry for landscape irrigation and toilet flushing (Li et al., 2020).

Public education campaigns that encourage water-saving behaviors such as turning off taps while brushing teeth, repairing leaks promptly, and running appliances only when fully loaded promote sustainable water management at household level (Willis et al., 2016). These collective technical and behavioral strategies promote a sustainable and resilient approach to domestic water consumption.

8.4.2 Sustainable Water Conservation for Agricultural Use

In order to maintain environmental balance and establish long-term food security in agriculture being one of the most consumers or freshwater, sustainable water management techniques need to be implemented. For instance, the delivery of water directly to the plant root zone through drip irrigation minimizes evaporation and runoff and has been shown to ameliorate water use efficiency by up to 60% compared to conventional irrigation (Kisekka et al., 2017).

Real-time soil moisture sensors, drone surveillance, and weather-based irrigation scheduling, which are accurate agriculture technologies permit farmers to optimize water use and maximize crop produce. These innovations notably batten decision-making related to resource allocation and irrigation (Zhou et al., 2021).

Drought-tolerant crop varieties, developed through traditional breeding or genetic engineering, ensure productivity under limited water accessibility. These cultivars offer an adaptive mechanism for sustaining agricultural output from increasing water stress (Mittler & Blumwald, 2015). Capturing seasonal rainfall and ensuring water availability during dry periods are achieved through water harvesting structures such as farm ponds and contour trenches (Nain et al., 2019). Collectively, these strategies enable an integrated, adaptive approach to sustainable water management in agriculture.

8.4.3 Sustainable Water Conservation for Industrial Use

Sustainable water management is essential for both environmental improvement and operational effectiveness for industries that use a lot of water such as food processing industries, textiles and so on. One of the best routines in industrial operations is reusing and recycling water. Closed-loop water recycling systems noticeably reduce the need for freshwater intake and reduces wastewater effluents by enabling reiterated treatment and reuse of process water (Qin et al., 2020).

Zero-liquid discharge (ZLD) technologies are being adopted by industries more and more due to the recovery of nearly all of the water from effluents using sophisticated treatment techniques including reverse osmosis, evaporation, and crystallization,. Moreover, industrial water consumption can be greatly reduced by process optimization, such as utilizing dry cleaning methods or moving from water-intensive cooling towers to air-cooled systems. Water footprint evaluations can aid industries in finding inefficiencies and places for conservation that calculate the overall amount of freshwater utilized in production. The practice of industries collaborating with agricultural or municipal stakeholders to reuse treated wastewater is known as Industrial symbiosis (Bai et al., 2024).

8.5 Comparison of Traditional and Microbial Methods of Water Conservation

8.5.1 *Integrated Approach (Combination of the Traditional and Microbial Methods for Water Conservation)*

Merging of conventional and microbiological techniques ameliorates water quality and availability. There are traditional practices such as watershed management and rainwater collection that are inexpensive and of great importance to the community to capture, store, and restore water supplies. These systems are sustainable and culturally acceptable due to the usage of local resources and modification to local requisites (Wani et al., 2016; Yannopoulos et al., 2019). However, aggrandized biological treatment for wastewater is attainable by microbial technologies, including biofilms, microbial fuel cells (MFCs), anaerobic digestion, and artificial wetlands. Water reuse is achieved by these techniques; this lowers the toxic burden and decompose both organic and inorganic impurities. In addition, systems, such as MFCs, put forth economic value through the generation of renewable energy while the treatment process is in motion (Manikandan et al., 2023; Montoya Vallejo et al., 2025).

Example: Water capture, purification, and energy production can all be done simultaneously through the fusion of rainwater harvesting with artificial wetlands and microbial fuel cells. Harvested rainwater is contained and conveyed into wetlands, where microbial populations make them pure, MFCs are integrated into the system to batten system efficiency and resilience by converting organic waste into electricity (Jacobs et al., 2024; Logan et al., 2019).

8.6 Challenges

Water conservation using microbial technologies is highly beneficial and promising to the environment however, it is faced with a number of challenges:

- **Infrastructure and Economical Barriers**: A challenge facing the use of microbial systems is the expensiveness of the deployment in large-scale process, especially those utilizing synthetic biology in the Global South. According to Tzouvelekis et al. (2019), lack of technical and financial resources can hinder the implementation of the research.
- **Gaps in Regulation and Policy**: Obsolete legal frameworks and contradictory regulations can hinder approval and safe implementation of the microbial systems. The defect of conspicuous regulations pertaining to the use of synthetic biology and genetically modified organisms (GMOs) prevents innovation and cross-border collaboration (OECD, 2020).

- **Biosafety and Environmental Issues**: Ecosystems can be disturbed through the induction of non-indigenous microorganisms through horizontal gene transfer or interference with native microbial populations. Adequate biosafety evaluations and environmental monitoring can be put in order to reduce these risks (de Lorenzo, 2020; Timmis et al., 2019).
- **Inequity in Technology and Knowledge**: Areas or countries with limited resources and water do not have access to advanced technologies like transcriptomics and metagenomics unlike wealthy nations and some research institute. This challenge prevents the development of appropriate microbiological solutions (UNESCO, 2023a, 2023b; WHO, 2022).
- **Social Acceptance and Public Trust**: Public acceptance of technology is solely dependent on successes of the technology in history and versatility of the technology. Inadequate information or delayed sensitization to the public can decrease trust and acceptance of the microbial innovations resulting in the rejection (Simis et al., 2016; Vraga & Bode, 2020). Without proper sensitization of the community, even a successful scientific approach can be opposed (Scheufele & Krause, 2019; Thomas et al., 2017).

8.7 Conclusion

In the real world, microbial technologies have already demonstrated efficacy in successful bioremediation initiatives across different regions of India and Israel's sophisticated wastewater recycling systems, where favorable policy frameworks, supportive infrastructure, and active public engagement have come together to address global water challenges through innovative, decentralized, and environmentally sustainable solutions. However, a wide range of intricate obstacles limits the wider use and scalability of microbial treatments. These include urgent socio-political, legal, and environmental issues in addition to technical constraints and economic viability. A globally coordinated approach that is morally good, considerate of local settings, and inclusive of the opinions of all stakeholders is necessary to realize their full potential. To achieve this goal, a multifaceted strategy is needed which is access to microbial technologies which can be enhanced through strengthening marginalized groups, robust and flexible legislative frameworks, long-term sustainability and public trust. The United Nations Sustainable Development Goals, especially SDG 6 (Clean Water and Sanitation) and SDG 13 (Climate Action) can be greatly aided by the successful integration of microbial water conservation technologies. In order to achieve a shared vision of water security, climate resilience, and environmental stewardship for future generations, we must all work together to close the gaps between science, governance, and societal values.

References

Abdulsalam, M., Aliyu, A., Musa, M., & Ibrahim, U. (2022). Assessment of fecal contamination in surface water sources in Abuja, Nigeria. *Journal of Water and Health, 20*(3), 345–356. https://doi.org/10.2166/wh.2022.183

Adelekan, I. O. (2020). Urban dynamics and water services in Nigerian cities: Challenges and opportunities. *African Journal of Environmental Management, 24*(2), 134–148.

Adeniran, A. E., Agbede, O. A., & Jimoh, M. A. (2022). Urban water supply and infrastructure challenges in Nigeria. *Journal of Water Resources and Environmental Engineering, 14*(4), 101–112.

Bai, X., Zhang, Y., & Chen, H. (2024). Industrial symbiosis and water reuse: Towards zero-liquid discharge systems. *Journal of Environmental Engineering, 150*(1), 107–119.

Chellaraj, G., Adeoti, A. I., & Margolis, D. (2018). *Nigeria's water infrastructure and household water access: A review of trends and challenges. The World Bank.*

de Lorenzo, V. (2020). Beware of metaphors: Time for a paradigm shift in synthetic biology. *BioDesign Research, 2020,* Article 9898296. https://doi.org/10.34133/2020/9898296

Domínguez-Chávez, C., Marrón-Marchante, J. R., & Muñoz-García, M. A. (2016). Rainwater harvesting and greywater reuse in urban areas: A study in sustainable water management practices. *Water Resources Management, 30*(5), 1671–1685. https://doi.org/10.1007/s11269-016-1234-7

EPA. (2023). *WaterSense: Saving water indoors. U.S. Environmental Protection Agency.* https://www.epa.gov/watersense

FAO. (2017). *The future of food and agriculture—Trends and challenges. Food and Agriculture Organization of the United Nations.* https://www.fao.org/publications

FAO. (2020). *FAO water report: Irrigation in Africa in figures. Food and Agriculture Organization.* https://www.fao.org

FAO. (2021a). *Water security for all: Case study on Lake Chad. Food and Agriculture Organization of the United Nations.*

FAO. (2021b). *Global water use distribution by sector. Food and Agriculture Organization of the United Nations.*

Gleick, P. H., & Palaniappan, M. (2023). Peak water limits and sustainable water management. *Environmental Research Letters, 18*(1), Article 014002.

Howell, T. A. (2001). Enhancing water use efficiency in irrigated agriculture. *Agronomy Journal, 93*(2), 281–289.

Igbinosa, I. H., & Okoh, A. I. (2012). Impact of water quality on childhood infections in sub-Saharan Africa: Evidence from Nigeria. *African Journal of Microbiology Research, 6*(2), 456–465.

IPCC. (2021). *Climate change 2021: The physical science basis. Intergovernmental Panel on Climate Change.* https://www.ipcc.ch

Jacobs, B., Torres, M., & Rahman, M. (2024). Integrated constructed wetlands and microbial fuel cells for wastewater treatment. *Environmental Science & Technology Letters, 11*(2), 88–94.

Joutey, N. T., Bahafid, W., Sayel, H., & El Ghachtouli, N. (2013). Biodegradation: Involvement of microbial consortia and potential for environmental remediation. *Indian Journal of Microbiology, 53*(3), 407–420.

Khan, S., Cao, Q., Zheng, Y. M., Huang, Y. Z., & Zhu, Y. G. (2020). Reuse of treated wastewater in China: Opportunities and challenges. *Environmental Science & Technology, 54*(7), 3518–3526.

Kisekka, I., Aguilar, J. P., & Rogers, D. H. (2017). Water use efficiency of drip versus conventional irrigation systems. *Irrigation Science, 35*(4), 309–321.

Kumar, M., Ahmaruzzaman, M., & Sinha, R. (2016). Microbial immobilization techniques for wastewater treatment. *Biotechnology Advances, 34*(5), 559–579.

Kwadzah, D. D., & Iorhemen, O. T. (2015). Impact of slaughterhouse waste on surface water in Benue State, Nigeria. *Journal of Environmental Protection, 6*(10), 980–988.

Li, F., Wichmann, K., & Otterpohl, R. (2020). Review of greywater treatment and reuse. *Environmental Science and Technology, 54*(1), 1–10.

Logan, B. E., Rossi, R., & Saikaly, P. E. (2019). Microbial electrochemical technologies for sustainable energy and water. *Nature Reviews Microbiology, 17*, 307–319.

Manikandan, N., Subramani, T., & Venkatesan, P. (2023). Biofilm-based treatment strategies for micropollutant removal. *Journal of Environmental Chemical Engineering, 11*(3), 109–118.

Maron, M., & Shafran, A. (2019). Urban wastewater reuse for agriculture in arid regions. *Environmental Economics and Policy Studies, 21*(3), 475–491.

Mittler, R., & Blumwald, E. (2015). Genetic engineering for modern agriculture. *Science, 350*(6256), 915–916.

Mohammed, R. A., Ibrahim, I. Y., & Ahmed, A. (2015). Textile effluent impact on Kaduna River water quality. *International Journal of Environmental Studies, 72*(3), 479–490.

Montoya Vallejo, A., Zhang, Q., & Li, X. (2025). Advances in microbial fuel cells for wastewater treatment and energy recovery. *Bioelectrochemistry, 155*, 108–115.

Nain, L., Verma, A., & Kumar, R. (2019). On-farm rainwater harvesting structures for agricultural sustainability. *Journal of Water Resource Engineering, 14*(3), 221–230.

Nnamani, N., Okafor, C., & Eze, J. (2021). Treated wastewater reuse in Nigerian cities: Challenges and prospects. *Water Policy, 23*(6), 1120–1133.

Odjugo, P. A. O. (2010). General overview of climate change impacts in Nigeria. *Journal of Human Ecology, 29*(1), 47–55.

OECD. (2020). *Environmental biotechnology: Regulation and innovation*. OECD Publishing. https://doi.org/10.1787/9789264309417-en

Orebiyi, E. O., Awomeso, J. A., Idowu, O. A., Martins, O., Oguntoke, O., & Taiwo, A. M. (2010). Assessment of pollution hazards of shallow well water in Abeokuta and environs, southwest Nigeria. *American Journal of Environmental Sciences, 6*(1), 50–56.

Qin, J. J., Oo, M. H., & Lee, H. K. (2020). Industrial wastewater recycling using zero-liquid discharge systems. *Water Practice and Technology, 15*(2), 502–509.

Rajagopalan, P., Mukherjee, M., & Abraham, S. (2020). Rainwater harvesting in urban planning: Challenges and solutions. *Sustainable Cities and Society, 61*, Article 102307.

Scheufele, D. A., & Krause, N. M. (2019). Science audiences, misinformation, and fake news. *Proceedings of the National Academy of Sciences, 116*(16), 7662–7669. https://doi.org/10.1073/pnas.1805871115

Sharma, S. K., Kennedy, M. D., & Schippers, J. C. (2023). Wastewater reuse for agriculture and cities: Opportunities and challenges. *Journal of Water Process Engineering, 55*, Article 102784.

Simis, M. J., Madden, H., Cacciatore, M. A., & Yeo, S. K. (2016). The lure of rationality: Why does the deficit model persist in science communication? *Public Understanding of Science, 25*(4), 400–414. https://doi.org/10.1177/0963662516629749

Singh, R., Paul, D., & Jain, R. K. (2019). Bioaugmentation in bioremediation: Recent advances. *Environmental Science and Biotechnology, 18*(1), 99–124.

String, G. M., Lantagne, D. S., & Webster, J. (2021). Decentralized water disinfection technologies in low-resource settings: A review. *Environmental Science & Technology, 55*(3), 1754–1771. https://doi.org/10.1021/acs.est.0c06097

Tabe, C., Okeke, C., & Okwori, F. (2021). Waterborne pathogens in Nigeria: Prevalence and control strategies. *Journal of Infectious Diseases and Public Health, 14*(5), 625–631.

Thomas, K., Parrott, L., & Lucy, M. (2017). Public engagement with emerging biotechnologies: Foundations for responsible research and innovation. *Science and Engineering Ethics, 23*(2), 447–468. https://doi.org/10.1007/s11948-016-9775-x

Timmis, K., de Lorenzo, V., Verstraete, W., Ramos, J. L., Danchin, A., Brüssow, H., & Singh, B. K. (2019). The contribution of microbial biotechnology to sustainable development goals. *Microbial Biotechnology, 12*(5), 995–1000. https://doi.org/10.1111/1751-7915.13475

Trevathan-Tackett, S. M., Kelleway, J. J., Macreadie, P. I., Sanderman, J., Baldock, J., & Ralph, P. J. (2020). Global-change impacts on water quality and microbial communities in aquatic ecosystems. *Nature Communications, 11*, 4566.

Tzouvelekis, A., Popova, A., Ferlampin-Acher, M., & Gorgulho, A. (2019). *Strengthening water and sanitation in developing countries through public-private partnerships. World Bank Group.* https://openknowledge.worldbank.org/handle/10986/31460

UN-Habitat. (2022). *State of African cities 2022: Urban water and sanitation. United Nations Human Settlements Programme.* https://unhabitat.org

UN OCHA. (2021). *Nigeria humanitarian needs overview. United Nations Office for the Coordination of Humanitarian Affairs.* https://reliefweb.int

UN Water. (2020). *Water and climate change: 2020 report. United Nations Water.* https://www.unwater.org

UNEP. (2022). *Water for environmental sustainability. United Nations Environment Programme.*

UNESCO. (2018). *Traditional knowledge and water management practices. United Nations Educational, Scientific and Cultural Organization.*

UNESCO. (2021). *Water security and sustainable development. United Nations Educational, Scientific and Cultural Organization.*

UNESCO. (2022a). *Reviving traditional water conservation systems. United Nations Educational, Scientific and Cultural Organization.*

UNESCO. (2022b). *The United Nations world water development report 2022: Groundwater—Making the invisible visible. United Nations Educational, Scientific and Cultural Organization.* https://www.unesco.org

UNESCO. (2023a). *Circular water economy: Innovations in reuse and recovery. United Nations Educational, Scientific and Cultural Organization.*

UNESCO. (2023b). *Science report: The race against time for smarter development. United Nations Educational, Scientific and Cultural Organization.* https://unesdoc.unesco.org/ark:/48223/pf0000384812

UNESCO and UN-Water. (2023). *Water for sustainable development: 2023 world water report.* https://www.unwater.org/publications

UNICEF. (2020). *WASH programme progress in Nigeria: Annual update. United Nations Children's Fund.* https://www.unicef.org/nigeria

Vraga, E. K., & Bode, L. (2020). Defining misinformation and understanding its bounded nature: Using expertise and evidence for describing misinformation. *Political Communication, 37*(1), 136–144. https://doi.org/10.1080/10584609.2020.1716500

Wani, S. P., Sreedevi, T. K., & Rockström, J. (2016). Integrated watershed management for resilience in semi-arid tropics. *Agricultural Water Management, 86*(1–2), 146–162.

WHO. (2022). *Global strategy on digital health 2020–2025. World Health Organization.* https://www.who.int/publications/i/item/9789240020924

Willis, R. M., Stewart, R. A., Panuwatwanich, K., Williams, P. R., & Hollingsworth, A. L. (2016). Residential water end use efficiency. *Journal of Environmental Management, 90*(3), 1448–1458.

World Bank. (2021). *World development report 2021: Data for better lives.* https://www.worldbank.org

World Health Organization (WHO). (2017). *Guidelines for drinking-water quality: Fourth edition incorporating the first addendum.* https://www.who.int/publications/i/item/9789241549950

WWAP (UNESCO World Water Assessment Programme). (2021). *The United Nations world water development report 2021: Valuing water.* https://unesdoc.unesco.org

Yannopoulos, S., Giannakis, G., & Manoli, E. (2019). Rainwater harvesting as a sustainable water management practice in Greece. *Water, 11*(6), 1110.

Yisa, J., & Jimoh, T. (2010). Analytical studies on water quality index of river Landzu. *American Journal of Applied Sciences, 7*(4), 453–458.

Zhou, X., Wang, Q., & Zhang, Y. (2021). Smart irrigation technologies for precision agriculture. *Sensors, 21*(9), 3171.

Chapter 9
Community Participation in Water Management: Pathways to Sustainable Water Governance in Nigeria

Opeyemi Caleb Fatayo and Lawrence Enejeta Akpoterai

Abstract Access to safe and reliable water services in Nigeria remains undermined by fragmented governance and weak community systems. While previous research largely emphasized infrastructure delivery, the social and institutional dimensions that determine long-term functionality have received less attention. This chapter addresses that gap through an integrative review of academic literature, policy documents, and empirical case evidence drawn from community-managed and donor-managed water schemes across Nigeria from 2010 to 2025. The analysis synthesizes theoretical perspectives from Arnstein's Ladder of Participation, Ostrom's Common-Pool Resource Principles, and Hardin's Tragedy of the Commons to evaluate how community participation depth, institutional clarity, and rule enforcement shape water service outcomes. We found that community-managed systems, supported by transparent financial mechanisms and state backstopping, achieve higher resilience and accountability than donor-managed projects that lack local ownership. We also identified emerging innovations such as digital monitoring tools, tariff-based maintenance, and gender-inclusive Water, Sanitation and Hygiene Committees (WASHCOMs) as means to strengthen collective governance and reduce system failures. By linking theoretical insights with Nigerian field evidence, the chapter contributes a context-specific framework explaining how participatory governance can improve sustainability and equity in water service delivery. This work extends existing WASH literature by articulating the institutional and gendered pathways through which community participation transforms short-term access into long-term service reliability.

Keywords Community participation · Governance · Water, sanitation and hygiene committees (WASHCOMs) · Service reliability · Nigeria

---

O. C. Fatayo (✉)
Pan African University Life and Earth Sciences Institute, Ibadan, Nigeria
e-mail: fatayoopeyemi@gmail.com

L. E. Akpoterai
Institute for Climate Change and Adaptation (ICCA), University of Nairobi, Nairobi, Kenya

M. I. Adeoba et al. (eds.), *Water Resources Management in Nigeria*, World Water Resources 10, https://doi.org/10.1007/978-3-032-22813-0_9

9.1 Introduction

Water use is of critical importance in Nigerian communities. However, access to this resource presents what we may call Nigeria's water paradox—a country endowed with vast freshwater reserves yet confronted by chronic scarcity, inequality, and service unreliability (Irene et al., 2025). Water sustains life, health, agriculture, and cultural identity, yet millions of Nigerians still lack reliable access. Despite an estimated 215 billion cubic meters of surface water and 87 billion cubic meters of groundwater (estimates from UNEP GRID country data), only a minuscule fraction is effectively available for domestic and productive use due to poor governance, uneven distribution, inadequate infrastructure, and increasing ecological pressures (Ibrahim et al., 2023; Isukuru et al., 2024). The implications are widespread: persistent rural-urban disparities, gendered burdens in water collection, and growing exposure to waterborne diseases and climate-induced stressors (Ayanlade et al., 2022; Adeoti et al., 2023; Ite et al., 2018).

Moreover, catastrophic levels of contamination, occasional flooding, pollution, high-intensity heatwaves, and prolonged droughts further complicate water management, rendering traditional water sources unreliable (Anifowose & Oyebode, 2019; Ighalo & Adeniyi, 2020). Traditional patterns of water use have hence been disrupted, and management systems strained, thus reducing the reliability of supply and heightening competition among communities and sectors. Meanwhile, the regular centralized model of water governance has been criticized for bureaucratic rigidity, overlapping institutional mandates, and policy failures, which undermine effective and equitable water resource management (Ngene et al., 2021). This underscores the imperative for more inclusive and participatory approaches to water governance in Nigeria (Ugbah et al., 2017). In this context, strengthening community participation emerges as a critical pathway toward more equitable, sustainable, and context-sensitive water management.

However, while numerous studies emphasize technical and infrastructural dimensions of Nigeria's water crisis (Kponee et al., 2015), the social and governance aspects that determine whether water systems remain functional over time are less explored. Hence, this study focuses on community participation as a critical but underexamined determinant of water service sustainability. Across sub-Saharan Africa, evidence shows that when communities are involved in planning, financing, and maintenance, water facilities are more functional and equitable (Mangai & De Vries, 2018; Tsekleves et al., 2022). Also, where local cooperatives and water-user associations are meaningfully empowered, systems are more likely to remain functional, with more revenues and rapid conflict resolution (Aarnoudse et al., 2018; Madon et al., 2018). Yet, in Nigeria, community participation mechanisms remain inconsistent and poorly institutionalized.

Hence, the central question guiding this study is: *How does community participation influence the functionality, equity, and resilience of water management systems in Nigeria*? To address this, we examine the structures and processes that enable community engagement; analyze the policy and institutional frameworks supporting participation; and identify practical pathways toward sustainable, community-driven water governance. We situate community participation within the broader water governance landscape in Nigeria to contribute to understanding how local engagement interacts with policy frameworks and emerging innovations to build resilient, inclusive, and accountable services.

9.2 Conceptual Foundations for Community Participation

In the context of community water management, participation is perceived as a continuum of collective engagement and management of water resources, rather than a distinct, isolated single event. As established by the works of Grassini (2019) and Geagea (2025), participation encompasses exhaustive consultation with all stakeholders, co-production, self-management, and monitoring and evaluation, with the goal of building communal capacity leveraging institutional structures such as the Rural Water Supply and Sanitation Agencies (RUWASSAs), Partnership for Expanded Water Supply, Sanitation and Hygiene (PEWASH) committees, and community-based organizations (CBOs) usually registered under state laws. Community participation in water management is rooted in grassroots efforts that ensure that water systems are functional (sustained physical and managerial performance of water infrastructure over time), potable, used sustainably, accessible to all, and resilient (anticipate, absorb, and adapt to shocks while maintaining service delivery) to extreme social and environmental stressors (Table 9.1). In this sense, community participation in water management is the active engagement of all relevant local stakeholders who have distinct interests in communal water resources (Adom & Simatele, 2022; Ali & Kamraju, 2024; Lema, 2025). These stakeholders include households, local leaders, farmers, herders, religious groups, marginalized groups, and other water user associations actively involved in the conceptualization, design, performance, and monitoring of water systems.

9.2.1 Consultation

Consultation is a need-based approach involving a systematic collaboration among community members, where individual/group preferences and needs pertaining to water supply, infrastructure type, quality, cost, and administration are gathered (Tondera et al., 2023). They can be done via surveys and at community meetings. When and where done effectively, consultations facilitate the production of problem-solving water systems that reflect local realities, leading to optimal usage, less wastage, and lower abandonment rates, as exemplified by the work of Waarde et al. (n.d.) in Cameroon, who upheld that locals excluded from water management consultations may get upset and sabotage such efforts. Consultations provide platforms for marginalized groups, especially women, low-income households, and

Table 9.1 Participation-outcome framework for community water management (CWM)

Participation dimension	Functionality	Equity	Resilience	Evaluation
Consultation	Joint water solutions address local realities, improve usage, and reduce abandonment.	Marginalized and vulnerable groups are included in planning for fairer distribution of water resources	Locally informed designs better withstand natural and social fluctuations	Documented community priorities help to hold managers accountable
Co-production	Indigenous knowledge complements technical expertise to reduce failure rates	Elite capture is prevented while water is equitably distributed	Shared ownership enhances localized crisis management strategies	Facilitates compliance with rules between water users and providers
Self-management	Short repair turnaround time, uninterrupted service delivery, and low-cost maintenance	Locally set fees and subsidies for vulnerable groups promote affordability and wider access	Community-led management mechanisms ensure long-term sustainability	Users' rights are prioritized, and they have direct oversight of water infrastructure
Monitoring and evaluation	Early detection of downtimes and contaminants builds resilience and ensures a consistent water supply	Constant monitoring exposes inequalities in sharing and promotes fairness	Feedback mechanisms help communities to adapt to stressors (e.g., drought)	Community scorecards and conflict resolution boards prevent abuse and misuse

Note The matrix is a newly proposed participation-outcome framework for CWM

people with disabilities, to express their concerns over the design and pricing of water facilities to ensure fair access (Groce et al., 2011; Sandoval, 2022). In the Dutsin-ma, Mani and Faskari Local Government Areas of Katsina state, Aliyu et al. (2020) noted that the consultative strategy employed by RUWASSA has a significant positive impact on water availability and infrastructure maintenance, with a similar situation observed in the Shao community of Kwara state, where communal consultations facilitate effective operation and maintenance of water facilities, as studied by Sule et al. (2016).

9.2.2 Co-production

In their study, Birkinshaw et al. (2021) express co-production as a process that involves designing, decision-making, and implementation of water management systems by aggregating ideas from the community, government, non-governmental organizations, and other water stakeholders. This may involve shared budgeting, co-designing water infrastructure, and shared rulemaking about water use. Co-production incorporates local knowledge on seasonal water variations, groundwater recharge zone, and extant water-sharing strategies to increase the technical success rate while reducing failure rate (Barnhart et al., 2018). It also ensures joint decision-making in water management and sharing, promotes mutual responsibility, prevents elite capture of water systems, and ensures all segments of the stakeholders are well served, rather than a few privileged ones. Sule et al. (2016) posited that co-production was a success factor in Kwara state's Shao water scheme. They noted that tariff rates and collection strategies were jointly determined by community stakeholders and RUWASSA personnel, creating a sense of shared ownership and more equitable access. An identical success was recorded under the FADAMA III irrigation projects, where farmers and other relevant associations participated in cost-sharing and project design, which significantly improved maintenance culture (Adekunle, 2020).

9.2.3 Self-management

This entails community-led operation and sustenance of local water systems. Many communities have institutionalized local water user associations, fee collection schemes, and strict maintenance schedules, which directly put them in charge of operating, financing, and maintaining water infrastructure (Aarnoudse et al., 2018). This local arrangement minimizes downtimes as the end-users double as the managers. Thus, the turnaround time for repairs is shortened, and resolutions are made promptly without the need for input from external bodies. Self-management also fosters affordability as water tariffs are set locally, with subsidy preferences given to vulnerable groups and households. In a case where external funding or NGOs withdraw from local water management, the community-led operation and maintenance framework ensures long-term sustainability of the local water systems (Lema, 2025). Empirical studies from rural areas of Jigawa and Katsina states showed that functional and accountable water, sanitation, and hygiene (WASH) committees facilitate a reduction of repair turnaround time (TAT) from months to weeks (UNICEF, 2020). The work of Olajuyigbe (2016) in the Ikaram cluster of Ondo state also demonstrated that self-management nurtures accountability and affordability as locally trained mechanics ensured continued water availability even after donor exit.

9.2.4 Monitoring and Evaluation

Local communities and committees are saddled with the responsibility of actively keeping track of water quality and quantity, the reliability of water service, the management of finances, and the administration of water. To promote effective monitoring and evaluation, WASH Committees (WASHCOMs) in Benue State that kept maintenance logs and household satisfaction reports were prioritized for state support and funding, according to the Federal Ministry of Water Resources (FMWR, 2018). Current evaluation frameworks may include household tracking of water usage using scorecards and water meters to promote responsible use of water and prevent wastage (Liu et al., 2015; Loteni, 2017; Richards et al., 2023). It may also include public water audits and a conflict resolution board, which addresses social and cultural issues arising from water sharing and pricing within local communities. Regular monitoring guarantees early detection of system breakdowns and contaminations and allows for proactive water management while exposing injustices in water distribution to interested stakeholders.

9.3 Theoretical Frameworks for Community Participation

Many classical theories explain individuals' participation in the responsible use and sustainability efforts of shared resources. These theories include Arnstein's Ladder theory (1969), Ostrom's common-pool resource principles (1990, revised 2002), and Garrett Hardin's Tragedy of the Commons (1968). Arnstein's citizen participation model follows a typical ladder pattern, with each rung of the ladder ranging from non-participation to tokenism to citizen power (Mellink, 2025). This is applicable in communal water management efforts because it showcases how participation can either empower or neutralize citizens' concerns and interests. *Non-participation* covers rungs 1–2 and gives no credence to community participation. This often leads to the provision of infrastructures and pricing mechanisms that are unsuitable for the end-users and puts the water system at risk of low patronage and failure (Lloyd et al., 2019). *Tokenism* extends between 3 and 5, giving room for communities to be heard without a guarantee of these inputs shaping policy outcomes (Nunes Silva, 2020). In tokenism, public opinions are usually sought via surveys, meetings, and focus groups, but it gives no room for negotiations, while promoting bias (internal exclusion) and elite capture. *Citizen power,* which covers rungs 6–8, is characterized by true participation in which local communities exercise significant responsibility over water management. This promotes shared ownership, inclusive administration, mutual accountability, and higher control of water systems by local communities (Mellink, 2025). However, this control is not absolute as power is often negotiated and shared between the locals and the state. Nigerian evidence suggests that many WASH initiatives still operate between tokenism and partial partnership. For instance, early PEWASH projects implemented by WaterAid Nigeria welcomed community input during consultations but retained decision-making power at the state level (Yusuf et al., 2022). However, functional WASHCOMs under RUWASSA, such as in Benue communities where locals control tariffs, select caretakers, and maintain logs, indicate higher rungs of participation (FMWR, 2018).

Elinor Ostrom's eight principles also provide clarity on how community participation should be structured to promote equitable distribution, sustainability efforts, and transparency. The theory advocates for a clearly defined limit for both the resource, water, for instance, and the users of the resource (Ostrom 1990, 1994, 2002). In essence, the principle is based on rational use of resources, which specifies what water resources should be used by a group of stakeholders to prevent outsider abuse. It seeks to prevent overexploitation, clarify water rights, and facilitate accountability among water users. This principle posits that successful water management only comes from organized community participation rather than emergency interventions. It is a model for a paradigm shift from externally imposed water projects towards community-oriented and community-sustained systems. In practice, Nigeria's FADAMA III project (Adekunle, 2020) and the Ikaram cluster water scheme (Olajuyigbe, 2016) that specify user boundaries and adopt local maintenance rules exemplify these principles. Ostrom's common-pool resource principle is the direct opposite of the Tragedy of the Commons, which explains how individuals selfishly overuse resources for short-term gains rather than responsibly, jeopardizing group interests (Frischmann et al., 2019; Hardin, 1968, 2013; Ostrom, 2008) and undermining the concept of community participation. Hardin's Tragedy of the Commons is perceived as a cautionary theory in resource management, the effects of which can be mitigated when robust local institutions, tariffs, and user associations enforce rules and stewardship (Frischmann et al., 2019).

Arnstein's Ladder provides a lens to evaluate the depth of participation achieved in community-managed water schemes rather than technical inputs, while Ostrom's principles illuminate the conditions under which common-pool resource institutions succeed. The Tragedy of the Commons, conversely, highlights the risks when governance is weak or absent. Together, these frameworks inform our interpretation of community practices and the institutional environments in which they operate (Table 9.2). They shape the lens with which we examine the literature, policy frameworks, and case studies, and inform the way our findings are interpreted.

9.4 Methods

This study uses a narrative review design, given the paucity of peer-reviewed studies directly addressing community participation in water management in Nigeria identified through the systematic search. We adopted this methodological approach (Fig. 9.1) to allow for a more flexible synthesis of diverse sources to synthesize empirical, policy, and programmatic evidence on community participation in water management in Nigeria, thereby providing a broader and more nuanced understanding of the subject. The review aimed to consolidate existing knowledge on the structures, processes, and outcomes of community involvement, while identifying gaps and pathways toward sustainable water governance.

Table 9.2 Comparative synthesis of theoretical frameworks for community participation in water management

Theory	Participation perspective	Participation outcome	Limitation
Arnstein's ladder of citizen participation	Evaluates the degree of empowerment and inclusiveness in decision-making	Helps assess whether WASH committees or RUWASSA schemes represent true empowerment (shared power) or tokenism	Focuses on power dynamics but overlooks institutional sustainability
Ostrom's common-pool resource (CPR) principles	Views participation as institutionalized cooperation through rules, boundaries, and monitoring	Explains why locally governed water systems endure when they are rule-based and supported by state agencies	Assumes cooperative norms exist with less attention to power dynamics
Hardin's tragedy of the commons	Frames participation failure as lack of governance	Useful in diagnosing collapse of donor-led water projects lacking defined ownership	Extremely pessimistic, neglects social norms and the possibility of collective stewardship

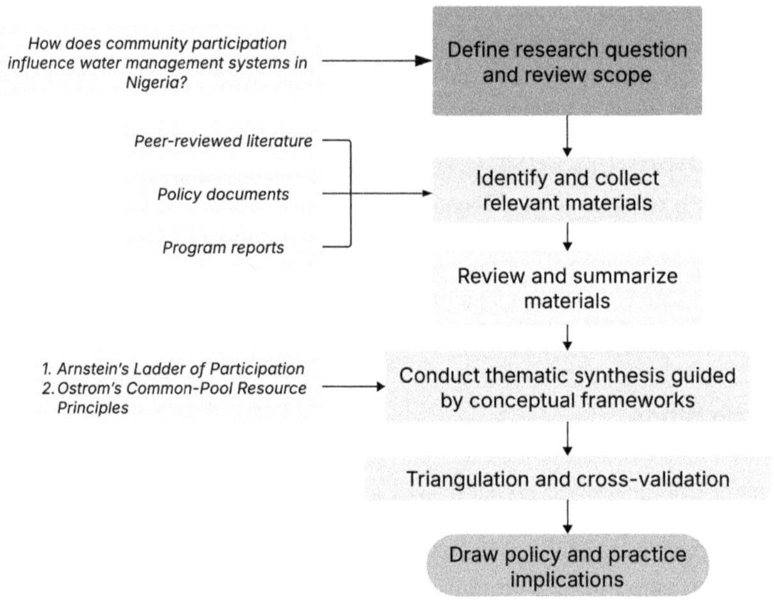

Fig. 9.1 Methodological framework showing the sequential process adopted in the narrative review

Table 9.3 Summary of selected materials used in the review

Type of source	Documents
Academic papers	Peer-reviewed journal articles
Policy and institutional documents	PEWASH Strategy (2016–2030), National Action Plan for WASH, SURWASH Program document, Federal Ministry of Water Resources (FMWR), UNEP GRID country data
Program and evaluation reports	UNICEF WASH evaluation; World Bank policy papers; FADAMA/FUAs program evaluations (2015–2022); JMP (2024) WASH monitoring data, World Bank ESS Assessments, World Bank program evaluations

9.4.1 Data Sources and Selection Criteria

Materials used, excluding the theoretical backgrounds, were published between 2010 and 2025, with data drawn from three primary sources: peer-reviewed academic literature, policy and institutional documents, and program and evaluation reports (Table 9.3). Sources were retrieved from Scopus, Google Scholar, Web of Science, and institutional repositories, using search terms such as "*community participation*," "*water management*," "*WASH*," "*Nigeria*," "*collective action*," and "*rural water governance*." Inclusion criteria required that each selected material: addressed community roles in planning, operation, financing, or governance of water services; provided empirical or policy-relevant evidence from Nigeria; and was published in English within the target period.

9.4.2 Analysis and Synthesis

The selected materials were analyzed thematically to identify recurring patterns and insights. The analysis was on theoretical frameworks informed by Arnstein's Ladder of Participation, Ostrom's Common-Pool Resource Principles, and Hardin's Tragedy of the Commons, focusing on forms and depth of participation, success and failure factors, and challenges to local governance. We present our findings under four major themes:

1. policy and institutional frameworks,
2. current state of water service and participation,
3. practical lessons for community-managed water services, and
4. emerging innovations and good practices.

Themes were developed iteratively as materials were reviewed, and attention was given to both consistencies and contradictions in the literature, allowing a balanced and interpretive synthesis. Based on our evaluation of selected materials used, we propose practical principles to strengthen community participation. Here, due to the limited availability of empirical studies specifically focused on community participation in water management within Nigeria, some evidence from comparable contexts outside the country was drawn upon to substantiate the discussion.

9.4.3 Triangulation and Validation

Key insights across different source types were cross-checked and validated. Findings from peer-reviewed studies were compared with program evaluations and policy reports for validity. Similar findings from at least two independent sources strengthened our argument, while divergent evidence was critically examined to contextualize differences, ensuring analytical transparency and minimizing interpretation bias. This interpretive triangulation strengthened the reliability of conclusions drawn from diverse evidence bases.

9.5 Policy and Institutional Framework for Community Participation in Water Management in Nigeria

Through an analysis of Nigeria's policy and institutional landscape for water management (Table 9.4), we observe, at least in policy, emphasis on community participation and empowerment, but in practice, performance remains mixed. While successive policy instruments, such as the Partnership for Expanded Water Supply, Sanitation, and Hygiene (PEWASH), the National Action Plan (NAP) for WASH, the Sustainable Urban and Rural WASH (SURWASH) program, and central actors like the River Basin Development Authorities (RBDAs) emphasize decentralization, genuine empowerment of communities has been constrained by weak coordination, limited financing, and fragmented institutional mandates.

9.5.1 Policy Landscape and Practical Outcomes

The PEWASH Strategy (2016–2030) marked a significant policy shift toward multi-stakeholder collaboration, assigning roles to federal, state, and local governments alongside communities. As a multi-stakeholder framework that nominally advocates decentralization, it assigns roles to communities, Local Government Areas (LGAs), and State Rural Water Supply and Sanitation Agencies (RUWASSAs) (FMWR, 2016). However, evidence from program assessments (Yusuf et al., 2022; Andres et al., 2018a, 2018b) shows that implementation remains largely top-down, with state-level agencies often bypassing community structures during project design and budgeting. As a result, while communities are nominally recognized, their influence

Table 9.4 Key provisions for community participation in Nigerian water policies

Framework	Stated goal for community participation	Key mechanisms	Implementing bodies	Implementation challenges
PEWASH strategy (2016–2030)	Increase access to rural water supply and sanitation through a multi-stakeholder strategy that decentralizes responsibilities	Decentralization of roles and responsibilities from federal to state and local governments	LGAs, state RUWASSAs, and communities	Persistent centralization of decision-making; inconsistent funding from LGAs; weak accountability for community engagement
NAP (2018)	Institutionalize community participation to improve sustainability and reduce high failure rates of water projects	Standardization of community engagement and the institutionalization of VLOM	WASHCOMs at the community level, with VLOM units at the state and LGA levels	Poor operational funding; limited capacity-building; unclear lines of responsibility between tiers
SURWASH program (2021)	Increase access to services and strengthen sector institutions by addressing social and institutional barriers like "limited community participation"	Performance-based funding (program-for-results) that incentivizes the achievement of specific, community-level outcomes	State program implementation units (SPIUs), LGAs, CBOs	Community data is not directly integrated into DLI verification; limited awareness of DLIs at local levels
River basin development authorities (RBDAs)	Harness water resources to accelerate agricultural and rural development, providing water from reservoirs for farmers and communities	Establishment of WUAs to manage irrigation and drainage systems	WUAs, farmers, local communities	Political interference; weak legal empowerment of WUAs; inadequate technical and financial autonomy
Legal/policy recognition of CBOs	Provide a legal basis for voluntary associations to manage water resources	Constitutional right to associate; legal provisions for customary water use	Unincorporated associations, CBOs, Project Management committees (PMCs)	Lack of formal registration frameworks in some states; poor linkage with government agencies

over operational decisions, especially tariff setting, maintenance funding, and service monitoring, remains minimal.

The National Action Plan for WASH (2018), launched after the federal declaration of a "state of emergency" in the sector, sought to institutionalize participation through the establishment of Village Level Operation and Maintenance (VLOM) and WASH Committees (WASHCOMs) (FMWR, 2018). However, the operationalization of these mechanisms is inconsistent across states. Limited fiscal autonomy of LGAs, poor capacity within RUWASSAs, and overlapping mandates between federal and state actors have weakened sustainability. In theory, the SURWASH Program (2021) represents a stronger results-based model, trying to fund to verified performance indicators known as Disbursement-Linked Indicators (DLIs). For instance, DLI 6 measures the number of communities having achieved community-wide sanitation status to reward genuine community integration. Yet, these incentives depend on state reporting accuracy and are rarely extended to community institutions. Under SURWASH, community inputs often stop at project inception, with little involvement in monitoring or maintenance planning (World Bank, 2021a, 2021b).

The Water Resources Act of 1993, which centralizes water resource ownership under federal control, remains a structural constraint to decentralized governance (Adeoti, 2010). Although it aims to regulate equitable access, its lack of clear enforcement at subnational levels and weak inter-agency coordination have perpetuated duplication of roles and accountability gaps. RBDAs have also gradually been part of efforts to promote participatory irrigation management through water-user associations (WUAs) and partial delegation of management responsibilities, however, federal–state overlaps are particularly pronounced, where political interference and poor legal empowerment of WUAs have stifled participatory management (Ogundele, 2019; Odjegba et al., 2023). Notably, operational water service delivery at state and local levels is increasingly institutionalized through RUWASSAs and water boards (Khan et al., 2018), illustrating agency-led support for communities (Aliyu et al., 2020; Ekpo & Umoh, 2022). State capacities and governance, however, differ-while some states integrate community engagement mechanisms, others are strengthening corporate governance (Adeoti et al., 2023; Ngene et al., 2021).

9.5.2 *Coordination and Implementation Gaps*

Nigeria's multi-tiered governance structure presents persistent coordination failures. Funding mechanisms are poorly aligned, and maintenance responsibilities are often undefined. The Federal Ministry of Water Resources designs policies and donor programs, yet RUWASSAs and LGAs, who support WASHCOMs on the ground, frequently lack the technical or financial capacity to follow through. This disconnect has led to weak oversight, limited accountability, and inconsistent service outcomes across states. Functionality data from Andres et al. (2018a, 2018b) and JMP (2024) indicate that only about 67% of rural water points are functional, and 30–38% fail within the first year of operation, primarily due to poor maintenance and institutional fragmentation. These figures underscore the gap between policy rhetoric and real-world outcomes, where sustainability often depends more on local initiative than on formal policy provisions. Despite these shortcomings, the recognition of community-based organizations (CBOs) and WASHCOMs remains a key institutional advancement (John & Ajibade, 2024). When linked effectively to RUWASSAs and LGAs through transparent funding and capacity-building, these entities demonstrate improved functionality, accountability, and inclusivity. Bridging the divide between policy intent and practice, however, requires consistent devolution of authority, predictable funding flows, and a monitoring framework for accountability of participatory outcomes.

9.6 Current State of Water Services and Participation

Water service delivery in Nigeria reflects two main governance types, each with unique merits and demerits. These include donor-managed systems and community-managed systems. Community-managed systems are locally-run arrangements where water users, through committees like WASHCOMs, take charge of daily supervision, small repairs, and tariff collection (UNICEF, 2020). Internal accountability, rule enforcement, and ongoing relationships with official agencies like RUWASSAs are all necessary for their performance. Community-managed programs frequently exhibit higher resilience, user satisfaction and cost recovery when these institutional ties are operational (UNICEF, 2020; Adeoti et al., 2023), while systemic failure may occur where they are not operational. Donor-managed systems, on the other hand, have little local authority over their design and initial operations and are funded and executed externally, usually by philanthropies, NGOs, or international development partners. Despite producing quick infrastructural improvements, if finance, oversight, and spare-part supply chains are removed, they are vulnerable to post-donor collapse (Khan et al., 2018). It is, therefore, noteworthy that Nigeria's uneven track record in accomplishing Sustainable Development Goal 6 is more a result of the governance model (which informs coverage, functionality, and inclusion trends) used to provide services than it is of technological shortcomings.

9.6.1 Coverage and Functionality

Nigeria has made unequal progress in improving household water access, falling well short of international standards. National monitoring reveals wide disparities in coverage between urban and rural areas, with the hope of achieving the 2030 targets for safely managed water services currently in jeopardy (JMP, 2024). According to recent studies, a significant portion of Nigerians continue to rely on outdated or sporadically operational sources, with metropolitan areas receiving significantly better service than rural ones (Olalekan et al., 2019; JMP, 2024). These nationwide patterns indicate ongoing deficiencies in service coverage and quality, particularly outside of major cities. Another trend in water resources in Nigeria is not only whether infrastructure exists, but also its consistent functionality. Where local communities maintain active WASHCOMs, such as the Katsina state VLOM units, average repair turnaround is under two weeks, in contrast to several months in donor-driven systems (UNICEF, 2020). When and where community management mechanisms are inadequate or insufficient, numerous field studies of Nigerian rural boreholes and piped projects reveal significant breakdown rates, little preventative maintenance, and brief operating lifespans (Adeniran, 2022; Adeoti et al., 2024). Moreover, more than 33% of rural water points in Nigeria may not be operational owing to a lack of maintenance and deficiencies in regular financial and technical capabilities (Enang, 2020; Imonikhe, 2020; Adeniran, 2022). Such is the case in Bida, Niger State, where more than 30% non-functional rural water schemes were attributed to poor community maintenance culture and wrong siting of water infrastructures, as studied by Mohammed (2021). This claim is also supported by the work of Andres et al. (2018b) who synthesized data from the 2015 Nigeria National Water and Sanitation Survey and observed that about 38% of all water points in Nigeria are non-functional while projecting that almost 40% water points may become non-functional over a few years post-construction.

9.6.2 Equity and Inclusion

Gender and social inclusion in water management remain critical concerns in Nigeria. With respect to equal access, rural areas normally depend on low-quality and distant water sources, whereas wealthy households usually have more access to high-quality water sources. In a 2018 report published by the Nigeria Demographic and Health Survey (NDHS, 2018), more than 74% of urban households have access to improved water sources, a sharp contrast to just about 50% reported in rural areas. These discrepancies are worsened by gendered dynamics. Women and girls usually handle much of the water collection work and are disproportionately impacted by unsafe sources, lengthy collection periods, and, in certain communities, exclusion from decision-making (NDHS, 2018; Adeoti et al., 2023). Recent empirical works, however, show that gender inclusion directly improves maintenance and accountability in community-managed water systems. According to UNICEF (2020), rural WASH committees with up to 40% female membership reported significantly higher rates of fee collection, record-keeping compliance, and stronger linkages to RUWASSA support schemes than those dominated by men. Also, women's regular interaction with household water users promotes early fault detection, raising functional water points to above 80%, while their participation in financial committees promotes transparent fund utilization and timely maintenance decisions. Haddout et al. (2024) argued that women's active engagement transforms water management from a technical routine into a socially responsive service, embedding trust and long-term stewardship. This implies that inclusion is not merely a fairness agenda but a sustainable ideology that promotes resilience of water facilities.

9.6.3 Institutional Capacity

Community structures do not operate in isolation. Rather, their success depends on technical and institutional backstopping from RUWASSAs and local governments. Community institutions serve as the primary means for grassroots involvement in water service provision and management. WASHCOMs and water-user associations significantly improve repair responsiveness and ownership when they are duly established and legally recognized. They also manage fare collection, conduct periodic maintenance, and mobilize local labour. Effective WASHCOMs encourage women's leadership in water management and connect communities to external aid when necessary, according to UNICEF program assessments and field case studies from states including Katsina, Benue, and Jigawa (UNICEF, 2020). Nevertheless, success is not universal. In cases where local committees lack technological support or transparent accounting, systems frequently collapse or are taken over by local elites. This is typical of donor-managed projects such as the Tulsi Chanrai Foundation/UNICEF partnership (Olawade et al., 2020), which demonstrated high initial success but poor post-donor performance because handover strategies to local and state institutions were weak (Khan et al., 2018).

9.7 Practical Lessons for Community-Managed Water Services in Nigeria

Communal water management services in Nigeria have produced ambivalent outcomes. Where local institutions are transparent, empowered, and connected to supply chains and higher-level agencies, water systems can continue to function and provide services. However, in the absence of these favourable circumstances, water infrastructures frequently fail shortly after their launch or after donors leave. Several key drivers are consistently identified by empirical investigations as success factors for community participation in water management. These drivers include distinct grassroots organizations (such as water-user associations and WASH committees) with accountable finance structures, training (technical and accounting), dependable supply-chain links for technicians and spare parts, reasonable water pricing, and legal recognition and connections to state agencies (for technical backup and auditing). When these factors are present, community ownership gains significance, and operational efficiency is enhanced. On the other hand, a lack of technical assistance and infrastructure spare parts, superficial community involvement, elite capture, low local ability or willingness to pay for operations and maintenance (O&M), and sudden donor departure without a transition strategy are all common reasons for breakdowns (Andres et al., 2018a, 2018b; Adeniran et al., 2021).

9.7.1 Cases of Community Participation in Water Management in Nigeria

1. *A rural handpump scheme with functional WASHCOM sustained for more than 5 years*: The Ikaram Millennium Village Project (MVP) exemplifies sustained grant-funded water and sanitation initiatives throughout the Ikaram cluster in Ondo state, Nigeria (Olajuyigbe, 2016). The project illustrates how community ownership creates a sense of collective responsibility and promotes indigenous repair capability through training and local skills, such that water facilities remain functional beyond donor exit. The Ikaram cluster WASHCOMs operate within the premise of Ostrom's principles with clear boundaries, local repair capacity and funding, and monitoring structures in place. It also illuminates the importance of long-term sustainability facilitated by official connections of local WASHCOMs with state authorities. This structured community participation ensures that rehabilitation interventions on water facilities are proactive rather than reactive.
2. *Small-town piped water under RUWASSA with tariff collection–Shao, Kwara State*: Piped water schemes can accomplish routine tariff collection and partial O&M cost recovery when a defined service provider exists and tariffs are viewed as legitimate. An instance is the Shao water tariff collection system in Kwara state, where effective tariff collection strategies form a critical part of communal water management plans and exemplify co-production (Sule et al., 2016). This corresponds to the partnership rung of Arnstein's Ladder, showing that negotiated authority can enhance accountability even without full citizen control. Nevertheless, the Shao community also reveals a high degree of service variability (intermittency in supply), suggesting that tariff revenue must complement uninterrupted supply chains, technical capability, and shared management, as is the case with the Ikaram cluster.
3. *Irrigation user associations—Nationwide FADAMA program*: In a bid to oversee irrigation infrastructure in low-lying plains along major river systems, Nigeria's World Bank-supported FADAMA programs specifically established farmer groups and FADAMA Users Associations (FUAs) (Kuza et al., 2018; Harmon, 2022). FUAs linked farmers to credit and input markets, facilitated cost-sharing for basic irrigation infrastructure, and enhanced water use coordination (Adekunle, 2020). However, the program faced challenges such as uneven institutional capacity, unequal membership practices, and poor scaling into long-term government support. This implies that when groups have clear membership rules, enforcement procedures, and connections to markets and extension,

collective irrigation management works well, thus validating Ostrom's emphasis on collective-choice arrangements. In a case where these components are absent, long-term performance suffers.

4. *Individual-led (influencer/celebrity) borehole projects*: Boreholes and water supply facilities have been financed as charitable initiatives by Nigerian socialites in recent years. For example, borehole donation in Amper, Kanke local government area of Jos, Plateau State (Nwahiri, 2024), and a solar-powered borehole on the University of Jos Village campus (Otse, 2025) are examples of such water projects. These initiatives result in high social visibility and instant access but expose persistent issues, such as occasional elite struggles over management of the new asset, a lack of formal succession plans, and a lack of training for local O&M. Therefore, while these initiatives may yield short-term benefits, they frequently lack the institutional underpinnings necessary for long-term success. Governance under celebrity projects ranks low on Arnstein's ladder (tokenism) and highly reflects Hardin's tragedy risk, as there is no clear ownership or accountability.
5. *Donor-exit and post-project non-functionality*: Following donor departure, many boreholes and handpumps become non-functional, a reflection of Hardin's tragedy of the commons. The UNICEF/Tulsi Chanrai partnerships provided water facilities in many parts of Nigeria but undermined local autonomy, failed to enforce effective tariff collection schemes and provided isolated O&M training. This frequently resulted in local communities' inability to pay for spare parts (for instance, in Shanga village, Yobe state) (UNICEF, 2023), lack of access to skilled technicians, or lack of established, accountable finance structures (Khan et al., 2018). Large percentages of rural water sites are measured by extensive empirical assessments as non-functional at any given time, confirming that donor investment without a practical transition to local O&M is a common source of failure.

Across all presented cases, a comparative synthesis shows that long-term functionality of water infrastructure depends on three interlocking conditions. These are institutional clarity (which defines community rights and responsibilities), financial accountability (transparent tariff and repair funds), and technical linkage (continuous support from RUWASSA and similar state agencies). While the Ikaram and FADAMA projects leveraged organized community participation and clearly defined water rights as success factors, the Shao community water scheme leveraged effective tariff collection and shared power between the locals and the state. However, individual-led water projects, which usually lack the institutional frameworks and technical support needed for long-term sustainability, risk failure in the short term.

9.7.2 Emerging Trends and Innovations

Recent developments in Nigeria point toward hybrid arrangements that combine community oversight with professional, data-driven, and incentive-aligned support. While many of these are lacking in the literature, we draw insights from relevant program reports that can inform future research directions. Four interlinked trends are particularly salient in the Nigerian context: (1) professionalisation and clustered maintenance support; (2) results-based financing through SURWASH DLIs; (3) citizen monitoring and digital reporting; and (4) linking ecosystem protection to local incentives. Each offers potential gains in reliability and accountability, but each also faces significant feasibility constraints that must be addressed in policy design.

1. *Professionalisation and clustered maintenance support*

There is a growing consensus that the traditional, fully community-managed model (where volunteer committees alone run systems) is insufficient for durable services at scale (Machado et al., 2022). Policy documents and reports for the SURWASH program emphasize the need to move toward more professionalized, state-supported service delivery and maintenance arrangements. The operational guidance and appendices discuss training, institutional roles for RUWASSAs and SPIUs, and the importance of linking communities to reliable technical support and spare-parts supply chains as part of program design. Evaluations of state projects (e.g., in Katsina and Bauchi) also document RUWASSA's role in regular equipment maintenance and capacity gaps that professionalisation seeks to address (JICA, 2017). Although evidence of large-scale, standardized "maintenance hub" rollouts remains limited in the public record, professionalisation (e.g., regional mechanics, LGA-level maintenance contracts) can reduce downtime and dependence on donor-funded repairs. It, however, requires clear cost-recovery models, procurement channels for spare parts, and safeguards against service capture by private contractors. The sector capacity assessments also emphasize substantial skills gaps in plant operation and maintenance that must be filled for professionalised models to function. (NEPAD, 2018)

2. *Results-based financing (SURWASH DLIs)*

Results-based instruments that tie disbursement to verified service outcomes are being used to strengthen the link between community performance and state support (Marshall et al., 2023). For instance, SURWASH is explicitly designed as a Program-for-Results, with a set of DLIs that tie financial releases to verified outcomes (for example: establishing a National WASH Fund, state PIR plan adoption, community sanitation targets, and functionality targets for water systems). World Bank program documents and the SURWASH Program Operations Manual detail the DLIs, verification arrangements, and the incentives intended to align state performance with community-level outcomes (World Bank, 2021a, 2021b). Implementation Status Reports, however, show that several DLIs are off-track or only partially achieved in early years, reflecting challenges in meeting verification and institutional readiness criteria (World Bank, 2025). While DLIs provide a credible design for linking finance to performance, achieving the expected downstream effects on community management depends on robust verification systems and timely state responsiveness. Nevertheless, DLIs create accountability pathways from federal/donor finance to results on the ground. In practice, their effectiveness at strengthening community participation will depend on (a) whether community-level indicators include measures of participation quality (not merely the number of committees), and (b) whether verification processes incorporate community voice rather than only state reports.

3. *Citizen monitoring and digital reporting (opportunities and limits)*

Donors and implementing partners increasingly recommend participatory monitoring approaches (community scorecards, social audits) as complements to formal verification. Global guidance on community scorecards has been adapted in Nigerian programming, and SURWASH appendices reference community engagement and verification steps. National WASHNORM surveys and sector mid-term reports show growing interest in routine outcome mapping and user feedback mechanisms; some states have piloted digital reporting or strengthened reporting cultures through RUWASSA-led initiatives and NGO partnerships (FMWR, 2022). However, there is limited publicly available documentation of widescale, integrated mobile-app reporting systems (for example, community caretakers logging faults into state dashboards) across multiple states. Where digital pilots exist, they commonly face the digital-divide constraints of limited smartphone ownership, intermittent connectivity, and low data literacy among community committee members (WASH Nigeria, 2014). Digital and participatory monitoring, however, can shorten repair cycles and increase transparency where connectivity and capacity permit. To avoid exclusion, programs must incorporate offline reporting options, basic digital training, and accessible verification channels that do not privilege the digitally connected only.

4. *Ecosystem (source-protection) incentives*

SURWASH and state programs explicitly recognize source protection (catchment and watershed measures) in their environmental and social assessments and program designs; pilot activities and donor-supported projects have linked upstream conservation with downstream service quality in limited contexts (World Bank, 2021a, 2021b). While the policy frameworks encourage combining catchment protection with livelihood incentives, the public literature provides only scattered examples of small-scale pilots rather than national programs that systematically link payments for ecosystem services to community water outcomes. Monitoring and financing such incentives remain nascent and require sustained funding and credible ecological indicators (WASH Nigeria, 2018). Its recognition in state programs and adoption (Ngene et al., 2021) suggests that integrating catchment protection into community water governance can produce multiple benefits.

Across these innovations, possible constraints in the Nigerian context include the affordability of higher tariff levels for the poorest households without targeted subsidies, the lack of technical and institutional capacity, verification and reporting gaps, and the digital divide and inclusion constraints. However, taken together, these emerging trends and innovations shift the locus of community water management from an isolated, volunteer-dependent model to a hybrid co-management model, where: communities retain oversight, local legitimacy, and tariff-setting roles where appropriate; state agencies (RUWASSAs, SPIUs) provide technical backstopping, procurement, verification, and institutional continuity; and performance incentives and participatory monitoring create accountability loops linking finance to outcomes. This shift requires realistic cost-recovery strategies, investments in RUWASSA/LGA capacity, inclusion safeguards for the digitally and economically marginalized, and robust verification systems that incorporate community voice. Reviewed program documents and evaluations demonstrate the design and early pilots of these changes in Nigeria, but current public evidence indicates they are still maturing rather than fully operational at scale.

9.8 Practical Principles for Strengthening Community Participation

Based on our synthesis of existing knowledge on the subject matter and building on gaps identified in policy implementation and emerging innovations, we outline practical principles for enhancing community participation in Nigeria's water management. These principles are grouped into institutional, financial, technical, and social domains, collectively reinforcing sustainable and inclusive service delivery. We draw from international good practice and existing frameworks. We further develop a proposed framework, which serves as a participatory governance model for strengthening community participation in water management (Fig. 9.2).

1. *Institutional Principles: Clarify Roles and Strengthen Local Governance*

Effective participation requires a clear definition of responsibilities and accountability channels among federal, state, and local tiers, as well as between government agencies and community structures. Formalized roles reduce ambiguity and improve upward accountability (see model DLI expectations: SURWASH, 2022). RUWASSAs, LGAs, and WASHCOMs should operate within a transparent hierarchy of responsibilities that minimizes overlap and ensures that communities have legally recognized input into planning, tariff setting, and monitoring (WaterAid, 2018; Lockwood, 2019). State-level by-laws should formalize community management units (CMUs) as part of local government service structures, rather than informal volunteer bodies. Accountability mechanisms should include annual performance audits, for instance, linking WASHCOM activity reports to RUWASSA dashboards to institutionalize feedback loops. WaterAid (2018) highlights that durable systems depend on service authorities providing continuous post-construction support, not only initial project oversight.

2. *Financial Principles: Align Incentives and Ensure Predictable Support*

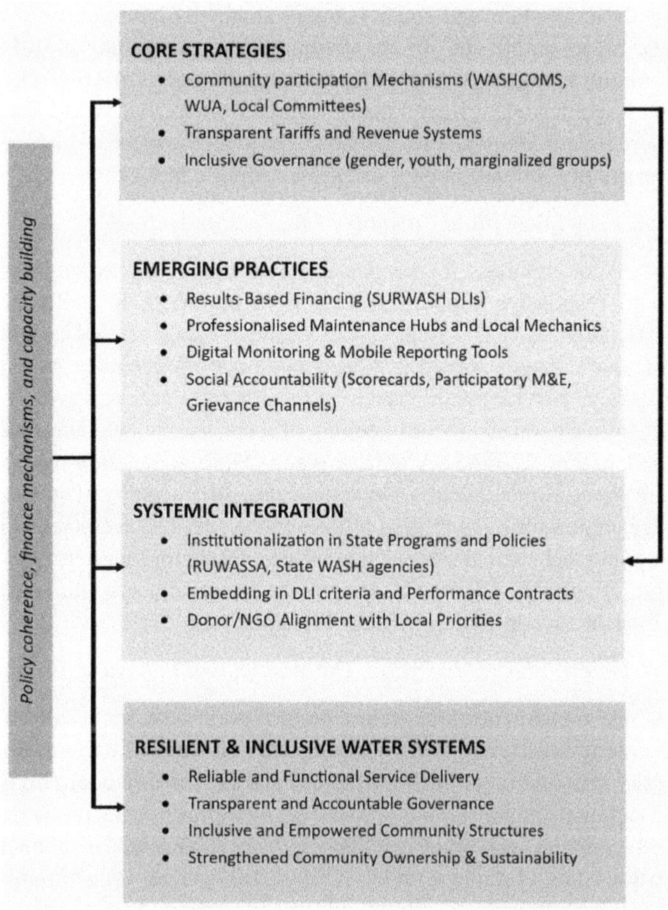

Fig. 9.2 Integrated framework for strengthening community participation and advancing sustainable water governance in Nigeria. This illustrates the interaction between enabling principles, emerging practices, and desired outcomes that collectively enhance community participation in Nigeria's water sector. The emerging practices and systemic integration represent strategies linking institutional reform with innovation. These practices contribute to resilient and inclusive water systems. Arrows indicate interactions and feedback loops through which performance data and community learning inform policy, institutional, and technical adjustments across all framework components

Sustainable participation depends on aligning community contributions with reliable external financing and performance incentives. In the co-management of water resources, there is a need to also ensure predictable, ring-fenced financing for operation and maintenance. Local cost-sharing models should clearly specify how user tariffs, LGA budgets, and program grants (e.g., DLIs) converge at the community level. Lockwood (2019) identifies "*support costs*" (technical advice, monitoring, training) as essential, yet often underfunded, components of rural water schemes.

Embedding these in state and LGA budgets stabilizes community engagement. Performance-based grants can reward communities maintaining ≥80% functionality rates or holding regular WASHCOM meetings, based on WASHNORM indicators (WASH Nigeria, 2018). Key evidence can be drawn from World Bank (2021a, 2021b), where functionality and maintenance frequency were tracked as proxies for participation quality, validating their use as measurable indicators.

3. *Technical Principles: Build Capacity and Integrate Technology*

The ability of communities to manage systems hinges on technical know-how and access to responsive support. Hence, it is important to build local spare-parts markets, strengthen technical capacity, and leverage digital monitoring and reporting of assets. States should institutionalize Area Mechanics Associations or similar clusters to provide maintenance and spare-parts access. Periodic refresher training for pump caretakers, record-keeping officers, and hygiene promoters should be embedded within RUWASSA annual plans. More so, mobile-based reporting tools (e.g., mWater, KoboCollect) (Almajano et al., 2013) can facilitate maintenance alerts and usage monitoring, provided offline options exist to overcome connectivity barriers. Ongoing technical assistance is a decisive factor for long-term functionality. For instance, Nigeria's SURWASH design documents embed routine monitoring within DLI verification systems (SURWASH, 2022).

4. *Social Principles: Foster Inclusion, Ownership, and Behavioural Change*

Beyond institutional structures, participation thrives on trust, inclusivity, and shared motivation. Strengthening community participation should involve measuring participation quality (beyond committee existence) and embedding community voice in verification and financial flows. Actionable steps include: ensuring that at least 40% of WASHCOM members are women and that youth are represented in maintenance and communication roles, aligning with UNICEF (2017) gender-transformative WASH guidelines; using community scorecards and public posting of financial and functionality data to strengthen transparency; and linking community meetings and hygiene campaigns to seasonal or local events (e.g., market days) to improve attendance and continuity. Participatory assessments demonstrate that gender-balanced committees and open information sharing correlate with improved system functionality and tariff compliance (FMWR, 2022).

In addition, there are relevant metrics for monitoring levels of community participation, including repair turnaround time and meeting frequency. Repair turnaround time (days between fault identification and repair) is adapted from WASHNORM functionality metrics (target < 7 days) (FMWR, 2022). Meeting frequency (number of WASHCOM meetings per quarter) derives from RUWASSA reporting templates and can be integrated into SURWASH's state monitoring dashboards. Another metric is the percentage of water points with a registered, trained caretaker and documented maintenance log. These indicators enable longitudinal tracking of community responsiveness and governance quality. Overall, these principles emphasize that community participation is not a stand-alone activity but an integrated system function, where institutional capacity, finance, technical competence, and social inclusivity are aligned.

9.9 Conclusion

This study contributes to understanding the current landscape of community involvement in water management in Nigeria. This reveals both promising opportunities and persistent challenges. Our findings demonstrate that while institutional frameworks (PEWASH, NAP, SURWASH) embed participation in principle, implementation remains constrained by institutional fragmentation and capacity deficits. Nevertheless, innovations such as professional maintenance networks, results-based financing, and digital accountability tools provide viable pathways to bridge the gap between policy and practice. Notably, there remains a limited body of empirical research that systematically assesses these emerging practices and models across diverse community contexts in Nigeria. To strengthen community participation towards sustainable water governance in Nigeria, three evidence-informed recommendations emerge: (1) clarify and legally anchor institutional roles to strengthen downward accountability and minimize duplication; (2) establish predictable funding for operations and maintenance, integrated into LGA and state budgets; and (3) institutionalize continuous community engagement and capacity development. These practical measures can consolidate emerging hybrid governance models that enhance functionality. Given the current gap in literature, future research should examine these governance arrangements through comparative field studies across Nigerian states, assessing how local institutional capacity and social dynamics influence the sustainability of participatory water systems. This empirical testing would deepen evidence on what works, where, and why, advancing the science and practice of sustainable water governance in Sub-Saharan Africa.

References

Aarnoudse, E., Closas, A., & Lefore, N. (2018). *Water user associations: A review of approaches and alternative management options for sub-Saharan Africa*. International Water Management Institute (IWMI), 43p (IWMI working paper 180). https://doi.org/10.5337/2018.210

Adekunle, S. B. (2020). Analysis of impact of Fadama III additional financing (AF1) on beneficiaries' income and poverty alleviation in Abuja, Federal Capital Territory, Nigeria. *International Journal of Development and Management Review, 15*(1), 212–223.

Adeniran, A. B. (2022). *New pipes, old ways: Water infrastructure failure, power, and the state in Nigeria* (Doctoral dissertation). The Australian National University, Australia.

Adeniran, A., Daniell, K. A., & Pittock, J. (2021). Water infrastructure development in Nigeria: Trend, size, and purpose. *Water, 13*(17), 2416. https://doi.org/10.3390/w13172416

Adeoti, O. (2010). Development of river basin organizations in Nigeria. *Research journal of soil and water management, 1*(3), 91–100.

Adeoti, O. S., Kandasamy, J., & Vigneswaran, S. (2023). Water infrastructure sustainability in Nigeria: A systematic review of challenges and sustainable solutions. *Water Policy, 25*(11), 1094–1111. https://doi.org/10.2166/wp.2023.173

Adeoti, O. S., Kandasamy, J., & Vigneswaran, S. (2024). Water infrastructure sustainability challenge in Nigeria: A detailed examination of infrastructure failures and potential solutions. *Water Supply, 24*(6), 2066–2076. https://doi.org/10.2166/ws.2024.127

Adom, R. K., & Simatele, M. D. (2022). The role of stakeholder engagement in sustainable water resource management in South Africa. *Natural Resources Forum, 46*(4), 410–427. https://doi.org/10.1111/1477-8947.12264

Ali, M.A., & Kamraju, M. (2024). The role of community participation in sustainable integrated water resources management: Challenges, opportunities, and current perspectives. In A. K. Yadav, K. Yadav, V. P. Singh (Eds.), *Integrated management of water resources in India: A computational approach.* Water science and technology library (Vol. 129, pp. 325–344). Springer. https://doi.org/10.1007/978-3-031-62079-9_18

Aliyu, M., Saleh, L., & Madu, A. Y. (2020). Assessment of the contributions of rural water supply and sanitation agency (RUWASSA) on provision of potable water supply in Katsina state, Nigeria. *Gusau International Journal of Management and Social Sciences, 3*(1), 182–196.

Almajano, P., Trescak, T., Esteva, M., Rodríguez, I., & López-Sánchez, M. (2013). V-mWater: An e-government application for water rights agreements. In S. Ossowski (Ed.), *Agreement technologies* (Law, governance and technology series) (Vol. 8). Springer. https://doi.org/10.1007/978-94-007-5583-3_34

Andres, L. A., Chellaraj, G., Dasgupta, B., & Grabinsky, J. (2018a). An evaluation of the contributing factors of water scheme failures in Nigeria. *World Bank Policy Research Working Paper, 8371.*

Andres, L. A., Chellaraj, G., Dasgupta, B., Grabinsky, J., & Joseph, G. (2018b). Why are so many water points in Nigeria non-functional? An empirical analysis of contributing factors. An empirical analysis of contributing factors (28 March 2018). *World Bank policy research working paper, (8388).* https://ssrn.com/abstract=3152088

Anifowose, A. J., & Oyebode, A. W. (2019). Studies on heavy metals contents of Osun River at the pre-urban settlement and across Osogbo City, Nigeria. *Journal of Taibah University for Science, 13*(1), 318–323. https://doi.org/10.1080/16583655.2019.1567899

Arnstein, S. R. (1969). A ladder of citizen participation. *Journal of the American Institute of Planners, 35*(4), 216–224. https://doi.org/10.1080/01944366908977225

Ayanlade, A., Oladimeji, A. A., Okegbola, O. M., Eludoyin, A. O., Eslamian, S., Ayinde, A. F., & Perkins, P. E. (2022). Effect of climate change on water availability and quality: An assessment of socio-resilience in Nigeria. In S. Eslamian & F. Eslamian (Eds.), *Disaster risk reduction for resilience* (pp. 245–262). Springer. https://doi.org/10.1007/978-3-030-99063-3_11

Barnhart, B. L., Golden, H. E., Kasprzyk, J. R., Pauer, J. J., Jones, C. E., Sawicz, K. A., Hoghooghi, N., Simon, M., McKane, R. B., Mayer, P. M., Piscopo, A. N., Ficklin, D. L., Halama, J. J., Pettus, P. B., & Rashleigh, B. (2018). Embedding co-production and addressing uncertainty in watershed modeling decision-support tools: Successes and challenges. *Environmental Modelling and Software, 109*, 368–379. https://doi.org/10.1016/j.envsoft.2018.08.025

Birkinshaw, M., Grieser, A., & Tan, J. (2021). How does community-managed infrastructure scale up from rural to urban? An example of co-production in community water projects in northern Pakistan. *Environment and Urbanization, 33*(2), 496–518. https://doi.org/10.1177/09562478211034

Ekpo, M. E., & Umoh, U. E. (2022). Corporate governance and the Management of Public Corporations: A study of Akwa Ibom water company limited. *AKSU Journal of Administration and Corporate Governance, 2*(4), 82–93.

Enang, R. (2020). *Towards effective asset management improving operation and maintenance of a water distribution system through better information management practices: Cases of Abuja, Nigeria and Kampala, Uganda* (Doctoral dissertation). Loughborough University.
FMWR. (2016). *Partnership for Expanded Water Supply, sanitation and hygiene (PEWASH) strategy 2016–2030*. Federal Ministry of Water Resources, Nigeria. Retrieved from https://policyvault.africa/policy/partnership-for-expanded-water-supply-sanitation-and-hygiene-pewash-programmed-strategy-2016-2030/
FMWR (2018). *National Action Plan for the revitalization of Nigeria's WASH sector (2018–2030)*. WaterAid/FMWR. Retrieved from https://washnigeria.com/wp-content/uploads/2022/10/Nigeria-National-Action-Plan-for-the-Revitalization-of-Nigerias-WASH-Sector.pdf
FMWR (2022). Federal Ministry of Water Resources (FMWR), Government of Nigeria, National Bureau of Statistics (NBS) and UNICEF. *Water, sanitation and hygiene: National outcome routine mapping (WASHNORM) 2021: A report of findings*. FCT Abuja, Nigeria. Retrieved from https://www.unicef.org/nigeria/media/5951/file/2021%20WASHNORM%20Report%20.pdf
Frischmann, B. M., Marciano, A., & Ramello, G. B. (2019). Retrospectives: Tragedy of the commons after 50 years. *Journal of Economic Perspectives, 33*(4), 211–228. https://doi.org/10.1257/jep.33.4.211
Geagea, D. H. (2025). *Recommoning water: Emerging water governance practices as a response to the austerity crisis* (Ph.D. thesis). Research and Graduation Internal, Vrije Universiteit Amsterdam. https://doi.org/10.5463/thesis.1003
Grassini, L. (2019). Participatory water governance between theories and practices: Learning from a community-based initiative in India. *International Journal of Water Resources Development, 35*(3), 404–429. https://doi.org/10.1080/07900627.2017.1354761
Groce, N., Bailey, N., Lang, R., Trani, J. F., & Kett, M. (2011). Water and sanitation issues for persons with disabilities in low-and middle-income countries: A literature review and discussion of implications for global health and international development. *Journal of Water and Health, 9*(4), 617–627. https://doi.org/10.2166/wh.2011.198
Haddout, S., Cardoso, L., Hoguane, A. M., Priya, K. L., C. Casila, J. C., Kurniawan, T. A., & Ljubenkov, I. (2024). Women's empowerment in water management across Africa. *Development in Practice, 34*(6), 810–823. https://doi.org/10.1080/09614524.2024.2338211
Hardin, G. (1968). The tragedy of the commons: The population problem has no technical solution; it requires a fundamental extension in morality. *Science, 162*(3859), 1243–1248. https://doi.org/10.1126/science.162.3859.1243
Hardin, G. (2013). The tragedy of the commons. In *Environmental ethics* (pp. 185–196). Routledge.
Harmon, G. M. (2022). *The policy Mobilities of farmer-led irrigation development in sub-Saharan Africa* (Doctoral dissertation). https://hdl.handle.net/1969.1/197830.
Ibrahim, U. A., Alkali, A. N., Sanyaolu, B. O., & Usman, B. (2023). Nigerian water resources management- an overview. *Arid Zone Journal of Engineering, Technology and Environment, 18*(1), 23–30.
Ighalo, J. O., & Adeniyi, A. G. (2020). A comprehensive review of water quality monitoring and assessment in Nigeria. *Chemosphere, 260, 127569*. https://doi.org/10.1016/j.chemosphere.2020.127569
Imonikhe, O. M. (2020). *Asset management: A way forward to improving the performance of urban water Utilities in Nigeria* (Doctoral dissertation). University of Leeds.
Irene, J., Irene, B. N., & Daniels, C. (2025). Not a drop to drink: Addressing Nigeria's deepening freshwater crisis. *Water, 17*(12), 1731. https://doi.org/10.3390/w17121731
Isukuru, E. J., Opha, J. O., Isaiah, O. W., Orovwighose, B., & Emmanuel, S. S. (2024). Nigeria's water crisis: Abundant water, polluted reality. *Cleaner Water, 2, 100026*. https://doi.org/10.1016/j.clwat.2024.100026
Ite, A. E., Harry, T. A., Obadimu, C. O., Asuaiko, E. R., & Inim, I. J. (2018). Petroleum hydrocarbons contamination of surface water and groundwater in The Niger Delta region of Nigeria. *Journal of Environment Pollution and Human Health, 6*(2), 51–61. https://doi.org/10.12691/jephh-6-2-2

JICA (2017). Internal ex-post evaluation for Grant aid project. The project for water supply in Bauchi and Katsina states. Retrieved from: https://www2.jica.go.jp/en/evaluation/pdf/2016_0961680_4_f.pdf

JMP (2024). Washdata.org. Retrieved from https://washdata.org/?utm. 10 September 2025

John, C. K., & Ajibade, F. O. (2024). Exploring the dynamics of WASH services: Challenges, enablers, and strategies for improvement. *Discover Civil Engineering, 1*(1), 79. https://doi.org/10.1007/s44290-024-00085-9

Khan, F. A., Niang, M., Jurji, Z., & Mahato, G. (2018). Ensuring sustainability and improving functionality of water supply facilities through VLOM in Nigeria—from pilot to national strategy. In *Transformation towards sustainable and resilient WASH services: Proceedings of the 41st WEDC international conference.*

Kponee, K. Z., Chiger, A., Kakulu, I. I., Vorhees, D., & Heiger-Bernays, W. (2015). Petroleum contaminated water and health symptoms: A cross-sectional pilot study in a rural Nigerian community. *Environmental Health, 14*(1), 86. https://doi.org/10.1186/s12940-015-0073-0

Kuza, Y., Okwoche, V. A., & Age, A. I. (2018). Assessment of the impact of Fadama III, development project on beneficiaries in Nasarawa state, Nigeria. *Greener journal of agricultural science, 8*(9), 197–202. https://doi.org/10.15580/GJAS.2018.9.091018133

Lema, M. W. (2025). Sustaining rural livelihoods through participatory water governance: A review of community-driven water resource management models in east and Central Africa. *Environmental Quality Management, 34*(3), *e70023*. https://doi.org/10.1002/tqem.70023

Liu, A., Giurco, D., & Mukheibir, P. (2015). Motivating metrics for household water-use feedback. *Resources, Conservation and Recycling, 103*, 29–46. https://doi.org/10.1016/j.resconrec.2015.05.008

Lloyd, C., Onyeabor, E., Nwafor, N., Alozie, O. J., Nwafor, M., Mahakweabba, U., & Adibe, E. (2019). Maritime transportation and the Nigerian economy: Matters arising. *Commonwealth Law Bulletin, 45*(3), 390–410. https://doi.org/10.1080/03050718.2019.1708426

Lockwood, H. (2019). *Sustaining rural water: A comparative study of maintenance models for community-managed schemes.* Sustainable WASH Systems Learning Partnership.

Loteni, N. S. (2017). *Strategy implementation of prepaid water meters in the City of Johannesburg* (Pro quest dissertations & theses). University of the Witwatersrand, Johannesburg (South Africa), 31797490.

Machado, A. V. M., Oliveira, P. A. D., & Matos, P. G. (2022). Review of community-managed water supply-factors affecting its long-term sustainability. *Water, 14*(14), 2209. https://doi.org/10.3390/w14142209

Madon, S., Malecela, M. N., Mashoto, K., Donohue, R., Mubyazi, G., & Michael, E. (2018). The role of community participation for sustainable integrated neglected tropical diseases and water, sanitation and hygiene intervention programs: A pilot project in Tanzania. *Social Science & Medicine, 202*, 28–37. https://doi.org/10.1016/j.socscimed.2018.02.016

Mangai, M. S., & De Vries, M. S. (2018). Co-production as deep engagement: Improving and sustaining access to clean water in Ghana and Nigeria. *International Journal of Public Sector Management, 31*(1), 81–96. https://doi.org/10.1108/IJPSM-03-2017-0084

Marshall, K., Setty, K., Mutegeki, A., Peletz, R., Khush, R., & Delaire, C. (2023). Water funds within village savings and loans associations: A promising solution to improve water user fee collection in rural Uganda. *PLOS Water, 2*(8), e0000159. https://doi.org/10.1371/journal.pwat.0000159

Mellink, E. (2025). *Participation or Tokenism? A Critical Analysis of Stakeholder Participation in Wadden Sea Governance* (Dissertation). Retrieved from https://urn.kb.se/resolve?urn=urn:nbn:se:hj:diva-68930 on 20 July 2025.

Mohammed, A. B. (2021). *Spatial impact analysis of community water boreholes in Bida Metropolis, Nigeria* (Doctoral dissertation).

NAP. (2018). National Action Plan for Revitalization of Nigeria's WASH Sector. Federal Ministry of Water Resources, Federal Secretariat, Area 1, Abuja, FCT. Retrieved from: https://washnigeria.com/wp-content/uploads/2022/10/Action-plan.pdf on 8 September 2025.

NDHS (2018). *Review of Nigeria demographic and health survey 2018*. Nigeria Demographic and Health Survey.

NEPAD (2018). West African water Centres of excellence. *Report on the National Strategy and Implementation on Skills Gap Assessment Framework for Human Resource Capacity Development in Nigerian Water Sector*. National Water Resources Institute (NWRI). Retrieved from: https://aquaknow.jrc.ec.europa.eu/system/files/Annex%20115_Nigeria_NWRI_HCD%20Framework.pdf

Ngene, B. U., Nwafor, C. O., Bamigboye, G. O., Ogbiye, A. S., Ogundare, J. O., & Akpan, V. E. (2021). Assessment of water resources development and exploitation in Nigeria: A review of integrated water resources management approach. *Heliyon, 7*(1), e05955. https://doi.org/10.1016/j.heliyon.2021.e05955

Nunes Silva, C. (2020). Citizen participation in spatial planning in Portugal 1920-2020 non-participation, tokenism and citizen power in local governance. In C. Nunes Silva (Ed.), *Contemporary trends in local governance: Reform, cooperation and citizen participation* (pp. 241–276). Springer. https://doi.org/10.1007/978-3-030-52516-3_12

Nwahiri, V. (2024, September 18). *"Asherkine for president": Content creator donates borehole to Jos community*. Retrieved from legit.Ng—Nigeria news. https://www.legit.ng/people/1614525-asherkine-donates-school-items-borehole-jos-communities-residents-celebrate/?utm. 10 September 2025.

Odjegba, E., Oluwasanya, G., Shittu, O., Idowu, O., & Brion, G. (2023). Service level indicators evaluation of water supply systems: Water users' perspective. *Urban Water Journal, 20*(10), 1532–1540. https://doi.org/10.1080/1573062X.2022.2134039

Ogundele, A. T. (2019). River basin development authorities in Nigeria: The neglected tools for national development. *International Journal of Advanced Academic Research, Sciences, Technology and Engineering, 5*(3), 39–49.

Olajuyigbe, A. E. (2016). Community participation and sustainability issue: An evaluation of a donor-driven water sector in Ikaram Millennium Village project, Nigeria. *Open Journal of Social Sciences, 04*(06), 90–103. https://doi.org/10.4236/jss.2016.46010

Olalekan, R. M., Adedoyin, O. O., Ayibatonbira, A., Anu, B., Emmanuel, O. O., & Sanchez, N. D. (2019). "Digging deeper" evidence on water crisis and its solution in Nigeria for Bayelsa state: A study of current scenario. *International Journal of Hydrology, 3*(4), *244–257*. https://doi.org/10.15406/ijh.2019.03.00187

Olawade, D. B., Wada, O. Z., Afolalu, T. D., Oladipo, T. C., & Asogbon, O. (2020). Assessment of rural water supply in selected communities in Osun state, Nigeria. *International Journal of Environmental Sciences & Natural Resources, 26*(1), 556177. https://doi.org/10.19080/IJESNR.2020.26.556177

Ostrom, E. (1990). *Governing the commons: The evolution of institutions for collective action.* Cambridge University Press.

Ostrom, E. (1994). Neither market nor state: Governance of common-pool resources in the twenty-first century.

Ostrom, E. (2002). Common-pool resources and institutions: Toward a revised theory. In *Handbook of Agricultural Economics* (Vol. 2, pp. 1315–1339).

Ostrom, E. (2008). Tragedy of the commons. In *The new Palgrave dictionary of economics* (pp. 1–5). Palgrave Macmillan.

Otse, V. (2025). Martins Vincent Otse initiative. Martins Vincent Otse initiative. Retrieved from https://mvoi4all.com/unijios-soloar-powered-boreholes. 10 September 2025.

Richards, C. E., Tzachor, A., Avin, S., & Fenner, R. (2023). Rewards, risks and responsible deployment of artificial intelligence in water systems. *Nature Water, 1*(5), 422–432. https://doi.org/10.1038/s44221-023-00069-6

Sandoval, D. (2022). Accesibility to services of water and sanitation. *Energy and Transportation for People with Disabilities in Latin America and the Caribbean.* https://doi.org/10.18235/0003613

Sule, B. F., Agunbiade, M., Adeogun, A. G., & Ihagh, G. (2016). Small town water supply: Situational assessment of shao, Kwara state, Nigeria. *Journal of Mechanical and Civil Engineering (IORS JMCE), 13*(4), 48–53. https://doi.org/10.9790/1684-1304034853

SURWASH Program. (2021). Sustainable Urban and Rural Water Supply Sanitation and Hygiene Program for Results (SURWASH). Environmental and Social System Assessment. World Bank. Retrieved from: https://ead.gov.ng/wp-content/uploads/2021/03/SURWASH-Environmental-and-Social-Systems-Assessment_Final_26APRIL2021.pdf on 10 September 2025.

SURWASH (2022). *Nigeria sustainable urban and rural water supply, sanitation and hygiene program (NG-SURWASH).* Program Operations Manual (POM). Retrieved from: https://surwash.ng/tools/SURWASH%20POM-12.01.22.pdf

Tondera, K., Brelot, E., Fontanel, F., Cherqui, F., Ellerbæk Nielsen, J., Brüggemann, T., Naismith, I., Goerke, M., Lopez, J. S., Rieckermann, J., Leitao, J. P., Clemens-Meyer, F. H. L. R., Moreno-Rodenas, A., Tait, S., & Anta, J. (2023). European stakeholders' visions and needs for stormwater in future urban drainage systems. *Urban Water Journal, 20*(7), 831–843. https://doi.org/10.1080/1573062X.2023.2211559

Tsekleves, E., Fonseca Braga, M., Abonge, C., Santana, M., Pickup, R., Yongabi Anchang, K., de Pippo, T., Sample, K., & Roy, M. (2022). Community engagement in water, sanitation and hygiene in sub-Saharan Africa: Does it WASH? *Journal of Water, Sanitation and Hygiene for Development, 12*(2), 143–156. https://doi.org/10.2166/washdev.2022.136

Ugbah, R., Meldrum, A., & Ehiwario, K. (2017). Water access and community engagement: Creating the right environment for maximizing the benefits of community engagement processes and increasing water participation in Nigeria. *WIT Transactions on Ecology and the Environment, 220*, 129–139.

UNEP GRID/Interactive Country Fiches (n.d.). Water/Nigeria. UNEP GRID. Retrieved from UNEP GRID: Nigeria country data, on 6 October 2025.

UNICEF (2017). *Gender-responsive water, sanitation and hygiene: Key elements for effective WASH programming.*

UNICEF (2020). *Evaluation of the UNICEF-supported Federal Government of Nigeria water, sanitation and hygiene Programme (2014–2017)*, Final report, May 2020.

UNICEF (2023). *Back to life.* Unicef.org. Retrieved from https://www.unicef.org/nigeria/stories/back-life on 12 October 2025.

Waarde, J. V. D., Musa, H. T., & Ischer, M. (n.d.). *Water catchment protection handbook* (pp. 1–32). Helvetas.

WASH Nigeria. (2014). *Mid-term review of the water supply and sanitation sector reform programme, Phase II (WSSSRP II), Nigeria.* Retrieved from: https://washnigeria.com/wp-content/uploads/2022/10/Mid-Term-Reportwssrp22.pdf

WASH Nigeria (2018). *National outcome routine mapping of water, sanitation, and hygiene service levels, Nigeria. Summary of survey findings.* Retrieved from: https://washnigeria.com/wp-content/uploads/2022/11/National-Outcome-Routine-Mapping-of-Water-Sanitation-and-Hygiene-Service-Levels-Nigeria.pdf

WaterAid. (2018). *Sustainability framework for WASH.* WaterAid. Retrieved from: https://washmatters.wateraid.org/sites/g/files/jkxoof256/files/sustainability%20framework.pdf

World Bank (2021a). *Draft environmental and social systems assessment (ESSA)—Nigeria sustainable urban and rural water supply, sanitation and hygiene program—P170734*, Washington, D.C. https://documents.worldbank.org/curated/en/453761619731259052

World Bank (2021b). *Nigeria sustainable urban and rural water supply, sanitation and hygiene program (P170734).* Retrieved from: https://documents1.worldbank.org/curated/en/498591620141029783/pdf/Final-Technical-Assessment-Nigeria-Sustainable-Urban-and-Rural-Water-Supply-Sanitation-and-Hygiene-Program-P170734.pdf

World Bank (2025). *Implementation status & results report. Nigeria Sustainable Urban and Rural Water Supply, Sanitation and Hygiene Program (P170734).* Retrieved from: https://documents1.worldbank.org/curated/en/099060325093521171/pdf/P170734-246b957a-2c9c-4afb-9dd3-0bcb5cf7de72.pdf

Yusuf, T. S., Murray, A., & Okereke, C. (2022). Working with local governments to increase access to WASH services: A case of WaterAid's participatory approaches in Nigeria. *H2Open Journal, 5*(3), 424–437. https://doi.org/10.2166/h2oj.2022.061

Part IV
Technological Innovations and Scientific Approaches

Chapter 10
Advanced Water Management Technologies

Nosa Idusuyi, Thomas Aghogho Orhadahwe, Adewuyi Felix Adedokun, and Victor Olakunle Ibidunmoye

Abstract There are over 8.2 billion people globally who depend on the limited water resources for domestic and industrial applications, putting a strain on existing water resources. This underscores the need to examine technologies that can boost water management. The different water management technologies and alternative water sources have been discussed in this chapter. Innovations in artificial intelligence (AI) and machine learning (ML) that can foster better water management were also examined. However, persistent gaps remain despite technological advancements in water management. These include high initial cost, energy demand, limited industrial adoption due to scalability issues and the perception of people towards some of these water sources. The chapter reviews advanced water management technologies and alternative sources to address global water scarcity and pollution. It discusses treatment methods like membrane bioreactors and nanofiltration, as well as smart technologies such as AI and IoT for monitoring. Additionally, it explores resource recovery from wastewater and alternative source strategies, including desalination and fog harvesting. By identifying technical gaps, financial constraints, and social acceptance issues, the chapter outlines future research priorities and policy directions for sustainable water management. The chapter recommends further research efforts to reduce the energy demand of these water technologies, as well as sustained awareness programmes to foster increased adoption among the global population.

Keywords Greywater · Wastewater treatment · Fog harvesting · Atmospheric water generation · AI and ML in water management

N. Idusuyi (✉) · T. A. Orhadahwe · A. F. Adedokun · V. O. Ibidunmoye
Department of Mechanical Engineering, University of Ibadan, Ibadan, Nigeria
e-mail: nosaidus@gmail.com

M. I. Adeoba et al. (eds.), *Water Resources Management in Nigeria*, World Water Resources 10, https://doi.org/10.1007/978-3-032-22813-0_10

10.1 Introduction

Water is one of the most vital resources and thus ensuring its availability in clean and safe form is crucial for health, agriculture, industry, and the environment. It is an important resource that shapes civilisation. Today, water infrastructure is exposed to severe strain due to the global rising population. The conventional water treatment plants produce basic and safe water but do not meet the requirements for recycling and reuse (Ngo et al., 2006; Miserli & Konstantinou, 2025). Although water makes up over 70% of the Earth's total surface, only 2% is available to support over 8 billion people (Ranade & Bhandari, 2014). In a developing country like Nigeria, improper wastewater treatment poses a serious risk to the environment. Wastewater contains harmful compounds often released untreated into the environment, thereby polluting rivers, lakes and oceans. This poses health risks to aquatic life and disrupts the ecosystem (Ranade & Bhandari, 2014; Chen et al., 2024; Jenila et al., 2023; Maldonado et al., 2023; Ren et al., 2023; Idusuyi et al., 2025). The conventional wastewater treatment methods include the use of activated sludge, coagulation and sedimentation. These methods are characterised by high cost and low efficiency. Meanwhile, advanced approaches like adsorption, photocatalysis, membranes, ozonation, and electrochemical treatment are more efficient. Millions of lives are lost yearly due to water-related illnesses (Wear et al., 2021). This underscores the need for climate-resilient and sustainable wastewater treatment solutions to address this global challenge. Industrial activities further contribute to water contamination. Over 70% of water at the point of consumption has been contaminated, mainly by industrial effluents (Lin et al., 2022). By 2050, it is projected that excessive biochemical oxygen demand will affect water quality for one in five people globally (UNESCO, 2016). These projections reveal the difficulty in the standard of living that is expected if infrastructural, regulatory, technological and social changes are not made. Water scarcity stems from various factors beyond population growth. The accelerating rate of urbanisation astronomically increases the demand for industrial operations and municipal consumption, depleting groundwater faster than natural replenishment rates (Behari et al., 2008). The ability of ecosystems and natural water cycles to meet expanding water demand has been compromised by pollution from economic and human activities (UNESCO, 2015). Financial constraints particularly affect developing regions, where limited infrastructure investment restricts access to advanced treatment technologies (Ewim et al., 2023). Traditional wastewater treatment methods used over the years to remove contaminants to prevent environmental damage have proved rather ineffective (Nwokediegwu et al., 2024). Advanced water management technologies are well-positioned to provide better wastewater recycling results, preparing water for agricultural use, industrial processes, or even direct potable reuse. For example, membrane bioreactors achieve clean effluent by combining biological treatment with physical filtration, producing effluent clean enough to require minimal additional processing (Tansel et al., 2006). These technologies combine the cost-effectiveness of biological processes with the reliability of physical separation (Ngo et al., 2006). The objective of this chapter is to provide a comprehensive review of advanced water management technologies and alternative water sources that can mitigate global water scarcity and pollution challenges. It seeks to examine conventional and emerging treatment methods such as membrane bioreactors, advanced oxidation processes, and nanofiltration, while exploring the integration of smart technologies like artificial intelligence, Internet of Things, and digital twins for efficient monitoring and conservation. Additionally, the chapter highlights opportunities for resource recovery from wastewater, including energy and nutrient reclamation, and evaluates alternative water sourcing strategies such as desalination, fog harvesting, and atmospheric water generation. By identifying key technical gaps, financial constraints, and social acceptance issues, and by summarizing comparative features of these technologies in Table 10.1, the chapter aims to outline future research priorities and policy directions that promote sustainable, climate-resilient water management practices.

10.2 Water Treatment Technologies

Water treatment technologies such as Membrane Bioreactors (MBRs), Advanced Oxidation Processes (AOPs), Nanofiltration (NF), etc., make up the advanced water treatment technologies. They effectively remove water pollutants that traditional treatments might not adequately eliminate. They ensure water purity and safety for non-potable use, such as industrial, agricultural purposes.

10.2.1 Membrane Bioreactors (MBRs)

With rising global water scarcity, wastewater treatment has increasingly been viewed as a part solution. Several technologies have been considered of which the Membrane Bioreactor (MBR) stands out for the treatment of industrial wastewater. This technology is particularly useful in developing countries facing rapid industrialisation and related environmental challenges. It combines membrane filtration and biological treatment, offering advantages such as high treatment efficiency, compact design, low chemical usage, and water reuse potential as illustrated in Fig. 10.1. Research has shown that MBRs can significantly reduce the levels of certain pollutants, such as diclofenac metabolite four hydroxy diclofenac, by 56% (Jijingi et al., 2024). The three basic membrane bioreactor configurations are the internal submerged membrane bioreactor, external submerged membrane bioreactor and the side stream membrane bioreactor (Al-Asheh et al., 2021).

Malamis et al. (2015) conducted a study at the world's largest petrochemical membrane bioreactor (MBR) plant to evaluate the efficiency of MBR technology in treating petrochemical wastewater before discharge into a sensitive water body with strict standards. A pilot-scale MBR, fed with the same effluent as the full-scale system, was used for optimisation. The pre-denitrification setup showed poor

Table 10.1 Summary comparison of the various wastewater management and resource recovery techniques

Parameter	Energy recovery	Nutrient recovery	Water reuse	Greywater
Type of waste stream	Sludge, high-strength wastewater	Secondary effluents, sludge liquors rich in N and P	Treated municipal effluent	Household water excluding toilet effluent
Technologies needed	Anaerobic digestion, co-digestion units	Struvite precipitation, ammonia stripping, EBPR, crystalisaton	MF/UF-RO membranes, UV/AOP, disinfection units	Settling chambers, grease traps, sand/biochar filters, membrane bioreactors, constructed wetlands
Final product(s)	Biogas, heat, electricity	Struvite, ammonium sulphate, fertilizer-grade products	High-quality reclaimed water	Non-portable quality water
Area of application of final product	Renewable energy for power/heat, grid injection	Agriculture, horticulture	Municipal supplies for households, industries and irrigation	Household reuse, small scale irrigation, landscaping
Technological requirements (implementation)	Medium to high—requires digesters, gas handling, CHP	Medium—Chemical dosing, reactor control, integration into WWTP	High—Advanced membrane systems, energy-intensive oxidation	Low-medium—Simple filters to wetlands; advanced if MBR is used
Financial implications* (pilot scale plant)	Moderate to high USD 0.1–1.5 million	Moderate USD 0.03–0.4 million	Moderate to high USD 0.15–1.2 million	Low USD 1000—60,000
Ease of adoption	Moderate—requires skilled operation, but proven technology	Moderate—promising but limited by market uptake of fertilisers	Low to moderate—Technically proven, but public acceptance and costs are barriers	High—simple, low-cost systems are widely adaptable at household or community scale

*Pilot/demonstration CAPEX ranges derived from peer-reviewed pilot and demonstration studies (Garcia-belinchón et al., 2013; Micolucci et al., 2018; Rodríguez et al., 2020; Bastos et al., 2021). Site factors and scope can cause large variations

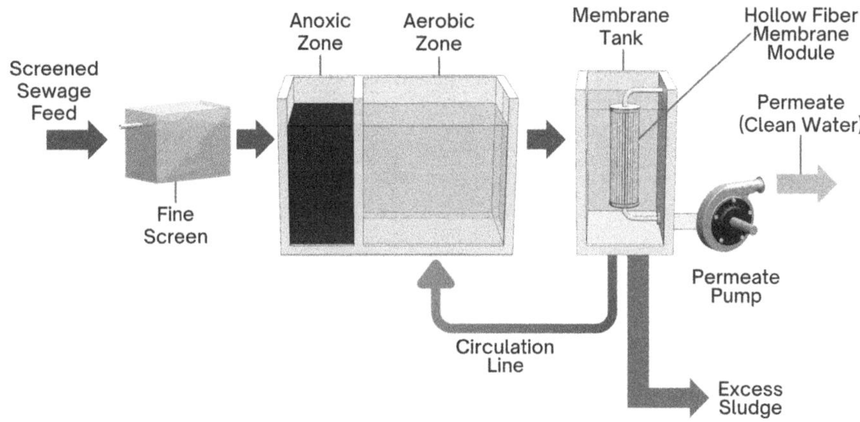

Fig. 10.1 Membrane Bioreactors (MBRs) process

ammonification performance. Heavy metal removal varied, with less than 40% removal for As, B, Ba, Mo, Al, Ni, Se, Sb, V, and Zn; 40–70% for Pb, Hg, Cu, Ag, Cr, Mn, and Co; and over 70% for Fe. Sludge clogging occurred in the membrane module, with higher accumulation of COD, nutrients, and metals (As, Zn, Mo, Ni, Cd, Sb, Fe, Se, and Co) compared to normal activated sludge. Combining MBR with suitable physicochemical pre-treatment ensured compliance with effluent quality standards. In another study, Hoinkis et al. (2012) reported two case studies where membrane bioreactor technology was applied because of its effectiveness for industrial water treatment: In a German laundry, a large-scale MBR and reverse osmosis system (200 m^3/day) achieved about 80% wastewater reuse and has operated successfully since 2007 and also in a Chinese textile factory, a small-scale MBR (0.4 m^3/day) achieved around 90% COD removal despite the presence of poorly biodegradable chemicals, though additional treatment such as nanofiltration or reverse osmosis was needed due to residual dyes.

10.2.2 Advanced Oxidation Processes (AOPS)

AOPs are oxidation processes which involve the generation of hydroxyl radicals (OH^-) in sufficient quantities to purify water as illustrated in Fig. 10.2. They offer great promise for disinfecting recycled water and removing micropollutants. AOPs are particularly useful for treating wastewater streams containing pharmaceuticals, pesticides and industrial chemicals (Ramm & Wojciechowska, 2025). Kumari and Kumar (2023) highlighted that Advanced Oxidation Processes (AOPs) are classified based on their mode of oxidation, which may occur through chemical (using reagents and catalysts), photochemical (using solar or UV energy), electrochemical (using electrical sources), or sonochemical (using ultrasound) methods. However,

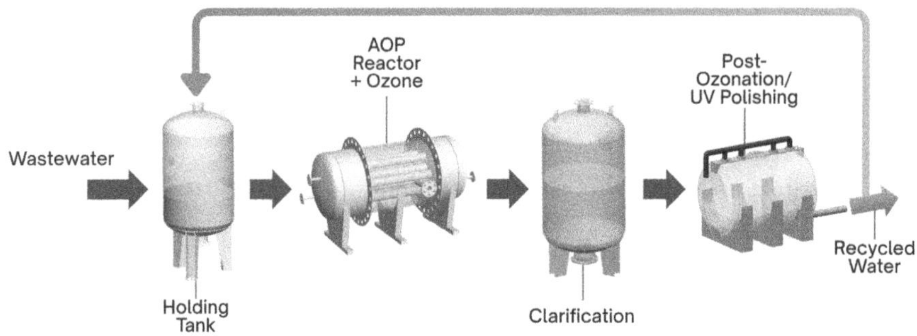

Fig. 10.2 Advanced oxidation processes (AOPS)

the selection of an AOP depends on factors such as pollutant type, resource availability, and cost, and is determined by the source of reactive oxygen species generation (Deng & Zhao, 2015). Recently, advanced variants like catalytic wet air oxidation, supercritical water oxidation, sulfate and carbonate radical-based AOPs, and combined AOP systems have shown great potential, achieving high pollutant degradation efficiencies across diverse wastewater types. Pharmaceuticals in wastewater pose risks to aquatic ecosystems and human health, prompting the EU's new directive for quaternary treatment. A local wastewater plant removed over 90% of key drugs like paracetamol and ibuprofen, but further treatment was needed. This prompted Skalska-Tuomi et al. (2025) to investigate the performance of three advanced oxidation processes, i.e. photocatalysis, ozonation and gas-phase pulsed corona discharge-based oxidation (PCD). Results showed that all three methods achieved more than 90% removal efficiency, with the PCD process having the best energy efficiency of 0.28 kWh m^{-3}, followed by ozonation with 0.55 kWh m^{-3}, while photocatalysis cannot compete with 47 kWh m^{-3} of energy required to obtain 90% removal. (Phan et al., 2023) explored advanced oxidation processes (AOPs)-O_3, O_3–H_2O_2, and O_3-UV combined with a biological activated carbon (BAC) process to remove natural organic matter (NOM), a disinfectant by-product (DBP) precursor. Among the tested methods, the O_3-UV process produced more biodegradable compounds, while the integrated O_3-UV-BAC system achieved the highest reductions in dissolved organic carbon (40%) and UV_{254} absorbance (52%). The study concluded that coupling AOPs with biological treatment could effectively minimise DBP precursors and organic micropollutants.

10.2.3 Nanofiltration (NF)

Nanofiltration is a widely used pressure-driven process with low cost yet high efficiency in separating monovalent and multivalent ions, which offers a promising solution to alleviate the issue of freshwater scarcity as illustrated in Fig. 10.3. Although

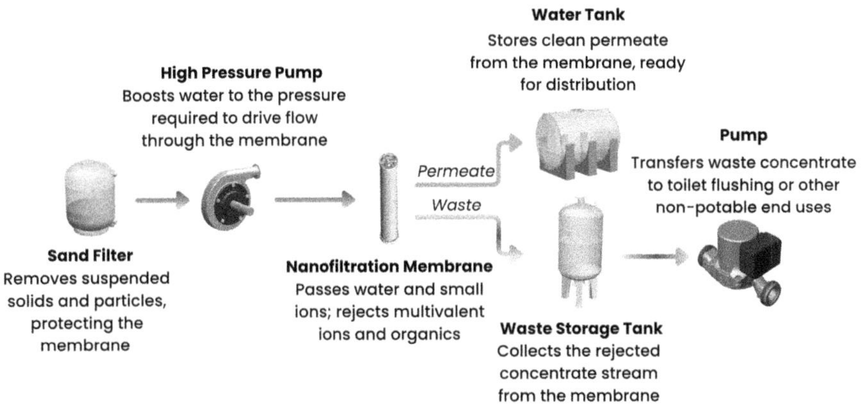

Fig. 10.3 Nanofiltration (NF) process

the practical performance of NF is regulated by the strong trade-off relationship between water permeance and selectivity (Zouhri et al., 2024).

Motta Cabrera et al. (2021) tested a commercial titania ceramic nanofiltration membrane system with a capacity of 20 m^3/h for nearly two years to treat recycled water from a Canadian oil sand mine. Operating at a 50% stage cut, the unit effectively reduced ion concentrations, Total Suspended Solids (TSS), and Total Organic Carbon (TOC). Results revealed a strong correlation between specific flux and rejection efficiency, with higher rejections occurring at lower specific fluxes, likely due to the formation of a cake layer on the membrane surface. Among the ions analysed, divalent cations exhibited the highest rejection, influenced by hydrated ion size and electrostatic effects. The membrane achieved 75–90% TOC and nearly 100% TSS removal, demonstrating its potential for improving water quality and reducing river water usage in oil sands operations. In another study, Tomaszewska et al. (2017) evaluated sulphate ion reduction in natural mineral water using a commercial Dow Filmtec NF-270 nanofiltration (NF) membrane operated at 0.8–2.0 MPa. The raw water had a Total Dissolved Solids (TDS) of 2176 mgL^{-1} and contained 798.7 mgL^{-1} of SO_4^{2-}, and the H_2S level was 2.0 mgL^{-1} on average. The NF process achieved over 98% SO_4^{2-} ions removal, reducing TDS by 61–64%, and significantly lowering calcium ion (82–84%) and magnesium (about 84%) levels. Monovalent ions have been reduced to a lesser extent, i.e., sodium by 42–46% and bicarbonates by 42–49%. Also, NF permeate filtration in the $CaCO_3$ bed showed increased calcium and bicarbonate ions concentration by 60.5% and 7%–8%, respectively. SEM analysis revealed gypsum crystal deposits on the NF-270 membrane surface.

10.2.4 Water Conservation Practices Using Smart Technologies

Smart water conservation technologies provide real-time monitoring and control of water use by integrating smart devices with building/structure management systems. These solutions track consumption patterns, improve efficiency, and support sustainability goals (Palermo et al., 2022) The integration of artificial intelligence (AI), internet of things (IoT), machine learning, and blockchain has transformed water management by enabling real-time monitoring, predictive maintenance, and decentralized control, thereby enhancing efficiency, sustainability, and resilience in water systems (Narendra et al., 2025).

10.2.5 Smart Water Meters

A smart water meter is an electronic device designed to record water consumption for utility companies and the users. Where traditional water meters read manually in monthly or yearly intervals, smart meters record consumption in real-time or near real-time, and communicate this information to the utility and consumer (Adams & Jokonya, 2021). Sønderlund et al. (2014) reported the effectiveness of consumption feedback in curbing water use with focus on smart water meters. This technology provides real-time, detailed information that educates consumers and encourages conservation. The enhanced effectiveness range from 3 to 53.4%, averaging 19.6%, showing strong potential for controlling domestic water consumption, hence improving conservation.

10.2.6 Leak Detection Systems

A leak detection system (LDS) is a collection of devices designed to find and alert users of the leaks in plumbing, pipelines, or other systems, often by detecting moisture or monitoring flow rates and pressure. Studies show that water loss due to leaks varies significantly across countries. In developed European nations, losses average around 15% with France at 27%, the UK at 21%, and the Netherlands at 5% while the United States records an estimated 13% loss (Fuentes & Mauricio, 2020). Several leak detection methods currently in use to detect and locate the leaks in water distribution networks include flow rate monitoring, pressure point analysis, negative pressure wave, distributed fibre optic sensing, acoustic sensing, real-time transient modelling, model simulation, infrared cameras, and lidar systems (Korlapati et al., 2022). In a study on LDS, Fang et al. (2023) experimented with a water distribution network and an acoustic emission detection system. A 42 mm diameter, 4.8 mm thick steel pipe with a controlled pressure of about 0.2 MPa was used, and leakage was simulated by opening a valve. The detection system included a YK-YD20 IEPE piezoelectric acceleration sensor (– 20–120 °C, 200 mV/g sensitivity, 0.5–5000 Hz range), a NI myDAQ data acquisition card, and NI LabVIEW software for data collection and analysis. The sensor, installed magnetically on the pipe, captured vibration signals produced when leakage occurred. In another study, (Islam et al., 2024) used machine learning to detect and locate underground water leaks, a major issue causing about $39 billion in annual losses. Two models were developed: a hybrid ML model using thermal imaging to identify leaks up to 750 mm deep with 95.2% accuracy, and an 1D-CNN model using binaural audio from geophones to localize leaks with 88.19% accuracy. Together, these models offer a fast and precise approach for large-scale leakage surveys, potentially reducing water loss significantly.

10.2.7 Artificial Intelligence and Machine Learning in Water Management

Artificial intelligence (AI) is transforming several aspects of science and more recently water management. AI has been used to forecast groundwater, flood risk management droughts in certain areas (Tanriverdi & Batmaz, 2025) and more. Ogunrinde et al. (2025) utilized Random Forest, Support Vector Machines, Gradient Boosting to forecast water demand. Long Short-Term Memory networks (LSTM) deep learning architecture was used for supply prediction under time-series conditions. These models were effective in predicting water demand and capture non-linear relationships and temporal dependencies in hydrological and consumption data. From their study, the LSTM model achieved the highest predictive accuracy for both demand and supply, with R^2 values above 0.93. In a related *study, Kim* et al. *(2023) developed an AI-based water demand prediction model for an intake station using deep neural* network (DNN) and long short-term memory (LSTM) models. The model was trained using data collected for about 10 years (2004–2015). The LSTM model achieved a correlation coefficient (CC) of 0.95 and normalized root mean square error (NRMSE) of 8.38, indicating excellent performance. Decision tree (DT) and random forest (RF) models, based on AI-based classification, were used to predict water demand at the Gurye intake station taking into consideration factors like maximum and minimum temperature, water demand and rainfall from the previous day. Model accuracy was evaluated using the F1-score, with the random forest model achieving a score of 0.88, indicating excellent performance.

Shu et al. *(2024) predicted water demand in different sectors in the Tuojiang River basin, China using Extreme* Learning Machine (ELM), Genetic Algorithm optimized Back Propagation Neural Network (GA-BP), Gaussian Process Regression (GPR), Support Vector Regression (SVR), and Random Forest (RF) using fifteen years data. Their study revealed that the GPR model better predicted water demand for the agricultural, domestic, and ecological sectors, with R^2 values of 0.9811, 0.9338, and 0.9142 respectively. Optimal predictions for industrial water demand were attained using the GA-BP model with an R^2 value of 0.8580. The results indicate that the prediction models would be useful long-term water demand forecasting tools, promoting sustainable water resource management in the basin studied and other areas.

*Kanyama et al. (*2025*) proposed an AI-driven anomaly detection framework for smart water* metering networks (SWMNs) using machine learning (ML) techniques and data resampling methods to enhance water conservation efforts. Support vector machine (SVM), decision tree (DT), random forest (RF), and k-nearest neighbors (kNN) models within ensemble learning strategies were used to model Six years monthly water consumption data from 1375 households. Data resampling techniques such as random under-sampling (RUS), SMOTE, and SMOTEENN were used to resolve class imbalance (a particular challenge with real world datasets). From the study, the SMOTEENN achieved the best overall performance for individual models, with the RF classifier reaching an accuracy of 99.5% and an AUC score of 0.998. Ensemble learning approaches also yielded strong results.

10.2.8 Application of IoT and Sensor Technologies in Monitoring Water Quality, Flow and Pressure

Water is a vital resource that requires continuous quality monitoring to ensure its safety and usability. Essamlali et al. (2024) highlighted that the Internet of Things (IoT) such as Low-Power Wide Area Networks (LPWAN), Wi-Fi, Zigbee and Radio Frequency Identification (RFID) has transformed water quality monitoring by enabling real-time data collection from various sources. When combined with Machine Learning (ML) algorithms, this data can be analyzed to predict water quality, support decision making, and prevent contamination.

Jabbar et al. (2024) designed and deployed a solar-powered, portable, IoT-based real-time water quality monitoring system for rural areas. The system used LoRaWAN technology and a LoRa node equipped with sensors to measure key parameters such as pH, total dissolved solids, turbidity, and temperature. Data collected via an Arduino-based controller is transmitted through a LoRaWAN gateway to The Things Network, ThingSpeak, and ThingView platforms for real-time online access. Powered by a solar cell and rechargeable battery, the system ensures continuous operation. The field test was performed at Gambang Lake in order to demonstrate its accuracy, durability, and waterproof ability. The results obtained closely matched the laboratory analyses, confirming the system's reliability and effectiveness in remote water quality monitoring.

In another study, Liu and Cheng (2025), proposed a difference analysis (DA) model based on the fuzzy neural network (FNN) and bi-directional long and short-term memory network (BiLSTM) attention mechanism to enhance water quality anomaly detection and alarm threshold setting. The model effectively minimizes the influence of outliers and missing data while considering seasonal and hydrological variations, improving the accuracy of water quality alerts. Using five key water quality indicators, the FNN extracts spatial features, while the BiLSTM captures time-series patterns. Experimental results show that the model achieves the lowest average mean absolute error (MAE) (0.174) and mean absolute percentage error (MAPE) (6.32%), and the highest correlation coefficient (0.936) on the YRB dataset, outperforming other models. Its robustness was further confirmed using an additional wastewater treatment plant dataset demonstrating strong generalization ability.

Lin et al. (2025) proposed a soft sensor designed to estimate sewer flow rates using water level measurements in manholes, offering a cost-effective alternative to conventional flow meters. The system employs a Multilayer Perceptron (MLP) model, which uses time-series water level data as input to predict corresponding flow rates. The model was trained with data from three real-life manholes and simulated data generated through 3D Computational Fluid Dynamics (CFD) modeling. Results demonstrated that the MLP accurately estimated flow rates and remained robust against noise and measurement bias. Overall, the proposed soft sensor is a low-cost, reliable, and low-maintenance solution for sewer system monitoring.

10.2.9 Digital Twin Technology in Water Management

Digital twin technology leverages digital technologies to create a real-time virtual model of water systems for simulating, monitoring, analysing, and optimizing water distribution network operations. Integrating sensor data, internet of things (IoT), AI, and other technologies reflects the real-time state and behaviour of the actual water network such as pressure, flow rate, and pipe status. The digital technology model enables dynamic management, predicts future trends, detects potential issues, and optimizes operational efficiency (Dui et al., 2025). Tavares et al. (2024) proposed a digital twin based multicriteria optimization approach for hybrid water supply systems within a multipurpose Pumped-Storage Hydropower (PSH) framework, utilizing the Generalized Reduced Gradient (GRG) method in Solver. The optimisation considers factors such as Net Present Value (NPV), renewable energy production, device count, reservoir capacity, source availability, topography, and energy tariffs. Results show that solar-only systems are inadequate for high hydropower scenarios, while wind integration boosts energy output, profitability, and NPV, which can exceed €180 million. The PSH system handled 130,000 m^3 of water per day, stored 25 MWh of energy, and cuts CO_2 emissions by over 18,000 tonnes annually. In a similar work by Syed et al. (2024), Digital Twin technology with a multimodal transformer-based model was deployed for accurate water usage prediction and leakage detection. The developed model synchronised real-time data from flow meters, pressure sensors, and thermal imaging devices with a digital twin of the water network. From their study the results showed the Informer model achieved an R^2 score and mean squared error of 0.9995 and 2.2 respectively indicating the effectiveness of the system. For leakage detection, the model attained 98.4% accuracy and precision, an F1 score of 97.6%.

Mohammed et al. (2024) proposed an Industrial Internet of Water Things (IIoWT) system based on blockchain and digital twin technologies, thereby enabling secure and standardized data exchange. A digital twin-enabled cross-platform environment using the Message Queuing Telemetry Transport (MQTT) protocol ensured smooth communication among heterogeneous systems. Additionally, a CNN-LSTM model combined with a Blockchain Data Transactional (BCDT) scheme was introduced to validate and process diverse sensor data efficiently. The simulation results that were obtained show that the proposed system reduced processing time and enhanced data standardization accuracy, providing a robust solution for modern water management systems. Sardar et al. (2025) highlighted that digital twins technologies have revolutionised water management by providing real-time operational insights, predictive maintenance, and improved monitoring for water quality, flood risks, and resource optimisation. They integrate advanced modelling, analytics, and real-time data to enhance efficiency, minimise downtime, and support proactive decision-making. Despite challenges like data complexity and cybersecurity, digital twins enable dynamic, data-driven responses to issues such as system failures and climate impacts. They also boost sustainability, resilience, and adaptability in the face of urbanisation and extreme weather.

10.3 Wastewater Management and Resource Recovery

For long, wastewater was regarded as a disposable commodity. This concept has changed over the last couple of decades. Wastewater treatment focuses on reducing public health risks and preventing pollution of rivers and coasts. In recent years, the discussion has shifted towards recognising wastewater as a resource that contains water, nutrients and even energy that could be recovered. Qadir et al. (2020) highlight that global wastewater production is close to 380 billion cubic metres each year, carrying large quantities of nitrogen and phosphorus. These nutrients, if recovered, could replace part of the fertiliser currently manufactured through energy-intensive industrial processes. The framing of wastewater as an "urban mine" reflects this wider interest in circular economy thinking. The potential, though, has not yet been realised. In many regions, wastewater treatment is still the exception rather than the rule. The European Investment Bank (2022) estimated that more than four-fifths of flows are discharged with little or no treatment, and this pattern is especially clear in developing countries. The implication is obvious: water that might have been reused is wasted, while rivers and coasts are left to absorb the impact. It is a frustrating contrast, because the technical possibilities are already known. Wang et al. (2023) argue that treatment plants should be thought of less as disposal facilities and more as recovery sites. In their view, the same infrastructure could provide multiple benefits: biogas for energy, clean water for reuse, and nutrients for agricultural purposes. Their analysis does not pretend that the shift will be simple or cheap. What it does suggest is that without such changes, the wider goals of sustainable water management will remain out of reach.

10.3.1 Energy Recovery

Wastewater treatment is well known for its high energy demand. Aeration, pumping and sludge handling together make up the bulk of operational costs, often more than sixty per cent of the total power bill (Lima et al., 2024). For many years, this was accepted as the unavoidable price of protecting rivers and public health. The plant was seen purely as a consumer of electricity, and the idea of it generating energy of its own seemed unrealistic. In recent times, there has been a shift from energy consumption to production using wastewater. Wastewater carries a significant organic load, and organic matter is, in essence, stored energy. When broken down in the right way, it can produce useful fuels. This realisation has shifted attention towards making treatment plants energy-neutral, and in some cases, even net energy producers. Anaerobic digestion is the most established pathway that can convert sewage sludge into biogas, which can be burned to generate heat or power. Huang (2024) noted that several Asian facilities have gone further by adding food waste to digesters, which increases the methane yield. There are other technologies for recovering bio-energy from waste streams. Cengiz et al. (2024) highlighted the role of anaerobic membrane bioreactors, which extend biomass retention and enhance gas output, although fouling and cost remain barriers. In Mexico, Ortiz-Sánchez and Cuevas-Rodriguez (2023) reported the successful application of up flow anaerobic sludge bed reactors, particularly in warmer regions where the process is more efficient. Full energy neutrality is rarely achieved, but these studies show that significant fractions of plant demand can be offset. The central point is that wastewater treatment facilities are beginning to be viewed not only as sanitation systems but also as small renewable energy plants in their own right.

Empirical evidence from recent field and experimental studies confirms that wastewater can yield substantial energy returns when technologies and operating practices are aligned.

Sakurai et al. (2025) presented a near full-scale comparison of a High-rate Contact-Stabilisation plus Activated Sludge configuration (HiCS–AS) against a conventional activated sludge baseline. The authors measured organic loading conversion under seasonal temperature variation and reported markedly higher methane potential from HiCS sludge. Quantitatively, methane yields per volatile solids rose from roughly 0.27–0.28 NL CH_4/g VS (conventional) to about 0.41–0.45 NL CH_4/g VS under the HiCS regime; overall methane recovery from influent COD increased by approximately 30–40%. Methodologically the study presents a robust solids characterisation, standard BMP (biochemical methane potential) assays and process monitoring that are the near full-scale setting, producing results that are practically relatable. Limitations include a limited geographical scope and the need for longer monitoring to capture fouling and seasonal operational issues.

Un (2025) took a different, systems-oriented approach. Using measured biogas production data from several wastewater treatment plants in Adana, Türkiye, the study evaluated the feasibility of upgrading biogas to biomethane for use as a transport fuel. Uncombined plant measurements with life cycle and emission calculations, reporting annual biogas volumes on the order of millions of normal cubic metres and estimated emission reductions of ~ 80–85% when biogas replaces diesel for municipal buses. The value of this work is twofold: it moved beyond on-site energy self-sufficiency and demonstrates cross-sector benefits (transport decarbonisation), while using real plant output data. Caveats are clear: upgrading and grid-injection infrastructure, gas quality standards and seasonal variability all affect feasibility and economics.

Rocha et al. (2024) presented controlled experiments on sludge pre-treatment and operational parameter optimisation (retention time, mixing intensity) and their effects on biogas yield. The paper took a pragmatic approach to show that modest changes in hydraulic retention time and mechanical mixing can increase cumulative methane yield by measurable margins and that simple thermal or chemical pre-treatments can further enhance biodegradability. Although the trials were smaller scale, the results provided actionable guidance for plant operators seeking incremental improvements without wholesale retrofits. The authors also provided energy balance calculations showing when interventions are net-beneficial versus when increased energy input for pre-treatment outweighs gains.

These studies show that co-digestion and targeted process upgrades reliably increase biogas yield. The yield depends on substrate mix and climate. It also shows that system boundaries matter. Assessing the CHP efficiencies and seasonal variations could provide a very different picture from single-point BMP yields. There are also operational constraints, such as membrane fouling, capital costs, gas-cleaning needs and market for biomethane. These constraints limit the proliferation of the implementation of energy recovery units in wastewater treatment plants.

10.3.2 Nutrient Recovery

Wastewater carries important nutrients like nitrogen and phosphorus, which, if mishandled, can lead to environmental pollution. The release of excessive nitrogen to the environment leads to an algal bloom, capable of upsetting aquatic ecosystems. These nutrients can be harnessed for reuse in agriculture. The scarcity of nutrients is a driver for the search for alternative sources. Wastewater streams can serve as reliable sources for recovering these nutrients. This can be achieved by adding resource recovery units to wastewater treatment plants. This will reduce over-dependence on fertiliser and promote the circular economy (Simoes et al., 2020).

Nutrients can be recovered from wastewater using several methods. One of the methods is struvite precipitation. It is used to produce crystalline fertiliser with slow dissolution rates when introduced to the soil. Another method is ammonia stripping, in which nitrogen is drawn off as ammonium salts. These nutrients can enhance biological phosphorus removal (EBPR) provides a biological route, often paired with chemical precipitation, to improve yields. The choice of process depends on wastewater composition, plant size and local demand. In practice, plants rarely rely on a single method but instead combine biological and chemical approaches (Kim et al., 2021). Evidence from operating facilities illustrates how these techniques work. Simoes et al. (2020) analysed a range of sludge dewatering liquors and characterised their phosphorus and ammonium contents to predict struvite formation potential under realistic operational conditions. Their work demonstrated that many sludge liquors have the requisite concentrations of PO_4–P and NH_4–N to form struvite with modest magnesium dosing, but also emphasised that variability across plants (liquor composition, dilution, and process configuration) strongly affects yields and the need for tailored control strategies. The study is valuable because it links liquor chemistry to expected recovery rates and highlights practical constraints such as contaminant content and variability in liquor flows. Similarly, Wang et al. (2022) reported an experimental work on producing high-purity struvite from hydrothermally-treated sludge supernatant. The study evaluates crystallisation conditions and product quality, showing that process adjustments (e.g., pH control and post-treatment) can significantly improve struvite purity and reduce co-precipitated contaminants. Their results support the feasibility of producing marketable fertiliser-grade struvite from sludge streams, provided adequate pre- and post-treatment steps are implemented. Mayor et al. (2023) presented a pilot-scale demonstration of hybrid systems (including membrane contactors) for recovering ammoniacal nitrogen from digestion liquors and other nitrogen-rich streams. The pilot results indicate that membrane contactors and related trapping strategies can recover ammonium into concentrated solutions suitable for use or further processing. The authors discuss operational issues (membrane fouling, trapping chemistry) and provide pilot data on recovery efficiency and energy inputs, offering a realistic assessment of scale-up prospects. Nutrient recovery presents a lot of benefits both to the environment and key stakeholders. Nutrient recovery limits eutrophication risks, thereby reducing greenhouse gas emissions from synthetic fertiliser production. It creates new revenue streams for utilities. Farmers are the direct beneficiaries, as they have access to low-cost but reliable fertilisers. At a strategic level, it strengthens phosphorus security, an issue made urgent by the geographical concentration of reserves (Witek-Krowiak et al., 2022).

However, recovery rates are sensitive to influent characteristics, process stability and reagent costs. Poorly controlled struvite precipitation can clog pipes. Market uptake is uneven: many farmers are cautious about recovered fertilisers. Economics are also uncertain where fertiliser prices are low or policies do not support resource recovery. Retrofitting existing plants requires capital and expertise. Kataki et al. (2016) noted that overcoming these hurdles, technical, financial and institutional, is essential if nutrient recovery is to move beyond promising pilots into widespread practice.

10.3.3 Water Reuse

Water reuse has moved from a niche application to a mainstream climate-resilience strategy in many water-stressed regions. Reuse ranges from non-potable applications (irrigation, industrial cooling) to indirect and increasingly direct potable reuse

(IPR and DPR). Modern potable-reuse trains typically combine membrane separation (micro/ultrafiltration and reverse osmosis) with advanced oxidation processes (AOPs) such as UV/H_2O_2 or UV/chlorine to remove pathogens and trace organic contaminants, and then add robust monitoring and redundancy to ensure safety (Lefebvre, 2018). The policy and engineering emphasis is now on process robustness, chemical challenge response, and energy–cost optimisation as reuse moves from demonstration to sustained operation.

Typical advanced potable-reuse trains follow a staged approach. After secondary treatment, effluents usually undergo extra polishing. This can involve filtration, coagulation or a short biological step, depending on the plant. Once stabilised, the water is pushed through membranes. Microfiltration or ultrafiltration acts first, and reverse osmosis follows, rejecting salts as well as most organics. Yet some trace contaminants remain. Compounds such as NDMA and 1,4-dioxane are well known for passing through RO. To deal with these, operators rely on UV-based advanced oxidation, most often UV/H_2O_2 or UV/chlorine. The final water is then charged into aquifers or blended with a portable water supply facility. The system requires continuous monitoring and frequent testing to maintain safety protocols (Tow et al., 2021; Kwon et al., 2020).

Kwon et al. (2020) developed a pilot comparison of UV-chlorine and UV/H_2O_2 AOPs on RO permeate for removal of challenging trace organics (including 1,4-dioxane, NDMA precursors). The study combined pilot reactor data with mechanistic kinetic modelling, reporting robust removal performance for both AOP variants under typical RO permeate matrices and identifying key design trade-offs (radical yields, by-product formation). The study is a blueprint that enables decision-making for the choice of DPR/IPR facilities in large-scale implementation. Tow et al. (2021) conducted an empirical modelling study that quantified specific energy consumption (SEEC) of UV-AOP and RO components at full-scale reuse facilities. The study combined measured pilot energy metrics with plant operations data to provide comparative SEEC estimates to create a balance between recovery, energy use and greenhouse gas implications. This highlights the importance of optimising AOP dosing and RO recovery to improve sustainability.

Recent studies have gone further than simply checking a few target contaminants. Using high-resolution mass spectrometry, Ishida et al. (2022) examined RO permeate before and after UV-based advanced oxidation. This approach, often called nontargeted chemical analysis, allows hundreds of trace organic compounds to be profiled at once. The results are revealing. Most of these organics are greatly reduced after treatment, and in many cases, entirely removed. At the same time, the method also provides evidence of transformation products, giving operators and regulators a clearer picture of the complete polishing step.

Advanced reuse trains deliver multiple benefits: they provide drought-proof water supplies, reduce pressure on freshwater sources, and, when properly designed, can meet drinking-water standards while controlling emerging contaminants. Reuse can decentralise supply risk and support resilience for utilities facing climate variability. Energy use is often raised as a concern for advanced reuse, yet there is growing evidence that smart process optimisation can ease this burden. For instance, higher recovery rates in reverse osmosis, as demonstrated in full-recovery pilot units, reduce the overall footprint. In parallel, UV-based advanced oxidation offers a flexible polishing step. It does not rely on heavy chemical dosing and has shown strong performance against a wide range of trace organics (Tow et al., 2021; Kwon et al., 2020).

Despite the promise, several hurdles remain for the wider adoption of potable reuse. Cost and energy demand still weigh heavily on utilities. Technically, it is also difficult to guarantee the removal of the many trace organic compounds present at very low concentrations. The choice between UV/H_2O_2 and UV/chlorine illustrates this dilemma. One produces high radical yields, while the other can generate halogenated by-products. Monitoring adds another layer of difficulty, since selecting the right surrogate indicators is still debated among practitioners and regulators. Public acceptance and rigorous regulatory frameworks (e.g., California DPR rule-making) are also essential. Finally, brine and concentrate management from high-recovery RO operations remains a pressing environmental and operational challenge (Ghimire et al., 2021).

10.3.4 Greywater Reuse Technologies

Greywater makes up the largest share of wastewater in most homes. It comes from showers, sinks, washing machines and kitchens, but not toilets. Greywater excludes sewage and is usually less polluted than blackwater. This is not surprising, since it mainly carries soaps, detergents and food particles rather than faecal matter. For that reason, it is considered easier to treat and reuse. In dry regions, especially, attention has turned to this stream as a substitute water source. With the right technology, it can supply irrigation, toilet flushing or even small-scale industrial needs (Diamantis, 2021).

Greywater can be treated in different ways. At the simplest level, the water passes through a settling tank or a grease trap to take out oils and larger solids. Some households add a sand or gravel filter, which helps to clear the water and remove odours. Biochar filters are also being tested for this purpose. Where higher quality is required, small membrane bioreactors are used, although they necessitate more maintenance. Another method is to construct wetlands where plants and microbes help to remove impurities. The methods to use depend on the available space, financial resources and the purpose of the final water product (Awasthi et al., 2024).

Empirical studies on greywater development include those conducted by Diamantis (2021), who worked with a very small membrane bioreactor, set up at a household scale. The unit handled greywater from kitchens and bathrooms. It gave effluent clear enough for garden use. Tests showed steady reductions in COD, BOD and turbidity. The plant's success was a result of combining biological treatment with a membrane barrier. However, the study reported that fouling frequently occurs, and membranes need regular cleaning. This is a major challenge for households with limited technical capacity to handle fouling. Hence, there is a limited adoption of greywater despite its ability to produce non-potable water.

In Chile, Rodríguez et al. (2022) studied greywater pilots in schools and rural communities. The Coquimbo Region faces frequent shortages, so local interest in reuse was high. The systems included filters, a biological step, and disinfection. Effluent from the pilots reached local standards for irrigation. There was a sharp drop in suspended solids and pathogens. The stability of greywater is season dependent. Its success also depends on the prevailing environmental conditions in the place of implementation and does not depend only on the plant design.

Parei et al. (2023) built a hybrid system with a vertical flow constructed wetland followed by a polishing column. Operating in a warm climate, the unit produced effluent with low BOD and low solids. It was reliable enough for reuse. Costs were also modest, as natural processes handled most of the work. Despite this, the study noted practical limits. Land was required, and clogging could occur if the loading was not carefully managed. This example shows both sides of nature-based solutions: low energy and low cost, but space and attention are essential.

10.4 Alternative Water Sources

A growing global population leads to a continuous rise in water demand. Rivers and aquifers are being stretched, and water scarcity intensifies in many regions. To cushion this effect, several communities have explored getting water from other sources other than the usual rivers and aquifers. These other sources are now referred to as "alternative water sources." One method of getting alternative water is by desalination, whereby seawater, or brackish water, is processed and made drinkable. Another source is Fog harvesting. It is the use of nets or surfaces to catch droplets from the air. In humid climates, machines can even pull moisture straight out of the atmosphere. Rainwater collection, long practised, also fits here, as does stormwater storage. None of these methods is without cost. Some need advanced plants, others need only simple structures. Alternate water sources reduce dependence on rivers and groundwater. They also give communities more resilience. In drought or heat, having more than one source often makes the difference between shortage and stability (Asad et al., 2022).

10.4.1 Desalination

In many dry regions, rivers and aquifers no longer meet demand. Attention has turned to desalination. Desalination is the removal of salt from seawater or brackish groundwater. Early plants were costly and consumed large amounts of energy. Reverse osmosis units now dominate, replacing older thermal plants in many areas. They are smaller, cheaper and more efficient than the first designs. Today, large coastal cities rely on desalination for daily water needs. For countries with limited freshwater, it has become a basic part of town planning. Desalination removes salt from seawater for consumption as drinkable water or other domestic usage. This is achieved using a thermal or membrane plant. A thermal plant produces steam through multi-stage or multi-effect distillation. Such plants perform well in excess availability of energy. In membrane plants, pressure is used to push water through the membrane, thereby eliminating the salt. Usually, a pretreatment process is done to prevent fouling by filtration or chemical treatment. In recent times, modern membrane desalination plants are equipped with energy recovery systems, thereby reducing the cost of the process. The most widely reported membrane system today is the reverse osmosis technology (Tayeh, 2024). There are many research efforts on the desalination process. For instance, Ruan and Wang (2021) embarked on a large-scale desalination project in China that used reverse osmosis (SWRO) and multi-effect distillation (MED) to remove salt from seawater. The results of their investigation indicated that a typical RO plant consumes about 4 kWh/m^3. On the other hand, MED plants operate at about 1.5 kWh/m^3, particularly when they are integrated with available waste heat. The two systems are able to deliver water quality that meets drinking standards. However, the cost of energy remains a significant barrier. The authors concluded that the choice of technology depends strongly on local energy prices and integration potential, with hybrid approaches offering promise for cost reduction.

Patel et al. (2021) did a comparison of the reverse osmosis and electrodialysis for the desalination of brackish water. Their study tested pilot configurations under varying salinity levels and recovery rates. The findings of the study revealed that electrodialysis produced higher energy efficiency in a moderate saline medium, while RO was more energy efficient at higher saline concentrations. The energy consumption of the two systems was below older benchmarks. This was a result of the improvement in the membranes and optimisation of the desalination process. The findings of the study suggest that technology selection should be based on site-specific parameters, such as water composition and utility costs, to ensure sustainability. This work underlines the importance of tailoring desalination solutions to local resource conditions.

A recent analysis by Alnajdi et al. (2024) conducted a study to understand the minimum practically required energy to complete the desalination of seawater by RO. The authors reviewed data from existing RO desalination facilities and reported that the specific energy consumption is between 2.5 and 4.0 kWh/m^3. This also depends on the system design and recovery rates. The values obtained for energy consumption for the systems considered were far lower than those of first-generation RO plants. This improvement is due to the incorporation of energy recovery devices in modern RO plants. However, further reductions toward the thermodynamic limit are unlikely without major technological breakthroughs. The authors concluded that future gains will depend more on reducing capital costs and improving operational reliability than on large energy savings.

The desalination process is limited by energy use. In spite of recent progress to lower the energy consumption, it remains considerably high when compared with traditional water sources. Disposal of brine remains a serious concern. In semi-enclosed seas or bays, salt levels can rise and harm marine life. Membranes also foul and scale, resulting in regular cleaning and increased costs. Building a plant requires a large upfront investment. This makes the technology unattainable for economically restricted regions. The environmental impact of such interventions leads to community resistance. These environmental issues are exacerbated when fossil fuel is used as the energy source (Shahid et al., 2023).

10.4.2 Fog Harvesting

Fog harvesting taps atmospheric moisture where clouds and wind meet elevated sites. It is mostly valuable in coastal hills or mountains where conventional water is scarce. Communities deploy mesh or structured surfaces to intercept fog droplets and channel them into storage. The method is low-energy and low-cost, if the right conditions exist. Nikbakht and Heyhat (2025) demonstrated that folded metal mesh outperforms flat mesh by doubling collection efficiency. This and other innovations are reviving interest in fog as a viable water source in arid areas. A fog-harvesting unit is usually a vertical mesh stretched between support poles. When wind drives fog through the mesh, tiny droplets hit the fibres, merge, and grow heavy enough to drip down. Channels at the bottom guide the collected water into storage tanks or barrels. The success of a typical fog harvesting unit depends on the mesh design, surface coating, wind speed and local topography (Jiang et al., 2023). Research has shown that fog capture and can be improved by the redesign of the hydrophilic coating and mesh patterns (Huan et al., 2023; Yu et al., 2024).

Empirical studies show that meshes can be used to harvest and treat fog water in areas prone to pollution. This was examined by Ghosh et al. (2023) who developed a polymer mesh that is coated with photocatalytic TiO_2. Disinfection occurs during sunshine, as the TiO_2 is activated to remove the microbial contaminants from the water film. The authors conducted a field test using the photocatalytic mesh. The results showed that pathogen counts in the water film decreased by more than 90% when compared to uncoated mesh. This dual function reduces the need for separate disinfection steps, which will be particularly useful for off-grid rural communities. This indicates that advances in materials science through material design can be leveraged to improve the performance of fog nets.

Nikbakht and Heyhat (2025) compared traditional flat mesh with a folded stainless-steel mesh design in a pilot fog-harvesting field. The folded geometry increased the active surface area while improving drainage, nearly doubling water yield under the same fog conditions. The study reported better wind resistance and lower clogging from airborne dust. Although folding raised fabrication cost slightly, the added efficiency made it cost-efficient, especially for remote dryland sites. The study showed that simple structural redesigns can substantially boost fog-harvesting performance without additional energy inputs.

Also, Kang et al. (2021) explored the influence of mesh wettability on the yield of fog harvesting systems. The researchers developed a scalable, small-scale mesh inspired by the structure of beetles and cacti with patterned hydrophilic-hydrophobic surfaces that guide droplets more effectively towards storage channels. This increased the collection efficiency of the modified meshes when compared with plain meshes. It was reported that when surface micro-features aligned with airflow, droplet coalescence improved and drainage losses declined. The approach is still at a prototype scale but demonstrates that surface patterning can enhance capture even without changing mesh size or geometry. Such biomimetic concepts could lead to lighter, more compact fog collectors.

Fog harvesting systems are heavily reliant on weather and geography. The yield drops significantly when the weather is dry and windless. Sites need regular fog with sufficient wind to push droplets through the mesh. Nets can clog with dust, insects or airborne salt, which reduces capture and increases the need for frequent cleaning. There is also rapid material degradation in heavily sunny and windy conditions, thereby increasing the system's maintenance costs over time. It is difficult to estimate the annual yield of fog harvesting systems due to the unavailability of reliable climate data in certain regions. These constraints have limited the large-scale adoption of fog harvesting.

10.4.3 Atmospheric Water Generation (AWG)

Atmospheric water generation (AWG) draws water vapour directly from the air and converts it into liquid water. The concept is not new, but it is gaining attention due to the depletion of surface and groundwater. AWG technologies either cool air below the dew point to form droplets through vapour compression refrigeration units or use absorbent beds that take up vapour and release it when heated. The absorbent beds use hygroscopic materials such as metal organic frameworks (MoFs) and polymer sorbents or photothermal materials that enable water release under sunlight. In either case, AWG systems give better yield in warm, moist air, but the output drops sharply in dry season. Research efforts are focused on producing sorbents with higher uptake at low humidity and lower regeneration energy. Hybrid systems that combine due harvesting, sorption and passive radiative cooling are being proposed for off-grid water production (Shafeian et al., 2022; Cattani et al., 2024).

Empirical studies on AWG technologies include Shafeian et al. (2022) who designed a portable hybrid solar AWG system combining thermoelectric cooling with a desiccant bed. The unit was field-tested in semi-arid Iran. Peak production reached 2.1 L/m^2/day under optimal sun and humidity, with an energy efficiency of approximately 0.19 mL/Wh. Performance declined during low-humidity afternoons, highlighting the importance of diurnal cycling between sorption at night and solar-thermal regeneration by day. The portability of the prototype and low electrical demand suggest promise for small off-grid households, although long-term reliability under dust and temperature extremes has not yet been assessed.

In a related study, Hu et al. (2022) reported a core-shell MOF supraparticle (MOF-801@MIL-101) designed for water capture at low relative humidity. Laboratory tests at 10–20% relative humidity showed water uptake of about 0.25 L/kg/day, outperforming conventional silica gel under the same conditions. The composite maintained structural stability over repeated adsorption-desorption cycles. Similarly, Song et al. (2023) tested a passive MOF-303 water harvester in outdoor trials at Berkeley (humid temperate) and Death Valley (hot-arid). The device required no external power beyond sunlight for desorption. Daily yields ranged from 114 g H_2O/kg MOF under very dry desert air (9–12% relative humidity) up to 210 g/Kg MOF under more favourable humidity. The MOF maintained sorption capacity over repeated cycles without structural loss. These results provide real-world evidence that MOF-based passive systems can deliver usable water in extremely arid climates, though present yields remain modest for community-scale supply.

Also, Sadowski et al. (2023) analysed performance data from commercial AWG units deployed at multiple sites across the United States, comparing refrigeration-based, sorption-assisted, and hybrid configurations. Their assessment highlighted strong climate sensitivity defined by production decline in regions where mean daily proposed benchmark metrics for water yield per kilowatt-hour and per unit cost, providing guidance for planners evaluating AWG as part of decentralised water supply portfolios. The study also identified vulnerability to power interruptions, underscoring the value of integrating AWG with renewable or backup energy systems.

AWG output depends heavily on local weather. Production significantly drops at humidity levels below 30%. Another prevalent issue is the high energy demand of refrigeration models, with huge power costs. This also makes it difficult to integrate them into off-grid power sources. To avoid this cost, the sorbent types do not use compressors. However, they still require a huge amount of energy for heating during the regeneration process. Furthermore, the efficiency of the system could be affected by dust and salt in the air that clog the filters or coat cold surfaces. Poor maintenance of the storage tanks can lead to the growth of microbial activities, thereby affecting the integrity of the water output from the system. These limitations combined with the high initial capital for setting up large units, have negatively impacted the uptake of AWG in many regions where it is needed.

10.5 Case Studies on Advanced Water Management Technologies

The chapter has examined several technological innovations that can improve water sourcing, treatment and reuse. These technologies underscore the potency of science and engineering in providing solutions to global problems. To further demonstrate how such concepts perform outside the laboratory, this section presents real-world examples. Each case highlights key technical achievements and limitations in implementation. The case studies show that water management can be driven by localised technologies and persistent community engagement.

Case 1: Singapore NEWater and Tuas Nexus facility (Asztalos et al., 2009; Lee & Tan, 2016; Bai et al., 2020; Kog, 2020)

Singapore is a small island in Asia with limited capacity in terms of land for freshwater storage. The country also has a few natural rivers. The country had to rely on rainfall or imports to meet its water demand for several decades. The country witnessed a rapid urbanisation in the 1980s and 1990s that led to an upsurge in water demand that exceeded the limits of these water sources. Another problem was the discharge of large volumes of untreated water into the few available water bodies by industries. This further compounded the problem of access to a reliable water supply. Policy makers had to consider wastewater as a recoverable resource rather than waste. This decision opened the way for the NEWater programme launched in 2003.

The NEWater scheme takes secondary-treated municipal wastewater and runs it through three main steps. The first is micro-filtration, which removes suspended solids and most microbes. The next step is reverse osmosis, a pressure-driven membrane step that strips out dissolved salts, nutrients and many trace organics. The final step involves the use of ultraviolet light (UV) to inactivate any remaining pathogens. The largest facility, the Changi NEWater plant, came online in 2010 and produces more than 200,000 m^3 of high-grade water each day. Singapore is now building the Tuas Nexus, which combines a water-reclamation plant with a waste-to-energy unit and includes anaerobic sludge digestion to generate biogas for part of its power needs. NEWater has been repeatedly tested and found to meet the WHO drinking-water standards for safety (Lee & Tan, 2016). Reports from plant operators show that the combined membrane and UV steps remove over 99% of microbes and trace organics (Bai et al., 2020). Much of this water goes to industry, which frees up more reservoir supply for households and reduces the need for imported water (Lefebvre, 2018). The new Tuas Nexus is designed to run sludge digesters for biogas, and it is expected to generate about 40 MW of electricity, which is sufficient to cover its operation energy requirement and possibly send extra units of energy to the grid.

Lessons learnt: Singapore successfully implemented large-scale reuse technology due to its reliability and public trust. Public observation, regular briefings, and good pretreatment reduced hesitation. High upfront costs were managed through national planning and production scaling. Policy support and clear communication are crucial for public acceptance.

Case 2: Chile—Large-scale fog-harvesting for coastal villages (Fessehaye et al., 2014; Di Bitonto et al., 2020)

The Atacama coastal desert, located in northern Chile, is regarded as one of the driest inhabited regions in the world. Yet dense fog, locally known as 'camanchaca', drifts inland from the Pacific almost year-round. For decades, scattered fishing and farming settlements along the coastal hills depended on water delivered by truck or carried from distant wells. In the early 1990s, researchers and local non-governmental organisations (NGOs) began exploring ways to capture fog as a supplementary water source. These trials gradually evolved into community-scale fog-net projects that still provide water for domestic use and small-plot irrigation in several villages.

The early Chilean installations used simple polypropylene Raschel-mesh nets stretched between wooden poles on ridge tops where the fog is thickest. Wind drives the fog through the mesh, droplets form on the fibres, and gravity lets them run down into a gutter that feeds a storage tank at the foot of the slope. Over time, the projects upgraded to taller steel-frame collectors, often 30–40 m^2 each, fitted with improved mesh that sheds water more efficiently and resists UV damage. Sites such as Falcon Hill near Mejillones now operate dozens of these units linked to a covered tank and gravity-fed pipes supplying households and small gardens.

Monitoring of fog-collector fields at Falcon Hill and nearby Atacama sites has reported typical yields of 5–10 L m^{-2} day^{-1} during the main camanchaca season ((Morales Vergara et al., 2023). Peaks above these values occur when wind-driven fog is dense. Collected water is stored in covered tanks and, after basic filtration, has been shown to meet microbial safety levels for household use (Fessehaye et al., 2014). The supply now covers part of the drinking and kitchen demand and allows small-scale irrigation of home gardens. These systems have reduced the reliance on water delivered by truck and strengthened the resilience of several coastal villages.

Lessons learned: Experience from the Atacama projects shows that site selection is critical; ridge tops that face prevailing ocean winds gave the best yields. Consistent maintenance of mesh and gutters is also essential, as dust and salt reduce efficiency quickly. Community participation proved decisive. Villages that formed local committees to clean nets and manage the tanks achieved steadier water output. The projects also benefited from long-term monitoring partnerships with universities, which guided mesh upgrades and water-quality checks. These lessons underline that success depends on both technical design and strong local stewardship of the systems.

Case 3: Domestic wastewater reclamation in Abuja and Lagos, Nigeria (Olasehinde & Alabi, 2019)

Rapid urbanisation and population growth in Nigeria's major cities, such as Abuja and Lagos, have placed significant pressure on existing water infrastructure. This has led to widespread challenges in wastewater management and safe water reuse. In response, the Federal Ministry of Water Resources and several state water agencies, in collaboration with research institutions, conducted comprehensive assessments and pilot reclamation projects within selected urban treatment plants between 2018 and 2022. The initiative sought to evaluate the performance of existing conventional systems, primarily activated sludge, anaerobic digestion, and oxidation ponds. It also involved the introduction of membrane-based technologies as tertiary treatment units. Membrane filtration systems, including ultrafiltration and microfiltration, were retrofitted to enhance effluent quality for non-portable reuse applications such as irrigation and construction. Results indicated substantial reductions in suspended solids, nutrients and microbial contaminants, producing effluent that met WHO standards for restricted reuse. Beneficiaries, including nearby agricultural cooperatives and construction firms, gained access to reliable alternative water sources, reducing freshwater demand and treatment costs. On a broader scale, this intervention underscores the potential of decentralised, hybrid systems in Nigeria's urban water management framework. This promotes a gradual transition from linear wastewater disposal to circular resource recovery, which aligns with sustainable development goals.

Case 4: Submerged membrane bioreactor for cassava processing wastewater in South-West Nigeria (Lawal et al., 2020)

Cassava processing generates large volumes of wastewater characterised by high organic load, suspended solids, and cyanogenic compounds, posing severe ecological threats to receiving water bodies. In a pilot study, conducted in 2021 by Lawal et al., an experimental submerged MBR system was designed to treat cassava effluent at laboratory and semi-field scale. The MBR integrated biological degradation with physical membrane separation using locally fabricated ceramic composite membranes derived from kaolin and rice husk ash to lower costs. The technology effectively achieved over 95% removal of chemical oxygen demand (COD) and total suspended solids (TSS), with significant detoxification of cyanide content. The system offered a compact, odour-free solution for local cassava millers that required minimal operation oversight. The treated effluent was found suitable for safe discharge or reuse in crop irrigation, demonstrating environmental and economic benefits. Although the case study is a pilot scale but it supports the local adaptation of the MBR technology as suitable decentralised treatment model that can work with locally sourced membrane materials.

10.6 Future Directions of Technological Innovation in Water Management

In the past decades, there has been tremendous progress in the reuse of wastewater, seawater desalination, fog and air-water harvesting. These have helped to reduce water stress in many regions. However, climate change, population rise and emergent pollutants create new challenges for water reuse. Several existing systems remain energy-intensive and too costly for poorer regions. Research is turning to lighter membranes, better sorbents, AI-driven sensors and hybrid recovery plants that treat water while recovering energy and nutrients (Ghimire et al., 2021). The next phase in water management aims for lower energy use, lower cost and resilience to extreme weather.

Emerging technological trends: Current research efforts are focused on developing advanced membranes to cut the energy cost of desalination and reuse. Graphene oxide and aquaporin-based membranes have shown lower salt passage and high flux in pilot trials (Jakowiecki et al., 2018). Utilities are adding IoT-linked sensors to monitor pressure, leakage and water quality in real time, allowing earlier detection of faults (Obunga et al., 2025). Large treatment facilities are shifting toward water-resource-recovery that also captures biogas and phosphorus (Qadir et al., 2020). For smaller settlements, nature-based wetlands and hybrid green-grey units are being tested to provide decentralised treatment with lower power demand (Sánchez et al., 2023). The trend across all these developments is to lower energy use and improve resilience to climate variability.

10.7 Key Challenges

Energy remains the main challenge of water management technologies. Many advanced processes draw significant power and the costs escalate when electricity prices spike. Brine and concentrate management is a particular problem for desalination plants, especially for inland plants with no easy discharge route. On the quality side, ultra-trace contaminants, such as PFAs, pharmaceuticals, and microplastics, demand monitoring and treatment that are not yet within established monitoring and treatment programmes. Operations depend on skilled staff and reliable sensors, which are typically scarce in smaller utilities. Another bottleneck is uneven financing, as water reuse projects compete with other public priorities with shorter payback periods. Critical to the success of the end-products is public acceptance. People support new water when it feels safe and affordable, and offer resistance when communication is inadequate. This challenge can be addressed through sustained sensitisation and awareness communications, as captured in Case 2. Furthermore, policies on wastewater reuse and alternative water sources can be enacted by the government or policymakers in water stressed regions and especially in developing countries such as Nigeria where awareness and technological advancement in water management is nascent. Finally, climate extremes stress assets, making resilience a design requirement.

10.8 Opportunities and Research Priorities

From the foregoing discussions and the abundance of literature consulted, three directions stand out for the future of innovative water management technologies. First, materials and process efficiency. There is a need to produce membranes and sorbents that foul less, last longer and run at lower pressure or temperature (Alluhaidan et al., 2024). Even small reductions in specific energy use can scale into major savings. Pairing treatment plans with renewable energy and heat recovery, such as solar PV for pumps, biogas from sludge, waste heat for thermal modules, can stabilise operating costs and cost emissions (Milani & Nabi Bidhendi, 2024). Second, digitally enabled operations, such as the use of simple, robust sensors tied to analytics for leak detection, early warning on fouling and dose control (Farah & Shahrour, 2024). The goal is to keep system surprises low and achieve faster recovery. Thirdly, there is a need to develop a circular design framework where plants that recover water, energy and nutrients are positioned together, with decentralised units that serve buildings or small clusters. On the governance side, clearer performance standards, outcome-based procurements, and community monitoring can speed up adoption. Priority research areas include brine minimisation, low-energy oxidation of organic traces, durable anti-scaling coatings and field-tested models for small utilities. The common goal is to achieve lower power, lower cost and higher reliability.

10.9 Conclusion

Technological innovations in water management are required to meet the rising demand for water due to population growth. These technologies include membrane bioreactors, advanced oxidation processes, nanofiltration and grey water systems. They are suitable for wastewater treatment and water reuse. To further boost the availability of water, especially in arid regions, alternative sources such as greywater systems, desalination, fog harvesting and atmospheric water generation. Typical practical examples of water reuse technologies include Singapore's NEWater and Tuas Nexus facilities. The country was able to meet a great portion of its water needs through these facilities. In another case study, Chile used a community-based approach to address the problem of water scarcity through fog-harvesting projects. Case studies from Nigeria showed that innovations in water management technologies are still at the pilot/laboratory scale. The high cost of these technologies should be a focus of further research. There is also a need to formulate policies that promote the use of reclaimed water in developing countries. Furthermore, the negative perception of people towards water reuse can be addressed by public education and awareness campaigns. This will ensure increased community participation in water management programmes. Researchers in developing countries like Nigeria can consider developing local alternatives to established water management technologies to foster water reuse. There is also a need for more investment in water management technologies and research from the relevant stakeholders and practitioners by developing pilot plants to demonstrate value, encourage research and provide the foundations for large scale plants in the future.

Glossary

AI Artificial Intelligence
AOPs Advanced Oxidation Processes
AWG Atmospheric Water Generation
BOD Biological Oxygen Demand
COD Chemical Oxygen Demand
DNN Deep Neural Network

DPR Direct Potable Reuse
DT Decision Tree
EBPR Enhanced Biological Phosphorus Removal
GA-BP Genetic Algorithm optimized Back Propagation Neural Network
GPR Gaussian Process Regression
IoT Internet of things
IPR Indirect Potable Reuse
LDS Leak Detection System
LSTM Long Short-Term Memory networks
MBRs Membrane Bioreactors
MED Multi-Effect Distillation
NF Nanofiltration
RF Random Forest
RO Reverse Osmosis
SVR Support Vector Regression
SWRO Seawater Reverse Osmosis
TOC Total Organic Carbon
TSS Total Suspended Solids

References

Abdel-Fatah, M. A. (2018). Nanofiltration systems and applications in wastewater treatment: Review article. *Ain Shams Engineering Journal, 9*(4), 3077–3092. https://doi.org/10.1016/j.asej.2018.08.001

Adams, M. N., & Jokonya, O. (2021). An investigation of smart water meter adoption factors at universities. *Procedia Computer Science, 196*, 324–331. https://doi.org/10.1016/j.procs.2021.12.020

Al-Asheh, S., Bagheri, M., & Aidan, A. (2021). Membrane bioreactor for wastewater treatment: A review. *Case Studies in Chemical and Environmental Engineering, 4*. https://doi.org/10.1016/j.cscee.2021.100109

Alluhaidan, A. S., Prabu, P., Basheer, S., Aziz, R., & Prakash, P. (2024). Integrated hybrid membrane system for enhanced water treatment and desalination for environmental preservation. *Desalination and Water Treatment, 320*(April), 100722. https://doi.org/10.1016/j.dwt.2024.100722

Alnajdi, S., Beni, A. N., Alsaati, A. A., Luhar, M., Childress, A. E., David, M., Alnajdi, S., Beni, A. N., Alsaati, A. A., Luhar, M., & Childress, A. E. (2024). Article practical minimum energy use of seawater reverse osmosis practical minimum energy use of seawater reverse osmosis. *Joule, 8*(11), 3088–3105. https://doi.org/10.1016/j.joule.2024.08.005

Asad, M., Kashif, M., Sheikh, U. A., Asif, M. U., George, S., & Khan, G. u. H. (2022). Synergetic effect of safety culture and safety climate on safety performance in SMEs: Does transformation leadership have a moderating role? *International Journal of Occupational Safety and Ergonomics, 28*(3), 1858–1864. https://doi.org/10.1080/10803548.2021.1942657

Asztalos, M., Wijndaele, K., De Bourdeaudhuij, I., Philippaerts, R., Matton, L., Duvigneaud, N., Thomis, M., Duquet, W., Lefevre, J., & Cardon, G. (2009). Specific associations between types of physical activity and components of mental health. *Journal of Science and Medicine in Sport, 12*(4), 468–474. https://doi.org/10.1016/j.jsams.2008.06.009

Awasthi, A., Gandhi, K., & Rayalu, S. (2024). Greywater treatment technologies: A comprehensive review. *International Journal of Environmental Science and Technology, 21*(1). https://doi.org/10.1007/s13762-023-04940-7

Bai, Y., Shan, F., Zhu, Y. Y., Xu, J. Y., Wu, Y. S., Luo, X. G., Wu, Y. H., Hu, H. Y., & Zhang, B. L. (2020). Long-term performance and economic evaluation of full-scale MF and RO process—A case study of the changi NEWater project phase 2 in Singapore. *Water Cycle, 1*(July), 128–135. https://doi.org/10.1016/j.watcyc.2020.09.001

Bastos, P. D. A., Santos, M. A., Carvalho, P. J., Velizarov, S., & Crespo, J. G. (2021). Pilot scale reverse osmosis refinery wastewater treatment-a techno-economical and sustainability assessment. *Environmental Science: Water Research & Technology, 7*(3), 549–561. https://doi.org/10.1039/d0ew00936a

Behari, J., Tiwari, D. K., & Sen, P. (2008). Application of nanoparticles in waste water treatment. *World Applied Sciences Journal, 3*(3), 417–433.

Cattani, L., Cattani, P., Figoni, R., & Magrini, A. (2024). Performance assessment of atmospheric water generators: A review of evaluation tools and proposal for a novel advanced global evaluation index for HVAC–AWG hybrid solutions. *Applied Sciences (Switzerland), 14*(24), 1–28. https://doi.org/10.3390/app142411793

Cengiz, A. I., Guven, H., Ozgun, H., & Ersahin, M. E. (2024). Enhanced energy recovery in municipal wastewater treatment plants through co-digestion by anaerobic membrane bioreactors: Current status and future perspectives. *Reviews in Environmental Science and Biotechnology, 23*(2), 385–410. https://doi.org/10.1007/s11157-024-09691-6

Chen, H., Cheng, Y., Wang, Y., Ding, Y., Wang, C., Feng, X., Fan, Q., Yuan, F., Fu, G., Gao, B., Liu, K., & Zou, X. (2024). Microplastics: A potential proxy for tracing extreme flood events in estuarine environments. *Science of the Total Environment, 918*, 170554. ISSN 0048-9697, https://doi.org/10.1016/j.scitotenv.2024.170554

Deng, Y., & Zhao, R. (2015). Advanced oxidation processes (AOPs) in wastewater treatment. *Current Pollution Reports, 1*(3), 167–176. https://doi.org/10.1007/s40726-015-0015-z

Di Bitonto, M. G., Angelotti, A., & Zanelli, A. (2020). Fog and dew harvesting: Italy and Chile in comparison. *International Journal of Innovative Research and Development, 9*(3). https://doi.org/10.24940/ijird/2020/v9/i3/mar20056

Diamantis, V. (2021). Performance of a micro-scale membrane reactor for greywater treatment at household level. *Membranes, 11*(1), 1–11. https://doi.org/10.3390/membranes11010063

Diawara, C. K. (2008). Nanofiltration process efficiency in water desalination. *Separation and Purification Reviews, 37*(3), 302–324. https://doi.org/10.1080/15422110802228770

Dui, H., Cao, T., & Wang, F. (2025). Digital twin-based resilience evaluation and intelligent strategies of smart urban water distribution networks for emergency management. *Resilient Cities and Structures, 4*(1), 41–52. https://doi.org/10.1016/j.rcns.2025.02.001

Essamlali, I., Nhaila, H., & El Khaili, M. (2024). Advances in machine learning and IoT for water quality monitoring: A comprehensive review. *Heliyon, 10*(6). https://doi.org/10.1016/j.heliyon.2024.e27920

European Investment Bank. (2022). Wastewater as a resource. *Water, 21*. https://doi.org/10.4324/noe0419229100.ch4

Ewim, D. R. E., Orikpete, O. F., Scott, T. O., Onyebuchi, C. N., Onukogu, A. O., Uzougbo, C. G., & Onunka, C. (2023). Survey of wastewater issues due to oil spills and pollution in The Niger Delta area of Nigeria: A secondary data analysis. *Bulletin of the National Research Centre, 47*(1). https://doi.org/10.1186/s42269-023-01090-1

Fang, Q., Wang, H., Xie, C., & Chen, J. (2023). Research on leak location method of water supply pipeline based on MVMD. *CMES - computer modeling in engineering and sciences, 134*(2), 1237–1250. https://doi.org/10.32604/cmes.2022.021131

Farah, E., & Shahrour, I. (2024). Water leak detection: A comprehensive review of methods, challenges, and future directions. *Water (Basel) (Switzerland), 16*(20). https://doi.org/10.3390/w16202975

Fessehaye, M., Abdul-Wahab, S. A., Savage, M. J., Kohler, T., Gherezghiher, T., & Hurni, H. (2014). Fog-water collection for community use. *Renewable and Sustainable Energy Reviews, 29*, 52–62. https://doi.org/10.1016/j.rser.2013.08.063

Fuentes, H., & Mauricio, D. (2020). Smart water consumption measurement system for houses using IoT and cloud computing. *Environmental Monitoring and Assessment, 192*(9). https://doi.org/10.1007/s10661-020-08535-4

Gao, T., Jin, Y., & Xiao, K. (2025). Techno–Economic and Environmental Impact Assessment of Membrane Bioreactors for Wastewater Treatment: A Review. *Engineering (London)*. https://doi.org/10.1016/j.eng.2025.05.001

Garcia-belinchón, C., Rieck, T., Bouchy, L., Galí, A., Rougé, P., & Fàbregas, C. (2013). Struvite recovery: pilot-scale results and economic assessment of different scenarios. *Water Practice and Technology, 8*(1), 119–130. https://doi.org/10.2166/wpt.2013.013

Ghimire, U., Sarpong, G., & Gude, V. G. (2021). Transitioning wastewater treatment plants toward circular economy and energy sustainability. *ACS Omega, 6*(18), 11794–11803. https://doi.org/10.1021/acsomega.0c05827

Ghosh, R., Baut, A., Belleri, G., Kappl, M., Butt, H. J., & Schutzius, T. M. (2023). Photocatalytically reactive surfaces for simultaneous water harvesting and treatment. *Nature Sustainability, 6*(12), 1663–1672. https://doi.org/10.1038/s41893-023-01159-9

Gopalakrishnan, G., Jeyakumar, R. B., & Somanathan, A. (2023). Challenges and emerging trends in advanced oxidation technologies and integration of advanced oxidation processes with biological processes for wastewater treatment. *Sustainability (Switzerland), 15*(issue 5). https://doi.org/10.3390/su15054235

Hoinkis, J., Deowan, S. A., Panten, V., Figoli, A., Huang, R. R., & Drioli, E. (2012). Membrane bioreactor (MBR) technology—A promising approach for industrial water reuse. *Procedia Engineering, 33*, 234–241. https://doi.org/10.1016/j.proeng.2012.01.1199

Hu, Y., Wang, Y., Fang, Z., Wan, X., Dong, M., Ye, Z., & Peng, X. (2022). MOF supraparticles for atmosphere water harvesting at low humidity. *Journal of Materials Chemistry A, 10*(28), 15116–15126. https://doi.org/10.1039/d2ta02026b

Huan, J., He, Y., Geng, X., Hou, Y., & Zheng, Y. (2023). Enhanced fog harvesting through programmable droplet movement via bidirectional wettable gradient and microchannel-connected pattern gradient. *Colloids and Surfaces A: Physicochemical and Engineering Aspects, 659*. https://doi.org/10.1016/j.colsurfa.2022.130780

Huang, X. (2024). The promotion of anaerobic digestion technology upgrades in waste stream treatment plants for circular economy in the context of "dual carbon": Global status, development trend, and future challenges. *Water (Basel) (Switzerland), 16*(24). https://doi.org/10.3390/w16243718

Hübner, U., Spahr, S., Lutze, H., Wieland, A., Rüting, S., Gernjak, W., & Wenk, J. (2024). Advanced oxidation processes for water and wastewater treatment—Guidance for systematic future research. *Heliyon, 10*(9). https://doi.org/10.1016/j.heliyon.2024.e30402

Idusuyi, N., Adebayo, M. A., Igwegbe, C. A., Aghogho, O. T., James, A., & Kazeem, R. A. (2025). A sustainable approach to dairy wastewater treatment through electrocoagulation: From beverage cans to clean water. *Waste Management Bulletin, 3*(1), 96–106. https://doi.org/10.1016/j.wmb.2024.12.012

Ishida, K. P., Luna, R. F., Richardot, W. H., Lopez-Galvez, N., Plumlee, M. H., Dodder, N. G., & Hoh, E. (2022). Nontargeted analysis of trace organic constituents in reverse osmosis and UV-AOP product waters of a potable reuse facility. *ACS Environmental Science & Technology Water, 2*(1), 96–105. https://doi.org/10.1021/acsestwater.1c00265

Islam, M. R., Azam, S., Shanmugam, B., & Mathur, D. (2024). Leak detection and localization in underground water supply system using thermal imaging and geophone signals through machine learning. *Intelligent Systems with Applications, 23*. https://doi.org/10.1016/j.iswa.2024.200404

Jabbar, W. A., Mei Ting, T., I. Hamidun, M. F., Che Kamarudin, A. H., Wu, W., Sultan, J., Alsewari, A. R. A., & Ali, M. A. H. (2024). Development of LoRaWAN-based IoT system for water quality monitoring in rural areas. *Expert Systems with Applications, 242*. https://doi.org/10.1016/j.eswa.2023.122862

Jakowiecki, J., Sztyler, A., Filipek, S., Li, P., Raman, K., Barathiraja, N., Ramakrishna, S., Eswara, J. R., Altaee, A., Sharif, A. O., Ajayan, P. M., & Renugopalakrishnan, V. (2018). Aquaporin–graphene interface: Relevance to point-of-care device for renal cell carcinoma and desalination. *Interface Focus, 8*(3). https://doi.org/10.1098/rsfs.2017.0066

Jamaaluddin, A. A., & Khoiri. (2023). Ultrasonic flow meters and microcontrollers for precise water management with 6.45% error margin. *IOP Conference Series: Earth and Environmental Science, 1242*(1). https://doi.org/10.1088/1755-1315/1242/1/012017

Jenila, J. S., Issac, P.K., Lam, S. S., Christina Oviya, J., Jones, S., Munusamy-Ramanujam, G., Chang, S. W., Ravindran, B., Mannacharaju, M., Ghotekar, S., Khoo, K. S. (2023). Deleterious effect of gestagens from wastewater effluent on fish reproduction in aquatic environment: A review. *Environmental Research, 236*(Part 2), 116810. ISSN 0013-9351. https://doi.org/10.1016/j.envres.2023.116810.

Jiang, Y., Machado, C., & Park, K. C. K. (2023). From capture to transport: A review of engineered surfaces for fog collection. *Droplet, 2*(2). https://doi.org/10.1002/dro2.55

Jijingi, H. E., Yazdi, S. K., Abakar, Y. A., & Etim, E. (2024). Evaluation of membrane bioreactor (MBR) technology for industrial wastewater treatment and its application in developing countries: A review. *Case Studies in Chemical and Environmental Engineering, 10*. https://doi.org/10.1016/j.cscee.2024.100886

Kang, J. H., Lee, J. W., Kim, J. Y., Moon, J. W., Jang, H. S., & Jung, S. Y. (2021). Effect of mesh wettability modification on atmospheric and industrial fog harvesting. *Frontiers in Physics, 9*(May), 1–11. https://doi.org/10.3389/fphy.2021.680641

Kanyama, M. N., Bhunu Shava, F., Gamundani, A. M., & Hartmann, A. (2025). AI-driven anomaly detection in smart water metering systems using ensemble learning. *Water, 17*, 1933. https://doi.org/10.3390/w17131933

Kataki, S., West, H., Clarke, M., & Baruah, D. C. (2016). Phosphorus recovery as struvite from farm, municipal and industrial waste: Feedstock suitability, methods and pre-treatments. *Waste Management, 49*, 437–454. https://doi.org/10.1016/j.wasman.2016.01.003

Kim, D., Choi, S., Kang, S., & Noh, H. (2023). A study on developing an AI-based water demand prediction and classification model for Gurye Intake Station. *Water, 15*, 4160. https://doi.org/10.3390/w15234160

Kim, E. J., Kim, H., & Lee, E. (2021). Influence of ammonia stripping parameters on the efficiency and mass transfer rate of ammonia removal. *Applied Sciences (Switzerland), 11*(1), 1–13. https://doi.org/10.3390/app11010441

Kog, Y. C. (2020). Water reclamation and reuse in Singapore. *Journal of Environmental Engineering, 146*(4), 1–8. https://doi.org/10.1061/(asce)ee.1943-7870.0001675

Korlapati, N. V. S., Khan, F., Noor, Q., Mirza, S., & Vaddiraju, S. (2022). Review and analysis of pipeline leak detection methods. *Journal of Pipeline Science and Engineering, 2*(4). https://doi.org/10.1016/j.jpse.2022.100074

Kumari, P., & Kumar, A. (2023). Advanced oxidation process: A remediation technique for organic and non-biodegradable pollutant. *Results in Surfaces and Interfaces, 11*. https://doi.org/10.1016/j.rsurfi.2023.100122

Kwon, M., Royce, A., Gong, Y., Ishida, K. P., & Stefan, M. I. (2020). UV/chlorine: Vs. UV/$H_2O_2$for water reuse at Orange County Water District, CA: A pilot study. *Environmental science: Water research and technology, 6*(9). https://doi.org/10.1039/d0ew00316f

Lawal, N. S., Ogedengbe, K., Ojo, O. O. S., & Odufowokan, A. A. (2020). Assessment of a submerged membrane bioreactor with composite ceramic filters for cassava wastewater treatment. *Research in Agricultural Engineering, 66*(2), 72–79. https://doi.org/10.17221/109/2018-RAE

Lee, H., & Tan, T. P. (2016). Singapore's experience with reclaimed water: NEWater. *International Journal of Water Resources Development, 32*(4), 611–621. https://doi.org/10.1080/07900627.2015.1120188

Lefebvre, O. (2018). Beyond NEWater: An insight into Singapore's water reuse prospects. *Current Opinion in Environmental Science & Health, 2*, 26–31. https://doi.org/10.1016/j.coesh.2017.12.001

Lima, D., Li, L., & Appleby, G. (2024). A review of renewable energy Technologies in Municipal Wastewater Treatment Plants (WWTPs). *Energies, 17*(23), 1–52. https://doi.org/10.3390/en17236084

Lin, L., Yang, H., & Xu, X. (2022). Effects of water pollution on human health and disease heterogeneity: A review. *Frontiers in Environmental Science, 10*. https://doi.org/10.3389/fenvs.2022.880246

Lin, R., Qiu, R., Hu, L., Ding, Y., & Yuan, Z. (2025). A low-cost soft sensor for sewer flow monitoring—Learning from water level measurements in manholes. *Water Research, 274*. https://doi.org/10.1016/j.watres.2025.123135

Liu, D., & Cheng, X. (2025). FNN-BiLSTM-attention-DA: A hybrid fuzzy neural network and BiLSTM with multi-sensor information fusion for water quality monitoring and warning. *Alexandria Engineering Journal, 124*, 624–639. https://doi.org/10.1016/j.aej.2025.04.011

Malamis, S., Katsou, E., Di Fabio, S., Frison, N., Cecchi, F., & Fatone, F. (2015). Treatment of petrochemical wastewater by employing membrane bioreactors: A case study of effluents discharged to a sensitive water recipient. *Desalination and Water Treatment, 53*(12), 3397–3406. https://doi.org/10.1080/19443994.2014.934112

Maldonado, I., Miranda-Mamani, J., & Paredes-Espinal, C. (2023). Heavy metals and ecological alterations resulting from wastewater discharge in inner Puno Bay, Lake Titicaca. *Environmental Nanotechnology, Monitoring & Management, 20*, 100903. https://doi.org/10.1016/j.enmm.2023.100903

Mayor, Á., Reig, M., Vecino, X., Cortina, J. L., & Valderrama, C. (2023). Advanced hybrid system for ammonium valorization as liquid fertilizer from treated urban wastewaters: Validation of natural zeolites pretreatment and liquid-liquid membrane contactors at pilot plant scale. *Membranes, 13*(6). https://doi.org/10.3390/membranes13060580

Micolucci, F., Gottardo, M., Pavan, P., Cavinato, C., & Bolzonella, D. (2018). Pilot scale comparison of single and double-stage thermophilic anaerobic digestion of food waste. *Journal of Cleaner Production, 171*, 1376–1385. https://doi.org/10.1016/j.jclepro.2017.10.080

Milani, S. J., & Nabi Bidhendi, G. (2024). Biogas and photovoltaic solar energy as renewable energy in wastewater treatment plants: A focus on energy recovery and greenhouse gas emission mitigation. *Water Science and Engineering, 17*(3), 283–291. https://doi.org/10.1016/j.wse.2023.11.003

Miserli, K., & Konstantinou, I. (2025). Determination of microplastics in municipal wastewater treatment plant effluents and sludge using micro-Raman spectroscopy. *Water Emerging Contaminants and Nanoplastics, 4*(1). https://doi.org/10.20517/wecn.2024.72

Mohammed, M. A., Lakhan, A., Abdulkareem, K. H., Abd Ghani, M. K., Marhoon, H. A., Kadry, S., Nedoma, J., Martinek, R., & Zapirain, B. G. (2024). Industrial internet of water things architecture for data standarization based on blockchain and digital twin technology☆. *Journal of Advanced Research, 66*, 1–14. https://doi.org/10.1016/j.jare.2023.10.005

Morales Vergara, L. P., Lima, R. C., Bonnail, E., & Allende, C. G. (2023). Fog Water Traps As a Low-Cost Alternative Source of Water in Coastal Desert Areas of the Pacific. *Revista Geografica De Chile Terra Australis, 59*, 38–45. https://doi.org/10.23854/07199562.2023591.morales

Motta Cabrera, S., Winnubst, L., Richter, H., Voigt, I., & Nijmeijer, A. (2021). Industrial application of ceramic nanofiltration membranes for water treatment in oil sands mines. *Separation and Purification Technology, 256*. https://doi.org/10.1016/j.seppur.2020.117821

Mukherjee, J., Lodh, B. K., Sharma, R., Mahata, N., Shah, M. P., Mandal, S., Ghanta, S., & Bhunia, B. (2023). Advanced oxidation process for the treatment of industrial wastewater: A review on strategies, mechanisms, bottlenecks and prospects. *Chemosphere, 345*. https://doi.org/10.1016/j.chemosphere.2023.140473.

Narendra, K., Manjushree, M., Adur, A. J., Raajasubramanian, D., Srinivasan, S., & Murali, R. (2025). Self-healing water systems: The future of smart, adaptive, and regenerative water networks. *Cleaner water, 4*. https://doi.org/10.1016/j.clwat.2025.100112.

Ngo, H. H., Nguyen, M. C., Sangvikar, N. G., Hoang, T. T. L., & Guo, W. S. (2006, December 1). Simple approaches towards the design of an attached-growth sponge bioreactor (AGSB) for wastewater treatment and reuse. *Water Science and Technology, 54*(11–12), 191–197. https://doi.org/10.2166/wst.2006.727

Nikbakht, S., & Heyhat, M. M. (2025). Experimental study of folded metal mesh for efficient fog harvesting. *Scientific Reports, 15*(1), 1–13. https://doi.org/10.1038/s41598-025-04345-1

Nwokediegwu, Z. Q. S., Daraojimba, O. H., Oliha, J. S., Obaigbena, A., Dada, M. A., & Majemite, M. T. (2024). Review of emerging contaminants in water: USA and African perspectives. *International Journal of Science and Research Archive, 11*(1), 350–360. https://doi.org/10.30574/ijsra.2024.11.1.0073

Obunga, P. O., Rwanga, S. S., Dinka, M. O., & Otieno, B. O. (2025). Review of the emerging technologies in the water sector with a focus on the deployment of internet of things solutions. *NPJ Clean Water, 8*(1). https://doi.org/10.1038/s41545-025-00487-x

Ogunrinde, V., Fruchman, M., & Mark, B. (2025). *AI in water resource management: Predicting water demand, supply, and contamination risks with AI algorithms*. https://www.researchgate.net/publication/395017739.

Olasehinde, S. O., & Alabi, T. M. (2019). Domestic wastewater reclamation and reuse in Nigeria: A case study of some selected treatment plants in Abuja and Lagos. *I-Manager's Journal on Future Engineering and Technology, 15*(1), 1–10.

Ortiz-Sánchez, M. F., & Cuevas-Rodriguez, G. (2023). Energy recovery from wastewater in Mexico: A systematic review. *Frontiers in Environmental Science, 11*(February). https://doi.org/10.3389/fenvs.2023.1116053

Palermo, S. A., Maiolo, M., Brusco, A. C., Turco, M., Pirouz, B., Greco, E., Spezzano, G., & Piro, P. (2022). Smart Technologies for Water Resource Management: An overview. *Sensors, Sensors (Basel), 22*(16). https://doi.org/10.3390/s22166225

Parei, A. N., Naeeni, S. T. O., & Akbari, Z. (2023). Application of hybrid vertical flow constructed wetland systems to treatment of greywater for their use to irrigation in rural areas. *Journal of Cleaner Production, 412*(April), 137368. https://doi.org/10.1016/j.jclepro.2023.137368

Patel, S. K., Biesheuvel, P. M., & Elimelech, M. (2021). *Energy consumption of brackish water desalination: Identifying the sweet spots for electrodialysis and reverse osmosis*. https://doi.org/10.1021/acsestengg.0c00192.

Phan, V. T. H., Nguyen, P. D., Bui, X. T., Tan-Thi, P., & Truong, N. T. (2023). Evaluation of DOC removal by advanced oxidation processes as pre-treatment of the biological process using low ozone doses. *Case Studies in Chemical and Environmental Engineering, 8*. https://doi.org/10.1016/j.cscee.2023.100447

Qadir, M., Drechsel, P., Jiménez Cisneros, B., Kim, Y., Pramanik, A., Mehta, P., & Olaniyan, O. (2020). Global and regional potential of wastewater as a water, nutrient and energy source. *Natural Resources Forum, 44*(1), 40–51. https://doi.org/10.1111/1477-8947.12187

Ramm, K., & Wojciechowska, M. (2025). Analysis of the impact of the advanced oxidation processes on reclaimed water microbiological safety. *Desalination and Water Treatment, 324*. https://doi.org/10.1016/j.dwt.2025.101431

Ranade, V. & Bhandari, V. (2014). *Industrial wastewater treatment, recycling, and reuse—Past, present and future*. https://doi.org/10.1016/B978-0-08-099968-5.00014-3

Ren, X., Lunt, D. J., Hendy, E., von der Heydt, A., Abe-Ouchi, A., Otto-Bliesner, B. L., Williams, C. J. R., Stepanek, C., Guo, C., Chandan, D., Lohmann, G., Tindall, J. C., Sohl, L. E., Chandler, M. A., Kageyama, M., Baatsen, M. L. J., Tan, N., Zhang, Q., Feng, R., Chan, W.-L., Peltier, W. R., Li, X., Kamae, Y., Zhang, Z., & Haywood, A. M. (2023). The hydrological cycle and ocean circulation of the maritime continent in the mid-Pliocene: Results from PlioMIP2. *Climate of the Past, 19*(10), 20253–22077. https://doi.org/10.5194/cp-19-2053-2023

Rocha, M. E., Lazarino, T. C., Oliveira, G., Teixeira, L., Marques, M., & Mangiavacchi, N. (2024). Analysis of biogas production from sewage sludge combining BMP experimental assays and the ADM1 model. *PeerJ, 12*, 1–31. https://doi.org/10.7717/peerj.16720

Rodríguez, C., Carrasco, F., Sánchez, R., Rebolledo, N., Schneider, N., Serrano, J., & Leiva, E. (2022). Performance and treatment assessment of a pilot-scale decentralized greywater reuse system in rural schools of north-Central Chile. *Ecological Engineering, 174*. https://doi.org/10.1016/j.ecoleng.2021.106460

Rodríguez, C., Sánchez, R., Rebolledo, N., Schneider, N., Serrano, J., & Leiva, E. (2020). Cost–benefit evaluation of decentralized greywater reuse systems in rural public schools in Chile. *Water (Basel), 12*(12), 1–14. https://doi.org/10.3390/w12123468

Ruan, G., & Wang, M. (2021). Progress and perspectives of desalination in China. *Membranes, 11*.

Sadowski, E., Mbonimpa, E., & Chini, C. M. (2023). Benchmarks of production for atmospheric water generators in the United States. *PLOS Water, 2*(6), 1–15. https://doi.org/10.1371/journal.pwat.0000133

Sakurai, K., Okayasu, Y., & Abe, C. (2025). Energy recovery from organic matter in municipal wastewater using a two-stage system with high-rate contact stabilization and activated sludge processes under seasonal water temperature variations. *Environmental Science: Water Research & Technology, 11*(4), 1016–1025. https://doi.org/10.1039/d4ew00820k

Sánchez, M., Fernández, M. I., Ruiz, I., Canle, M., & Soto, M. (2023). Combining constructed wetlands and UV photolysis for the advanced removal of organic matter, nitrogen, and emerging pollutants from wastewater. *Environments, 10*(3). https://doi.org/10.3390/environments10030035

Sardar, A. A., Amir, M. A., Rehman, E. A., Malik, M. A., & Khalid, K. F. (2025). Integration of digital twin Technology for Water Resource Management of smart cities and communities: A narrative review. *International Journal of Advanced Natural Sciences and Engineering Researches, 2*(2).

Shafeian, N., Ranjbar, A. A., & Gorji, T. B. (2022). Design and implementation of a new portable hybrid solar atmospheric water-generation system. *Clean Energy, 6*(6), 879–892. https://doi.org/10.1093/ce/zkac061

Shahid, M. K., Mainali, B., Rout, P. R., Lim, J. W., Aslam, M., Al-Rawajfeh, A. E., & Choi, Y. (2023). A review of membrane-based desalination systems powered by renewable energy sources. *Water (Switzerland), 15*(3). https://doi.org/10.3390/w15030534

Shu, J., Xia, X., Han, S., He, Z., Pan, K., Liu, B. (2024, May 22). Long-term water demand forecasting using artificial intelligence models in the Tuojiang River basin, China. *PLoS One, 19*(5), e0302558. PMID: 38776352; PMCID: PMC11111086. https://doi.org/10.1371/journal.pone.0302558

Simoes, F., Colston, R., Rosa-Fernandes, C., Vale, P., Stephenson, T., & Soares, A. (2020). Predicting the potential of sludge dewatering liquors to recover nutrients as struvite biominerals. *Environmental Science and Ecotechnology, 3*, 100052. https://doi.org/10.1016/j.ese.2020.100052

Skalska-Tuomi, K., Kaijanen, L., Monteagudo, J. M., & Mänttäri, M. (2025). Efficient removal of pharmaceuticals from wastewater: Comparative study of three advanced oxidation processes. *Journal of Environmental Management, 375*. https://doi.org/10.1016/j.jenvman.2025.124276

Sønderlund, A. L., Smith, J. R., Hutton, C., & Kapelan, Z. (2014). Using smart meters for household water consumption feedback: Knowns and unknowns. *Procedia Engineering, 89*, 990–997. https://doi.org/10.1016/j.proeng.2014.11.216

Song, W., Zheng, Z., Alawadhi, A. H., & Yaghi, O. M. (2023). MOF water harvester produces water from death valley desert air in ambient sunlight. *Nature Water, 1*(7), 626–634. https://doi.org/10.1038/s44221-023-00103-7

Su, M., Zhang, Y., Liu, S., Wang, Y., & Li, T. (2025). Challenges and solutions for nanofiltration membranes in water treatment. *Frontiers in Chemical Engineering, 7*. https://doi.org/10.3389/fceng.2025.1695014

Syed, T. A., Muhammad, M. A., AlShahrani, A. A., Hammad, M., & Naqash, M. T. (2024). Smart water management with digital twins and multimodal transformers: A predictive approach to usage and leakage detection. *Water (Basel), 16*, 3410. https://doi.org/10.3390/w16233410

Tanriverdi, I., & Batmaz, İ. (2025). AI-driven U.S. drought prediction using machine learning and deep learning. *Climate Dynamics, 63*, 249. https://doi.org/10.1007/s00382-025-07720-w

Tansel, B., Sager, J., Garland, J., Xu, S., Levine, L., & Bisbee, P. (2006). Deposition of extracellular polymeric substances (EPS) and microtopographical changes on membrane surfaces during intermittent filtration conditions. *Journal of Membrane Science, 285*(1–2), 225–231. https://doi.org/10.1016/j.memsci.2006.08.031

Tavares, M., Pérez-Sánchez, M., Carravetta, A., Coronado-Hernández, O. E., López-Jiménez, P. A., & Ramos, H. M. (2024). Smart feasibility optimization of hybrid renewable water supply systems by digital twin technologies: A multicriteria approach applied to isolated cities. *Sustainable Cities and Society, 115*. https://doi.org/10.1016/j.scs.2024.105834

Tayeh, Y. A. (2024). A comprehensive review of reverse osmosis desalination: Technology, water sources, membrane processes, fouling, and cleaning. *Desalination and Water Treatment, 320*(August), 100882. https://doi.org/10.1016/j.dwt.2024.100882

Tomaszewska, B., Bodzek, M., & Rajca, M. (2017). Research on improving the composition of mineral water using nanofiltration. *Desalination and Water Treatment, 64*, 287–291. https://doi.org/10.5004/dwt.2017.11398

Tow, E. W., Hartman, A. L., Jaworowski, A., Zucker, I., Kum, S., AzadiAghdam, M., Blatchley, E. R., Achilli, A., Gu, H., Urper, G. M., & Warsinger, D. M. (2021). Modeling the energy consumption of potable water reuse schemes. *Water Research X, 13*, 100126. https://doi.org/10.1016/j.wroa.2021.100126

Un, C. (2025). Assessing biogas from wastewater treatment plants for sustainable transportation fuel: A detailed analysis of energy potential and emission reductions. *Gases, 5*(1). https://doi.org/10.3390/gases5010006

UNESCO (2015) *The United Nations world water development report 2015: Water for a sustainable world*. UNESCO, Paris. https://unesdoc.unesco.org/ark:/48223/pf0000232272.

Wang, C., Deng, S. H., You, N., Bai, Y., Jin, P., & Han, J. (2023). Pathways of wastewater treatment for resource recovery and energy minimization towards carbon neutrality and circular economy: Technological opinions. *Frontiers in Environmental Chemistry, 4*, 1–7. https://doi.org/10.3389/fenvc.2023.1255092

Wang, Z., Guan, S., Wang, Y., Li, W., Shi, K., Li, J., & Xu, Z. (2022). High purity struvite recovery from hydrothermally-treated sludge supernatant using magnetic zirconia adsorbent. *International Journal of Environmental Research and Public Health, 19*(20). https://doi.org/10.3390/ijerph192013156

Wear, S. L., Acuña, V., McDonald, R., & Font, C. (2021). Sewage pollution, declining ecosystem health, and cross-sector collaboration. *Biological Conservation, 255*. https://doi.org/10.1016/j.biocon.2021.109010

Witek-Krowiak, A., Gorazda, K., Szopa, D., Trzaska, K., Moustakas, K., & Chojnacka, K. (2022). Phosphorus recovery from wastewater and bio-based waste: An overview. *Bioengineered, 13*(5), 13474–13506. https://doi.org/10.1080/21655979.2022.2077894

WWAP (United Nations World Water Assessment Programme) 2016. *The United Nations World Water Development Report 2016: Water and Jobs*. UNESCO, Paris. https://www.unesco.org/en/wwap/wwdr/2016

Xing, S., Du, S., Huang, Y., Qi, X., & Sui, M. (2024). Advances in high-performance nanofiltration membranes facilitated by two-dimensional materials. *Water (Basel) (Switzerland), 16*(7). https://doi.org/10.3390/w16070988

Yu, B., Hu, H., Li, J., Ding, X., & Li, Z. (2024). Wetting-enabled microfluidic surface for fluid/droplet manipulation: Fabrication, strategies, and applications. *Advanced Engineering Materials*.

Zouhri, N., Addar, F. Z., Tahaikt, M., Elamrani, M., ELmidaoui, A., & Taky, M. (2024). Techno-economic study and optimization of the performance of nanofiltration and reverse osmosis membranes in reducing the salinity of M′rirt water city (Morocco). *Desalination and Water Treatment, 317*. https://doi.org/10.1016/j.dwt.2024.100042

Chapter 11
Microbial Technology in Water Purification

Joshua Ogboghena Akhigbe, Aisha Eniola Adegbite, Amina Yahaya, Sesan Abiodun Aransiola, and Naga Raju Maddela

Abstract The constant increase in global population, coupled with existing water pollution has put a strain on water resources globally. In some parts of the world, millions of people do not have access to clean and safe drinking water. Conventional water purification processes have their limitations and most of those processes cannot be employed for the removal of new and emerging water pollutants that have arisen in recent years from novel industrial processes and technologies. Most of those traditional processes are also not environmentally sustainable and contribute to environmental pollution in one way or another. This has led researchers to explore more efficient and environmentally sustainable methods of producing clean water. Various organic methods, both conventional and contemporary have been discussed and tested by scientists for applications in water purification as they pose little to no risk to the natural environment. Among these, the concept of utilizing microorganisms and their various metabolic capabilities for the breakdown of water pollutants is significant. Microorganisms through their metabolic activities can breakdown pollutants via bioaugmentation and biodegradation, or produce bioflocculants that bind suspended pollutant particles, making their removal easier. These organisms have also been applied in microbial fuel cells and microbial electro-deionization, systems that generate electricity while purifying water. These methods of using microbes in water treatment could be fairly easy to scale up and almost seamlessly integrated into existing water treatment processes. In comparison to conventional chemical treatments, they are more cost effective, environmentally sustainable and produce less secondary pollution. This technology is very promising and could have limitless applications with proper development.

Keywords Pollution · Microbes · Technology · Water · Purification

---

J. O. Akhigbe (✉) · A. E. Adegbite · A. Yahaya · S. A. Aransiola
Faculty of Science, Department of Microbiology, University of Abuja, Abuja, Nigeria
e-mail: akhigbe.joshua2020@uniabuja.edu.ng; xavierakhigbe@gmail.com

N. R. Maddela
Departamento de Ciencias Biológicas, Facultad la Ciencias de la Salud, Universidad Técnica de Manabí, Portoviejo, Manabí, Ecuador

M. I. Adeoba et al. (eds.), *Water Resources Management in Nigeria*, World Water Resources 10, https://doi.org/10.1007/978-3-032-22813-0_11

11.1 Introduction

The extent of environmental pollution is causing concern. Every aspect of the biosphere is impacted by environmental pollution. Particularly, the introduction of pollutants to water bodies is severely concerning, as water resources form the base of life and impact all aspects of anthropogenic activities (Somma et al., 2021). Global scarcity of water and water pollution hinder access to clean drinking water to a lot of people, although it is a basic human right. Only about 0.3% of clean freshwater is available for consumption to the 8.1 billion people in the world, and this statistic is especially striking, considering that approximately two-thirds of the earth's surface is covered by water (Stephens et al., 2020; Sadigov, 2022). The exhaustion of fresh water resources poses a threat to various human activities. 1.2 billion people have access to only basic water services and 2 billion people do not have access to safe water. The current demand for water is unable to be met by existing water treatment processes (Alemayehu et al., 2023). The increasing demand for clean water necessitates research and innovative approaches to the hurdles of modern water purification. This growing distress concerning increased water pollution and scarcity insists on the development of novel and viable water purification technologies with the ability to check for remove these contaminating agents (Barjasteh et al., 2021; Eshkoraev, 2024). Scientific and technological progress have facilitated and is still facilitating the advent of water purification processes that are efficient, centre environmental sustainability and are cost effective.

Microbial technology in water purification refers to the use of microorganisms to treat and enhance the quality of water, playing a pivotal role in modern water management systems (Qi and He, 2025). This clever approach to water purification has garnered much attention owing to its high pollutant removal efficiency, thus addressing the problem of water scarcity and contamination while promoting sustainability (Mukherjee et al., 2022). This technology leverages the natural abilities of microorganisms to break down pollutants through their metabolic processes. These processes have been applied to various treatment systems, including bioreactors and biofilters, which utilize microbial communities for effective biodegradation and pollutant removal (Yadu et al., 2021). Microbial technology can transform wastewater into safe, reusable water, contributing significantly to environmental health and community resilience by employing the natural metabolic activities of bacteria, algae, and fungi (Qi & He, 2025).

Advancements in microbiology in the late nineteenth century shed light on the role of microorganisms in various biological processes, including wastewater treatment led to the development of microbial technology (Varsha et al., 2025). Initial water purification methods dating back to ancient civilizations were reliant on physical processes such as sedimentation and filtration. Biofiltration is one of the oldest methods applied in water treatment. It encourages the growth of microorganisms on granular materials, leading to the biological breakdown of pollutants as polluted water goes through the filter (Sinha et al., 2022). The Egyptians used coarse sand for water filtration, and the Greeks and Romans developed aqueducts and settling basins to improve the quality of their water (Gobana et al., 2023). The understanding of microorganisms and their role in water purification, however, was limited until the advent of microbiology in the nineteenth century (Qi & He, 2025).

The discovery of microorganisms and their various abilities laid the foundation for microbial technology in water treatment. Scientists in the late nineteenth century demonstrated microorganisms' role in fermentation and disease, paving the way for understanding the potential of microbes in water treatment (Rani et al., 2019). This led researchers to investigate the ability of specific microorganisms to break down organic pollutants and remove contaminants from water, marking a significant change (Inbathamizh et al., 2023). By the mid-twentieth century significant advancements in the understanding of microbial processes had been made and the field of microbial technology began to flourish. Specialized microbial cultures tailored for specific contaminants have been developed and this enhances the efficiency of wastewater treatment processes (Kumar et al., 2021; Varsha et al., 2025). Bioremediation was introduced and this created an avenue for transforming harmful substances into less toxic forms, revolutionizing approaches to managing industrial wastewater (Shivalkar et al., 2020).

Employing microbial technology to water treatment does not only enhance pollutant removal, it also contributes to sustainability through natural processes like nutrient cycling and ecosystem regeneration (Gür et al., 2021). This integration of microbial technology into water purification systems also supports public health through the mitigation of waterborne diseases while fostering community engagement in sustainable water management practices. Impending advancements in predictive modeling and microbial engineering hold promise for the future of microbial technology with potential for improved water quality and an increased resilience against environmental stressors, making it a vital part of global water resource management strategies (Inbathamizh et al., 2023).

Despite its many advantages, the application of microbial technology to water treatment faces challenges. It may be difficult to maintain a stable and diverse microbial community and to navigate regulatory hurdles that can hinder the widespread adoption of the processes (Varsha et al., 2025). In addition, the initial investment costs for establishing large-scale microbial treatment facilities may be significant, posing barriers, particularly in resource-constrained regions.

In spite of these, current research is improving and innovating these microbial treatment processes, thus increasing their viability for application to current and emerging water quality issues (Qi & He, 2025). Recent decades have seen the combination of microbial technology and other water treatment methods like membrane filtration and chemical treatments which have led to an overall improvement of water management practices (Qi & He, 2025). Current research in the field continues to innovate microbial treatment processes as new technologies such as genetically modified organisms are developed for more efficient pollutant degradation (Chauhan et al., 2022).

11.2 Water and Water Pollution

11.2.1 Sources of Water

Clean and safe drinking water is an essential component of life. The first step in successful water purification is identifying the source (Adesina et al., 2024). The various sources of water for drinking and microbiological treatment each have their advantages, disadvantages, and contamination risks (which range from microbial pathogens to chemical pollutants) the sources include:

11.2.1.1 Surface Water

Surface water includes rivers, streams, lakes, and reservoirs, basically any body of water that can be seen flowing or sitting on the ground. These sources are often the first choice for drinking water as they are easy to access and can supply large amounts (Singh et al., 2024). However, these bodies of water are also exposed to all the anthropogenic and natural processes happening around them, including runoff from farms, sewage, industrial waste, and even natural debris. This makes surface water need more treatment before it is safe to drink in comparison to groundwater. Surface water includes;

Rivers and Streams These water bodies are among the most common sources of drinking water owing to their ease of accessibility especially in urban areas. However, they are easily polluted by various sources as they are openly exposed in the environment. Farm runoff, sewage, animal waste and industrial pollutants could easily get into these water bodies (Singh et al., 2024). The quality of water in these bodies are also impacted by seasonal changes. Heavy rain can bring with it more contaminants and dry periods could slow down the rate of flow (Singh et al., 2024). These water sources also contain various types of microorganisms, leading to an increased difficulty in their treatment. Most times, it is necessary for the water to undergo multiple treatment steps like settling, filtering, and disinfecting before it is considered safe to drink (Bai et al., 2022).

Lakes and Reservoirs Based on the research by Siddha & Sahu, (2022), lakes and man-made reservoirs are another popular option for drinking water. A major advantage of them is that they store large amounts of water, leading to stability in water supply throughout the year. However, as they hold water for longer periods, organic and inorganic matter like algae, leaves, or waste can collect and sink to the bottom. This may affect the smell, taste and overall safety of the water. During seasons where it is warmer than usual, algal blooms can even release toxins which can be detrimental to health (Feng et al., 2024). To deal with these issues, water from lakes and reservoirs often need more advanced treatment methods like special filters or processes to get rid of organic materials and harmful microbes (Siddha & Sahu, 2022).

Wetlands Wetlands are often referred to as nature's water filters because they slow down water flow and help trap dirt, break down waste, and also absorb excess nutrients (Balwan & Kour, 2021). Recently, more places are using constructed wetlands, which are basically, man-made versions of natural wetlands. These man-made wetlands are a cheap and eco-friendly way to clean water, especially in farms (Balwan & Kour, 2021). These systems use plants, soil, and microbes to get rid of things like nitrogen, phosphorus, and even some germs, without needing a lot of chemicals (Sharma & Naik, 2024). They may take up more space and need some upkeep, but they're a great option where traditional water treatment systems aren't practical.

11.2.1.2 Groundwater

Groundwater is water that sits below the earth's surface, usually stored in layers of rock or soil called aquifers and it makes up about 30% of all the freshwater on Earth (Islam, 2023), and because it's underground, it's often cleaner compared to surface water. In places where there's farming, factories, or a lot of human activity, it can get polluted by things like pesticides, heavy metals, or even sewage that seeps down into the ground (Yahaya et al., 2021). That's why testing and treating groundwater before drinking it is still important.

Aquifers Aquifers are underground layers of water-holding rock or gravel that store groundwater. These are great sources because the water tends to be protected from pollution on the surface (Islam, 2023). But again, in cities or industrial areas, things like fertilizers, oil, or waste can still manage to go in. Also, overusing aquifers can lead to them drying up or even sinking land above them. So, while aquifers are reliable, they need to be carefully managed and monitored as mentioned by Bai et al., (2022).

Springs and Seeps Springs and seeps are places where groundwater naturally comes up to the surface. They're often seen as clean and pure, especially in mountain areas. But just because water looks clear, it doesn't mean it's safe (Islam, 2023). These sources can still carry bacteria or pick up minerals from the ground, which might require treatment using filters or UV light before it's safe to drink.

11.2.1.3 Rainwater

Direct Rain Harvesting Catching rainwater from rooftops or specially built surfaces is a smart way to get clean water in places where other sources are limited (Judeh et al., 2022). Since rain doesn't touch the ground, it usually starts off pretty clean. But once it hits rooftops, it can pick up dirt, bacteria, or droppings from birds. That is why it should be filtered and disinfected before drinking (Fig. 11.1).

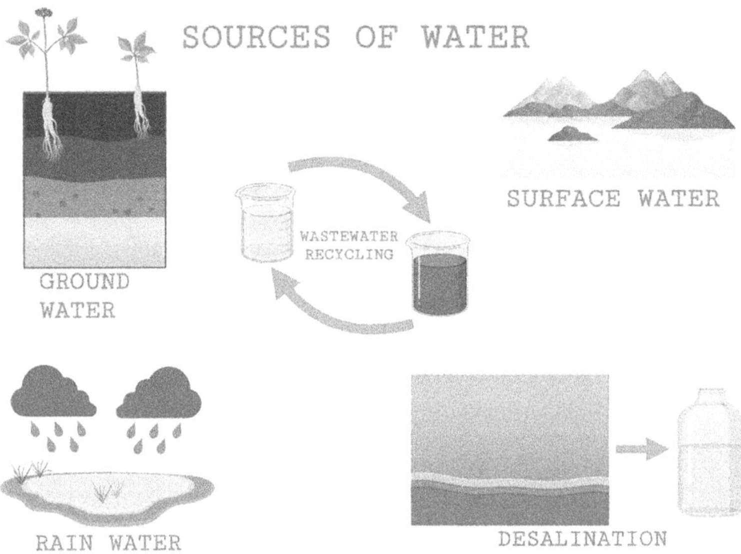

Fig. 11.1 Sources of water

Atmospheric Water Harvesting Based on the research by Vaghasiya & Pumera, (2024), this is a newer, high-tech approach where machines pull water straight out of the air, either from humidity, fog, or dew. These systems are still developing, but they could be a game-changer for dry areas where rain is rare. As the technology improves, simple treatment like UV light or filters will probably be enough to make this water safe to drink (Elhenawy et al., 2024).

11.2.1.4 Recycled Water

Wastewater Treatment & Potable Reuse Though it sounds incredulous, wastewater, including water from toilets can be purified enough to be considered safe for drinking. This has been successfully achieved in Orange County, California, where sewage water is cleaned through microfiltration, reverse osmosis, and UV disinfection before it gets reused (Zahmatkesh et al., 2022). Water reuse offers safety and sustainability, particularly in areas with water shortages.

Greywater Reuse Greywater comprises wastewater from showers, sinks and laundry. It is mostly water that has been used for domestic purposes, excluding sewage. Although it is not clean enough for drinking, it can be reused for basic tasks like flushing toilets or watering gardens after basic treatment and as it contains fewer germs than raw sewage, treatment processes are more simple. However, disinfection remains necessary in order to ensure safety during reuse (Van de Walle et al., 2023).

Table 11.1 Comparison of common water sources for microbial purification

Water source	Typical contaminants	Microbial risk	Treatment needed (conventional)	References
Surface water	Sediments, pathogens, fertilizers, waste runoff	High	Filtration, disinfection	Singh et al., 2024
Groundwater	Nitrates, heavy metals, coliforms	Medium	Filtration Sometimes, disinfection	Zahmatkesh et al., 2022
Rainwater	Dust, bacteria, bird droppings	Low to medium	First flush and disinfection Filtration Ultraviolet disinfection	Elhenawy et al., 2024
Recycled water	Organic matter, pathogens, chemicals	Very high	Multiple treatments like ultraviolet disinfection, reverse osmosis, advanced oxidation processes	Van de Walle et al., 2023

11.2.1.5 New and Emerging Sources

In trying to mitigate global water challenges, various regions across the globe are seeking alternative sources of water. Although most of them are still in development, these methods could serve as water sources to areas with limited access to freshwater. They include;

Desalination: Involves the removal of salt from brackish water to make it safe for drinking (Pearson et al., 2021).

Fog and Dew Collection: this method involves collecting moisture from the air via mesh nets, typically in humid or foggy regions (Giovanna et al., 2024).

11.2.2 Comparison of Common Water Sources for Microbial Purification

Table 11.1 presents the common water sources of water and the possible treatments.

11.2.3 Pollution of Water

Water pollution is the contamination of water bodies (lakes, rivers, oceans, aquifers, and groundwater) usually as a result of human activities. Pollutants are substances that make water unsafe for human use or harm aquatic ecosystems (U.S. Environmental Protection Agency [EPA], n.d.). Contamination is the introduction of pollutants like microorganisms and harmful chemical entities into bodies of water. Water pollution reduces the quality of water and access to clean water (Murugesan et al., 2020).

11.2.3.1 Sources of Water Pollution

Several anthropogenic and natural factors lead to water pollution. Human activities like urbanization and industrialization, population growth and climate change (Halder and Islam, 2015), improper sand, gravel and solid waste disposal (Ustaoğlu et al., 2020) and even religious activities (Dwivedi et al., 2018) directly affect water quality and contribute to water pollution. Some of the naturally occurring pollutants are nitrogenous compounds, heavy metals, oxides and radioactive substances (Evans et al., 2019). However, industrialization is the chief factor contributing to water pollution (Chen et al., 2019; Chowdhary et al., 2020). Food, iron, textile, steel, paper, tannery and other industries release several toxic chemicals, both organic and inorganic, which constitute the main water polluting agents. When these substances are released into the environment, mainly water systems, without proper treatment, they cause water pollution (Chowdhary et al., 2020). The increased rate of urbanization, technological advancement, and industrial production have collectively led to an increase in water pollution (Wu et al., 2020). The general publix has been exposed to higher levels of biological toxins and industrial chemicals and an increased occurrence of parasitic and infectious diseases has been recorded. This is owed to the combination of several factors including increased urbanization and industrialization processes and a lack of proper investment in water treatment and supply facilities (Lin et al., 2022).

Water pollution is also significantly influenced by agriculture and agricultural practices. Some of the most significant sources of water pollution are of agricultural origins like organic farm waste and inorganic compounds used for agricultural operations like pesticides and fertilizers. Nitrates, phosphorus, pathogens, soil sediments and other contaminants are introduced into water bodies through various agricultural operations (Parris, 2014). When these substances are applied in agricultural practices, the excesses are washed off by rainwater or irrigation and enter groundwater, eventually making their way to water bodies and causing pollution.

11.2.3.2 Types of Water Pollutants

The Environmental Protection Agency classified water pollutants into;

- Organic pollutants
- Inorganic pollutants
- Radionuclides
- Microorganisms
- Disinfectant by-products (Fig. 11.2) (U.S. Environmental Protection Agency, 2025)

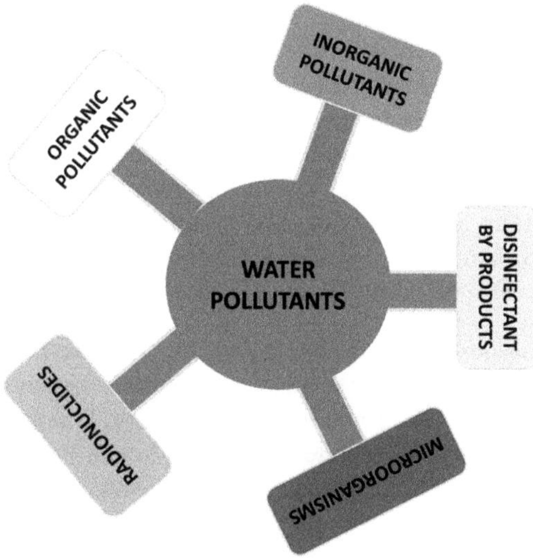

Fig. 11.2 Types of Water Pollutants

Organic pollutants Compounds containing carbon, hydrogen, oxygen, nitrogen, and sulphur are organic compounds. Organic pollutants comprise a group of non-natural chemical compounds that make their way into water bodies and drinking water as a consequence of their use in various industrial processes. They include compounds like hydrocarbon distillates, pesticides and fertilizers, paints, dyes, plastics, etc. (Meffe & de Bustamante, 2014). Waste products from different industries and industrial processes contain notable concentrations of these organic compounds in different forms. These pollutants, when present in water bodies completely metabolize the available oxygen, causing anaerobic conditions, ultimately throwing off the ecological balance (Kordbacheh & Heidari, 2023). Some of these organic pollutants like synthetic pesticides and fibers, dyes, volatile organic compounds, pesticides and insecticides, detergents and food additives are non-biodegradable and once introduced into the environment, water in this case, linger in the environment for long periods of time. They are called persistent organic pollutants (POPs) (Kordbacheh & Heidari, 2023). These pollutants are particularly harmful to human health, as they can affect food sources, like fish, and make their way up the food chain, potentially leading to health complications in humans (Elhenawy et al., 2024).

Inorganic pollutants This group of pollutants do not contain carbon. Inorganic pollutants include heavy metals (metallic elements having a greater atomic number than 20 and an elemental density greater than 5 g per cubic centimetre, and are toxic even in minute concentrations), nitrates, nitrites, ammonium, halogens, mineral phosphates, salts and others (Kordbacheh & Heidari, 2023). These non-natural contaminants often make their way in to drinking water supplies and are considered a threat to natural waters and human health (Mukherjee et al., 2021). Heavy metals are introduced into the environment by various industrial activities (Kapoor & Singh, 2021), and once introduced, they persist in the environment for long periods of time and bio accumulate. This way, their concentration in living (plant) tissues are increased and then they are introduced to higher trophic levels. Various adverse health conditions like developmental delays, cancer, reproductive problems, etc. can arise from exposure to heavy metals in drinking water and even small concentrations can cause high levels of exposure (Sankhla et al., 2016). High levels of other chemicals like mineral phosphates and nitrogenous compounds generated from agricultural operations could also potentially cause health problems when they make their way into drinking water. Nitrites are carcinogenic and can lead to increased bodily ammonium levels (World Health Organization, 2003).

Radionuclides Radionuclides are radioactive water contaminants. They are unstable isotopes of some elements that emit hazardous radiation. Elements like caesium, plutonium, and uranium are radioactive (Rajkhowa et al., 2021). In nature, radioactive materials are generated in the earth's crust and when they make their way to the earth's surface, they contaminate drinking water. Additionally, radioactive materials from industrial activities like medical research, remnants of uranium extraction, nuclear weapons tests, explosions in nuclear reactors, accidental leaks and effluents from nuclear facilities also contribute to water contamination by radionuclides (Verma, 2018). These radioactive materials can catalyse various health problems like damage to cells and DNA or genetic mutation, cancer, radiation sickness and a host of other problems (Rajkhowa et al., 2021).

Microorganisms Various microorganisms found in water are categorized as biological contaminants and it is essential that they are removed before water can be classified as safe for drinking. While some microbes found in water cause no harm, while others are extremely pathogenic and cause diseases when ingested by humans and other animals via drinking water. Some such pathogenic organisms include;

- Bacteria; *E. coli, Salmonella* sp., *Vibrio cholera, Campylobacter* sp., *Shigella* sp. (Olaolu et al., 2014; Kristanti et al., 2022)
- Viruses; hepatitis A virus, rotavirus, norovirus (Nwadike et al., 2024)
- Protozoa; *Giardia lamblia, Cryptosporidium parvum* (Omarova et al., 2018)
- Helminths; *Ascaris lumbricoides, Trichuris trichiura* (Kristanti et al., 2022).
- Fungi; *Alternaria* sp., *Cladosporium* sp., *Fusarium* sp., *Penicillum* sp., *Aspergillus* sp. (Novak Babič et al., 2017)

These disease vectors can get into drinking water via various means like sewage discharge and release of industrial wastewater, ingestion of which can cause waterborne disease (Olaolu et al., 2014), which can be fatal to humans. Cholera, typhoid, dysentery, polio, and hepatitis are the most common microbial-pollutant-causing waterborne diseases in humans. Notably, pathogenic organisms found in sewage could possibly cause severe gastrointestinal diseases (Ashbolt, 2015).

Disinfectant by-products Disinfectants have widespread application in pathogen elimination from drinking water due to their particular effectiveness in eliminating various harmful pathogens that propagate waterborne diseases. Chlorine, chloramine, and hydrogen peroxide are the most commonly used disinfectants in public drinking water systems (Ngwenya et al., 2012). However, disinfection byproducts (DBPs) are formed when chemical reactions occur between these disinfectants and some organic compounds found in the water. Entire populations are exposed in different degrees to DBPs through swimming pools and drinking water due to the frequency of disinfectant use in water treatment (Villanueva et al., 2015). DBPs have been linked to various health problems like cancer and fertility issues (Pandian et al., 2022; Zhang et al., 2023).

11.3 Purification of Water through Microbial Technology

Biological based water treatment techniques are known to have high efficiency in waste water treatment and removal of contaminants from water. They can also mitigate known setbacks of traditional water purification methods. This is achieved through a process called bioremediation. Bioremediation is a process that employs microbes in the removal of polluting species from contaminated sites. This is achieved by the microorganisms' ability to immobilize, oxidize and transform the polluting molecular species. Bioremediation is environmentally sustainable and cost effective (Wen et al., 2017). It depends on phenomena that exists naturally to degrade and remove contaminants by making use of microbes like fungi, bacteria and algae to catabolize harmful organic compounds and transform toxic substances into simple, non-toxic compounds (Eshkoraev, 2024). Due to the high amount of nutrients it contains, and its availability throughout the year, organic wastewater is perhaps the most suitable substrate for bioremediation (Tharali et al., 2016).

11.3.1 Constructed Wetlands (CWs)

A wetland is an area flooded with shallow water, containing soil saturated for a long enough period that a hydric, anaerobic environment is created capable of supporting macrophytes adapted to such conditions. Wetlands are generally divided into;

- Natural Wetlands, which are further classified into;
- Natural fresh water wetlands
- Natural salt water wetlands
- Constructed Wetlands (Omondi & Navalia, 2021).

For the longest time, natural wetlands have been employed in waste water remediation. The logic behind this, however, was not treatment, but discarding. The wetlands were a convenient dumping site, and were usually closer than the nearest water body and this uncontrolled wastewater disposal led to the pollution and eventual destruction of many naturally occurring wetlands across the globe (Vymazal, 2022). Attempts at the utilization of natural wetlands, under controlled conditions, for remediation to wastewater continued throughout the 70's and 80's in the United States. These experiments brought forth various unmanageable difficulties with system maintenance and it was quite difficult to predict the efficiency of the treatment process. This led to CWs being used to replace natural wetlands in the bioremediation process (Vymazal, 2022). A defining feature of CWs is shallow water depth and slow rate of water flow. This allows for long water retention time, ultimately increasing contact time between the polluted water and wetland component (Hassan et al., 2021).

CWs are engineered systems developed to mimic, under more controlled conditions, the processes of pollutant transformation and removal employed by natural wetlands. This allows for more a stringent control of processes, allowing for specifically designed substrate composition, flow patterns and vegetation type (Hassan et al., 2021). Typically, CWs consist of;

- Shallow channels, usually less than 1 m
- Substrates like sand, river rock or gravel
- Microorganisms
- Plants and Vegetation (Hassan, 2021; Maryjoseph & Ketheesan, 2020).

CWs combine intricate operations involving water, substrate (soil), plants, animals, microorganisms, and the environment to carry out their contaminant remediation. They also employ a variety of remediation processes biodegradation, phytoremediation, and natural attenuation (Truu et al., 2015). Vegetation native or non-native to the area are planted into the substrate, and organic matter in the form of contaminants in the wastewater is broken down by the action of microorganisms in the roots of the plants (Karungamye, 2022). Microbes play a key role in contaminant degradation in CWs and the eventual biodegradation of contaminants in the CW depends on the presence of microbes relevant to the degradation process and a suitable environment. Typically, a microbial consortium is involved in the degradation and the transformation processes of the contaminants. Usually, there are degradation pathways unique to a particular microbe for a particular contaminant (Azubuike et al., 2016). The pollutant degradation process in CWs also allow for harmful microorganisms like coliform bacteria to be expelled (Hdidou et al., 2021).

Contaminants are removed in CWs through:

- Chemical Processes which include; precipitation, sorption and complexation
- Physical Processes like filtration and sedimentation
- Biological Processes; microbe/plant uptake, microbial transformation and biodegradation (Biswal & Balasubramanian, 2022)

In comparison to natural wastelands, constructed wetlands prove to be more advantageous. It is possible to select the specific site, measure and execute the exact size and dimensions of the wetland, and perhaps most importantly, have control over water retention time and hydraulic pathways. CWs are also easily operated, maintained and are cost effective. They are also non-noise and non-odor generating, aesthetically pleasing, are environmentally sustainable as they do not require electricity and do not require a lot of man power to operate and also requires simple training (Rahman, 2020). In addition, they leave a low carbon footprint, do not require extra, toxic chemical additives, support changes in flow, hydraulic and organic load and also produces useful by-products (Masoud, 2022). As an added advantage, CWs can serve as habitats for wildlife (Zhang et al., 2020).

11.3.2 Microbial Fuel Cells (MFCs)

The bio electrochemical process is a green strategy that produces energy while simultaneously treating wastewater. The Microbial Fuel Cell is one of such strategies. MFC technology combines microbial metabolism and electrochemical reactions and is used for cheap electricity production and simultaneous wastewater treatment. The waste water acts as the substrate and can originate from various sources (Munoz-Cupa et al., 2021). A classic MFC is a variation of the traditional biological fuel cell and is a bio-electrochemical device capable of converting chemical energy, with microorganisms as biocatalysts, stored in substances like alcohols and sugars directly into bioelectric energy (Gupta et al., 2023). The MFC generates electricity by a biochemical conversion system from the potential energy present in the organic pollutants found in wastewater (Tatinclaux et al., 2018).

In general, the microbial fuel cell consists of an anode and cathode chamber with an ion-exchange membrane in between them. (Vishwanathan, 2021). The anode chamber is half biological while the cathode chamber is abiotic. The organic substrate (usually wastewater) used in the MFC is broken down by cellular respiration and microbial activity and carbon dioxide is released. As this substrate degradation happens, protons and electrons are generated in the anode chamber. Via an external circuit, the electrons produced in the anode chamber flow to the cathode and this flow generates energy due to potential difference (Logan, 2008). Microbial Fuel Cells can treat different types of wastewaters (domestic, industrial, municipal, etc.) and convert organic matter into electrical energy (Dwivedi et al., 2022). Due to MFCs ability to treat various kinds of groundwater, the technology is an appealing option for rural and urban areas. MFCs can also be integrated into existing wastewater treatment plants, which improves their efficiency and reduces their impact on the environment (Hamedani et al., 2024).

MFCs have wider applications then traditional wastewater treatment methods and produce less byproducts, thus improving their wastewater treatment potential (Ziara et al., 2018). Microbial Fuel Cells are an environmentally sustainable substitute to conventional wastewater treatment and bioelectricity generation, however, like most novel technologies, they have their setbacks. Scaling up this technology is a major challenge that could potentially increase the total cost and power use. Thus, it is imperative, that more research is carried out in this area to develop MFC-based technologies that would improve water treatment processes and contribute to environmental sustainability (Nawaz et al., 2020).

11.3.3 Biofiltration

Biofiltration, as a bio-based technique, is applied in the treatment of polluted water and air. Biological filters, known as biofilters are used in the biofiltration process (Boehm et al., 2020). The biofiltration process can be applied in the removal of organic pollutants like dissolved organic carbon from polluted media (Basu et al., 2016). Brown et al., (2015) stated that biofiltration is said to occur when media filters, commonly made up of anthracite, sand or granular activated carbon are used without exposure to disinfectants like ozone or chlorine, enabling microbial attachment and ultimately, degradation of the pollutant containing substrate. Biofilters employ naturally occurring microorganisms in the degradation of substrates like organic matter. Microbial colonization of the media filters occurs after attachment. These microbes then degrade and remove the pollutants present by oxidation from the substrate (water and air) (Ralebitso-Senior et al., 2012). It is applied as a tertiary treatment process for the removal of toxins, nutrients and recalcitrant compounds (García-Sánchez et al., 2019).

Biofiltration is an attached growth system and differs from suspended treatment process. The difference in the processes lies in the fact that in the biofiltration process, there exists a distinct separation between the biofilm (microbial biomass) and the treated effluent. The microorganisms, via immobilization, remain on the filter bed and the effluent flows through the filter media and this creates a separation between the microbial biomass and the effluent (Loh et al., 2021).

Steps in the Biofiltration Process Given by Schroeder (2002), the following steps are involved in the biofiltration process;

- First, contaminated substrate (water or air) is fed into the biofilter setup
- Contaminants are then absorbed in the aqueous phase on to the biofilm of the biofilter bed
- The contaminants act as a nutrient source and microbial proliferation and colonization occur, resulting in contaminant degradation
- Finally, the treated products are removed from the setup along with the end products (water, biomass and carbon dioxide)

Biofilters have widespread applications in the removal of various organic and nitrogenous pollutants from municipal waste water. The biofiltration process also has high efficiency in remediating various kinds of waste water. It has been shown to remediate domestic waste water (Cui et al., 2019), raw sewage (Gilbert et al., 2012), oil and gas produced water (Riley et al., 2018), groundwater (Zeng et al., 2019) and river water (McKie et al., 2016). There is a growing interest in identifying satisfactory substitutes to conventional filter materials in order to make the biofiltration process more cost effective. Wheat straws, woodchips and natural fibers have been brought up to replace conventional media like sand and anthracite (Sosa-Hernández et al., 2016). These suggested organic materials are readily available and contain similar physicochemical properties to conventional synthetic filter media, making

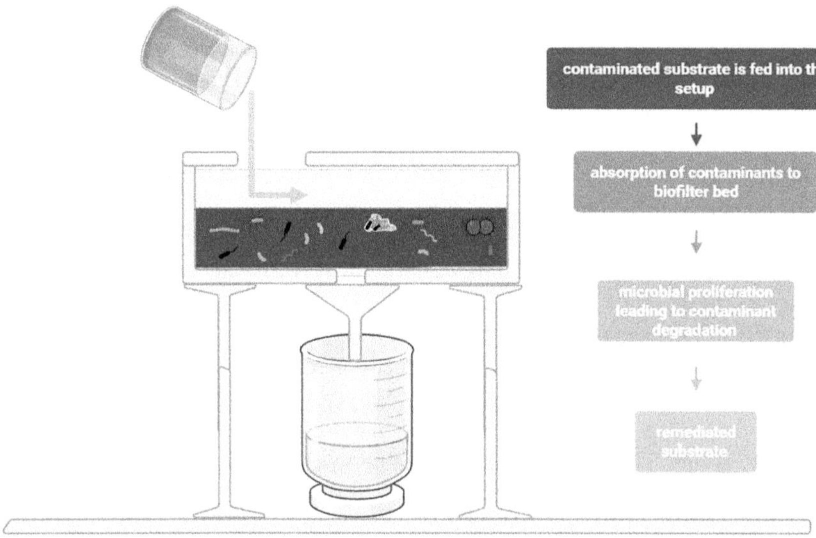

Fig. 11.3 Basic setup of biofiltration

these materials good candidates for potential conventional filter media replacement (Fig. 11.3) (Loh et al., 2021).

11.4 Other Recent Innovations in Water Purification

Millions of people in various parts of the world do not have access to clean water. Increase in global population and depletion of water resources has leas to research in new and improved methods of water purification. In addition, emerging water pollutants like pharmaceutical products, artificial sweeteners, hormones and steroids, surfactants, etc. (Ojha, 2020) are not able to be removed from water by traditional treatment methods. The development of novel technologies to effectively eliminate both traditional and emerging pollutants from unpurified water from natural sources is of the utmost importance (Berger et al., 2016). Global research into this issue has brought from some innovative technological advancements in recent years, some of which will be discussed in this section.

11.4.1 Non-Thermal Atmospheric Pressure Plasma (NTAPP)

Plasma is considered to be the fourth state of matter. It is applied in physics and chemistry to describe the condition of an ionized gas (Burm, 2012). Plasma is made up of electric fields, photons, ions, electrons and neutrals (López et al., 2019). Non-Thermal Atmospheric Pressure Plasma (NTAPP) is produced using gases like helium, argon, oxygen and nitrogen in ambient air at STP (Joshi & Thagard, 2013). NTAPP serves as a source of radicals, charged particles, shockwaves, reactive species, UV radiation and ultrasound. These entities are capable of degrading organic molecules and destroying bacterial cells and viral infections, without any harm to the environment or human health (Hara & Adachi, 2021; Mumtaz et al., 2023). Strong reactive species like H_2O_2, NO_2, NO and hydroxide radicals given off by plasma jets possess strong sterilization properties and thus, could contribute to water treatment processes (Suraj et al., 2018).

A variety of plasma containing different reactor parameters have been studied for applications in water purification and treatment in contribution to the development of improved water treatment processes in recent years. One study developed a water purifier by exposing a prototype to the plasma stream and its accompanying UV radiation to water. The study method was able to effectively eliminate pathogens from water. The authors were able to show that this new plasma–water treatment system was generally more efficient and economical in comparison with other water treatment options.

11.4.2 Nanofiltration/Use of Membranes

This involves the use of a physical separator to remove pollutants from water as the tiny openings on the surface of the membrane allow the tiny water molecules to pass through while holding back contaminant molecules. Nanofiltration can be applied to the purification of water from several pollutants (Elhenawy et al., 2024).

Various studies have been carried out to determine the efficiency of nanofilters to remove various pollutants from water;

- Nanofiltration and reverse osmosis was applied in the removal of trichloroethylene and tetrachloroethylene from water. This novel process was able to eliminate a very significant percentage of the pollutants, up to 93% under laboratory conditions and 100% of contaminants upon application to real groundwater samples. (Ainscough et al., 2021)
- Modi and Bellare (2019) successfully remediated Methylene Blue and Rhodamine B dye-polluted water using hollow fiber membranes (HFMs) containing zinc oxide nanoparticle-dispersed carboxylated graphene oxide nanosheets (ZnO/cGO nanohybrid). A removal efficiency of 98.6% and 98.5% was recorded for the pollutants, respectively.
- Thin film nanocomposite (TFN) membranes, made with graphene oxide quantum dots (GOQDs) dispersed within a tannic acid (TA) film were used to treat water polluted by methylene blue and Congo red dyes. The study was done on a feed solution with a starting concentration of 100 mg/L dyes and a removal rate of

97.6% and 99.8% respectively for each contaminant was recorded (Zhang et al., 2017)

- An attempt was made by Wei et al., (2016) to remediate water contaminated by various phthalates using hollow fiber nano filtration membranes. The removal efficiencies achieved were 82.3% for dimethyl phthalate (DMP), 86.7% for diethyl phthalate (DEP), 91.5% for dibutyl phthalate, 95.1% for di-n-octyl phthalate (DnOP) and 95.4% for diethylhexyl phthalate (DEHP).

The results of the above studies are positive with respect to the use of membranes to remove pollutants from wastewater, making them a promising opportunity for application in wastewater treatment processes.

11.4.3 AI & IoT-Driven Smart Water Systems

Traditional water treatment processes are dependent on conventional, standard practices which usually do not have control systems or real-time monitoring (Sun & Scanlon, 2019). This lapse in process efficiency led to waste of water and even, in some cases, a reduction in the quality of treated water. The coming together of recent technological advances like artificial intelligence (AI) and the Internet of Things (IoT) bears tremendous potential for implementation in the optimization of wastewater treatment processes (Rane, 2023). The seamless interaction and integration of these technologies is capable of revolutionizing the way water resources are managed.

IoT covers the entire interconnected network of physical devices equipped with sensors and other communication devices capable of collecting and transmitting real time data. The placement of these devices in strategic locations across to the entire water supply chain, ranging from water harvesting sources to the individual homes for domestic use, it becomes possible to harvest real time data on various water quality parameters which can be applied to unbiased anf informed decision making (Esmaeilian et al., 2020). A fundamental change in approach of water monitoring and management systems occurred with the advent of IoT. It became possible to collect large swaths of real time data from every point of the water supply chain, leading to never before seen changes in the process (Jan et al., 2021; Oberascher et al., 2022).

Artificial intelligence, with its vast processing capabilities, in conjunction with machine learning algorithms, analyze and interpret these large swaths of data gathered by IoT systems. AI technologies have had various real word applications over the last decade in mitigating environmental threats and ecological risks; in reducing flood risks and various other industries and wastewater management is not an exception (Zhong et al., 2021). Chowdhury and Karanfil, in their 2024 review, stated that the most commonly used AI techniques or drinking water treatment processes as; Decision Tree (DT), Support Vector Machine (SVM), Self-Organizing Map (SOM), Particle Swam Optimization (PSO), Random Forest (RF), Artificial Neural Network (ANN), Genetic Algorithm (GA), K-Nearest Neighbor (K-NN), Genetic Programming (GP), Fuzzy Logic (FL), Principal Component Analysis (PCA) and Machine

Learning (ML). AI equips smart water systems with advanced capabilities in analytics and decision making. AI and its various algorithms and systems uncover patterns, predict possible future events and convert data into applicable information (Himeur et al., 2023; Lowe et al., 2022).

This relationship between AI and IoT fundamentally improves water management systems, enabling operations optimization and swift adaptation to changing environmental conditions (Dada et al., 2024). AI complements the abilities of IoT by introducing advanced data analytics, predictive models and state of the art decision making models for the analysis of the large data IoT collects. AI will identify patterns and make predictions on system behavior, making way for increased efficiency in the water resource management processes. The coming together of these technologies improves the overall process and creates smart, dynamic systems capable of self-improvement and performance optimization over time (Kamyab et al., 2023). This is where the actual potential of integrating IoT and AI into water resources management lies. The interconnectedness of these technologies reveals their complementary nature. The coaction of these technologies working hand-in-hand allows for the transformation of water management into an intelligent and dynamic process with the capacity to manage modern issues quickly and accurately (Adeyemi et al., 2017).

11.5 Conclusion

The issue of water pollution and access to clean and safe drinking water is a pertinent problem and is shaping up to be one of the biggest global challenges in recent times. Novel water pollutants arising from industrial processes invented and perfected in recent years may not be easily identified and treated by traditional water treatment processes. For this reason, contemporary treatment processes are being researched to aid in combating these emerging pollutants. Among these, developments in the field of microbiology and advancements in harnessing the metabolic power of microbes for applications in water treatment processes have been made. Microorganisms can be applied in different ways including but not limited to; deployment of constructed wetlands, improvements to the processes of biofiltration, harnessing the bioelectrochemical power of microbial fuel cells and the further combination of artificial intelligence and microbiological processes. It is extremely vital to carry out proper research in line with and environmental compliance laws as technologies evolve so as to stay ahead of these emerging pollutants and limit their entry into the environment.

References

Adesina, O. B., William, C., & Oke, E. I. (2024). Evolution in water treatment: Exploring traditional self-purification methods and emerging Technologies for Drinking Water and Wastewater Treatment: A review. *World News of Natural Sciences, 53*, 169–185.

Adeyemi, O., Grove, I., Peets, S., & Norton, T. (2017). Advanced monitoring and management systems for improving sustainability in precision irrigation. *Sustainability, 9*(3), 353.

Ainscough, T. J., Oatley-Radcliffe, D. L., & Barron, A. R. (2021). Groundwater remediation of volatile organic compounds using nanofiltration and reverse osmosis membranes—A field study. *Membranes, 11*(1), 61.

Alemayehu, M. A., Agimas, M. C., Shewaye, D. A., Derseh, N. M., & Aragaw, F. M. (2023). Spatial distribution and determinants of limited access to improved drinking water service among households in Ethiopia based on the 2019 Ethiopian mini demographic and health survey: Spatial and multilevel analyses. *Frontiers in Water, 5*, 1166733.

Ashbolt, N. J. (2015). Microbial contamination of drinking water and human health from community water systems. *Current Environmental Health Reports, 2*(1), 95–106.

Azubuike, C. C., Chikere, C. B., & Okpokwasili, G. C. (2016). Bioremediation techniques–classification based on site of application: Principles, advantages, limitations and prospects. *World Journal of Microbiology and Biotechnology, 32*, 1–18.

Bai, X., Dinkla, I. J. T., & Muyzer, G. (2022). Microbial ecology of biofiltration used for producing safe drinking water. *Applied Microbiology and Biotechnology, 106*(13–16), 4813–4829.

Balwan, W. K., & Kour, S. (2021). Wetland- an ecological boon for the environment. *East African Scholars Journal of Agriculture and Life Sciences, 4*(3), 38–48.

Barjasteh, A., Dehghani, Z., Lamichhane, P., Kaushik, N., Choi, E. H., & Kaushik, N. K. (2021). Recent progress in applications of non-thermal plasma for water purification, bio-sterilization, and decontamination. *Applied Sciences, 11*(8), 3372.

Basu, O. D., Dhawan, S., & Black, K. (2016). Applications of biofiltration in drinking water treatment–a review. *Journal of Chemical Technology & Biotechnology, 91*(3), 585–595.

Berger, M., Pfister, S., & Motoshita, M. (2016). Water footprinting in life cycle assessment: How to count the drops and assess the impacts. *Special Types of Life Cycle Assessment*, 73–114.

Biswal, B. K., & Balasubramanian, R. (2022). Constructed wetlands for reclamation and reuse of wastewater and urban stormwater: A review. *Frontiers in Environmental Science, 10*, 836289. https://doi.org/10.3389/fenvs.2022.836289

Boehm, A. B., Bell, C. D., Fitzgerald, N. J., Gallo, E., Higgins, C. P., Hogue, T. S., Luthy, R. G., Portmann, A. C., Ulrich, B. A., & Wolfand, J. M. (2020). Biochar-augmented biofilters to improve pollutant removal from stormwater–can they improve receiving water quality? *Environmental Science: Water Research & Technology, 6*(6), 1520–1537.

Brown, J., Summers, R. S., LeChevallier, M., Collins, H., Roberson, J. A., Hubbs, S., & Dickenson, E. (2015). Biological drinking water treatment? *Naturally. Journal-American Water Works Association, 107*(12), 20–30.

Burm, K. T. A. L. (2012). Plasma: The fourth state of matter. *Plasma Chemistry and Plasma Processing, 32*, 401–407.

Chauhan, A. S., Kumar, A., Parmar, K., & Kumar, V. (2022). Microbial ecology of wastewater treatment processes: Trends, challenges, and perspectives. In V. Kumar & I. S. Thakur (Eds.), *Omics insights in environmental bioremediation* (pp. 301–315). Springer. https://doi.org/10.1007/978-981-19-4320-1_13

Chen, B., Wang, M., Duan, M., Ma, X., Hong, J., Xie, F., Zhang, R., & Li, X. (2019). In search of key: Protecting human health and the ecosystem from water pollution in China. *Journal of Cleaner Production, 228*, 101–111.

Chowdhary, P., Bharagava, R. N., Mishra, S., & Khan, N. (2020). Role of industries in water scarcity and its adverse effects on environment and human health. In *Environmental concerns and sustainable development: Volume 1: Air, water and energy resources* (pp. 235–256).

Chowdhury, S., & Karanfil, T. (2024). Applications of artificial intelligence (AI) in drinking water treatment processes: Possibilities. *Chemosphere, 141958.*

Cui, B., Yang, Q., Zhang, Y., Liu, X., Wu, W., & Li, J. (2019). Improving nitrogen removal in biological aeration filter for domestic sewage treatment via adjusting microbial community structure. *Bioresource Technology, 293*, 122006.

Dada, M. A., Majemite, M. T., Obaigbena, A., Daraojimba, O. H., Oliha, J. S., & Nwokediegwu, Z. Q. S. (2024). Review of smart water management: IoT and AI in water and wastewater treatment. *World Journal of Advanced Research and Reviews, 21*(1), 1373–1382. https://doi.org/10.30574/wjarr.2024.21.1.0171

Dwivedi, K. A., Huang, S. J., Wang, C. T., & Kumar, S. (2022). Fundamental understanding of microbial fuel cell technology: Recent development and challenges. *Chemosphere, 288*, 132446. https://doi.org/10.1016/j.chemosphere.2021.132446

Dwivedi, S., Mishra, S., & Tripathi, R. D. (2018). Ganga water pollution: A potential health threat to inhabitants of ganga basin. *Environment International, 117*, 327–338.

Elhenawy, S., Khraisheh, M., AlMomani, F., Al-Ghouti, M., Selvaraj, R., & Al-Muhtaseb, A. A. (2024). Emerging nanomaterials for drinking water purification: A new era of water treatment technology. *Nanomaterials, 14*(21), 1707.

Eshkoraev, S. (2024). Innovative methods in water purification: Paving the way for sustainable clean water solutions. *Journal of universal science research, 2*(11), 458–463.

Esmaeilian, B., Sarkis, J., Lewis, K., & Behdad, S. (2020). Blockchain for the future of sustainable supply chain management in industry 4.0. *Resources, Conservation and Recycling, 163*, 105064.

Evans, A. E., Mateo-Sagasta, J., Qadir, M., Boelee, E., & Ippolito, A. (2019). Agricultural water pollution: Key knowledge gaps and research needs. *Current Opinion in Environmental Sustainability, 36*, 20–27.

Feng, L., Wang, Y., Hou, X., Qin, B., Kuster, T., Qu, F., Chen, N., Paerl, H. W., & Zheng, C. (2024). Harmful algal blooms in inland waters. *Nature Reviews Earth & Environment, 5*, 631–644.

García-Sánchez, L., Gutiérrez-Macías, T., & Estrada-Arriaga, E. B. (2019). Assessment of a Ficus benjamina wood chip-based aerated biofilter used for the removal of metformin and ciprofloxacin during domestic wastewater treatment. *Journal of Chemical Technology & Biotechnology, 94*(6), 1870–1879.

Gilbert, S., Gasperi, J., Rocher, V., Lorgeoux, C., & Chebbo, G. (2012). Removal of alkylphenols and polybromodiphenylethers by a biofiltration treatment plant during dry and wet-weather periods. *Water Science and Technology, 65*(9), 1591–1598.

Giovanna, M., Rodríguez, M., Caruana, R., Angelotti, A., Guilizzoni, M., & Zanelli, A. (2024). Experimental analysis on fog harvesting by means of bi- and three-dimensional meshes: A laboratory study. *Journal of the Textile Institute*, 1–13.

Gobana, M. K., Wakjira, B. A., Belihun, T. G., & Hailu, A. B. (2023). Effectiveness of ceramic filter water treatment method in improving drinking water quality and reducing water-related disease: Systematic review and meta-analysis. *Journal of Health and Environmental Research.* https://doi.org/10.11648/j.jher.20230901.11

Gupta, S., Patro, A., Mittal, Y., Dwivedi, S., Saket, P., Panja, R., Saeed, T., Martíne, F., & Yadav, A. K. (2023). The race between classical microbial fuel cells, sediment-microbial fuel cells, plant-microbial fuel cells, and constructed wetlands-microbial fuel cells: Applications and technology readiness level. *Science of the Total Environment, 879*, 162757.

Gür, S. D., Bakhshpour, M., & Denizli, A. (2021). Applications of microbes in bioremediation of water pollutants. In Inamuddin, M. I. Ahamed, & R. Prasad (Eds.), *Recent advances in microbial degradation. Environmental and microbial biotechnology* (pp. 465–483). Springer. https://doi.org/10.1007/978-981-16-0518-5_19

Halder, J. N., & Islam, M. N. (2015). Water pollution and its impact on the human health. *Journal of environment and human, 2*(1), 36–46.

Hamedani, E. A., Abasalt, A., & Talebi, S. (2024). Application of microbial fuel cells in wastewater treatment and green energy production: A comprehensive review of technology fundamentals and challenges. *Fuel, 370*, 131855. https://doi.org/10.1016/j.fuel.2024.131855

Hara, H., & Adachi, T. (2021). Molecular mechanisms of non-thermal atmospheric pressure plasma-induced cellular responses. *Japanese Journal of Applied Physics, 60*(2), 020501.

Hassan, I., Chowdhury, S. R., Prihartato, P. K., & Razzak, S. A. (2021). Wastewater treatment using constructed wetland: Current trends and future potential. *PRO, 9*(11), 1917. https://doi.org/10.3390/pr9111917

Hdidou, M., Necibi, M. C., Labille, J., El Hajjaji, S., Dhiba, D., Chehbouni, A., & Roche, N. (2021). Potential use of constructed wetland systems for rural sanitation and wastewater reuse in agriculture in the Moroccan context. *Energies, 15*(1), 156. https://doi.org/10.3390/en15010156

Himeur, Y., Elnour, M., Fadli, F., Meskin, N., Petri, I., Rezgui, Y., Bensaali, F., & Amira, A. (2023). AI-big data analytics for building automation and management systems: A survey, actual challenges and future perspectives. *Artificial Intelligence Review, 56*(6), 4929–5021.

Inbathamizh, L., Rakshi, A., Sandhya, S., Usha, N. S., & Sudha, S. (2023). Microbial Technologies in Wastewater Treatment: A positive step towards sustainable environment. *Research Journal of Chemistry and Environment, 27*(3), 100–107. https://doi.org/10.25303/2703rjce1000107

Islam, M. S. (2023). Groundwater: Sources, functions, and quality. *Springer Nature*, 17–36.

Jan, F., Min-Allah, N., & Düştegör, D. (2021). IOT based smart water quality monitoring: Recent techniques, trends and challenges for domestic applications. *Water, 13*(13), 1729.

Joshi, R. P., & Thagard, S. M. (2013). Streamer-like electrical discharges in water: Part I. Fundamental mechanisms. *Plasma Chemistry and Plasma Processing, 33*, 1–15.

Judeh, T., Shahrour, I., & Comair, F. (2022). Smart rainwater harvesting for sustainable potable water supply in arid and semi-arid areas. *Sustainability, 14*(15), 9271.

Kamyab, H., Khademi, T., Chelliapan, S., SaberiKamarposhti, M., Rezania, S., Yusuf, M., Farajnezhad, M., Abbas, M., Jeon, B. H., & Ahn, Y. (2023). The latest innovative avenues for the utilization of artificial intelligence and big data analytics in water resource management. *Results in Engineering, 20*, 101566.

Kapoor, D., & Singh, M. P. (2021). Heavy metal contamination in water and its possible sources. In *Heavy metals in the environment* (pp. 179–189). Elsevier.

Karungamye, P. N. (2022). Potential of Canna indica in constructed wetlands for wastewater treatment: A review. *Conservation, 2*(3), 499–513.

Kordbacheh, F., & Heidari, G. (2023). Water pollutants and approaches for their removal. *Materials Chemistry Horizons, 2*(2), 139–153.

Kristanti, R. A., Hadibarata, T., Syafrudin, M., Yılmaz, M., & Abdullah, S. (2022). Microbiological contaminants in drinking water: Current status and challenges. *Water, Air, & Soil Pollution, 233*(8), 299.

Kumar, R. R., Bhattacharya, C., & Vishwakarma, N. P. (2021). Application of microbial technology for waste removal. In P. Bhatt, S. Gangola, D. Udayanga, & G. Kumar (Eds.), *Microbial Technology for Sustainable Environment* (pp. 261–291). Springer. https://doi.org/10.1007/978-981-16-3840-4_16

Lin, L., Yang, H., & Xu, X. (2022). Effects of water pollution on human health and disease heterogeneity: A review. *Frontiers in Environmental Science, 10*, 880246.

Logan, B. E. (2008). *Microbial fuel cells*. John Wiley & Sons.

Loh, Z. Z., Zaidi, N. S., Syafiuddin, A., Yong, E. L., Boopathy, R., Hong Kueh, A. B., & Prastyo, D. D. (2021). Shifting from conventional to organic filter media in wastewater biofiltration treatment: A review. *Applied Sciences, 11*(18), 8650.

López, M., Calvo, T., Prieto, M., Múgica-Vidal, R., Muro-Fraguas, I., Alba-Elías, F., & Alvarez-Ordóñez, A. (2019). A review on non-thermal atmospheric plasma for food preservation: Mode of action, determinants of effectiveness, and applications. *Frontiers in Microbiology, 10*, 622.

Lowe, M., Qin, R., & Mao, X. (2022). A review on machine learning, artificial intelligence, and smart technology in water treatment and monitoring. *Water, 14*(9), 1384.

Maryjoseph, S., & Ketheesan, B. (2020). Microalgae based wastewater treatment for the removal of emerging contaminants: A review of challenges and opportunities. *Case Studies in Chemical and Environmental Engineering, 2*, 100046.

Masoud, A. M., Alfarra, A., & Sorlini, S. (2022). Constructed wetlands as a solution for sustainable sanitation: A comprehensive review on integrating climate change resilience and circular economy. *Water, 14*(20), 3232. https://doi.org/10.3390/w14203232

McKie, M. J., Andrews, S. A., & Andrews, R. C. (2016). Conventional drinking water treatment and direct biofiltration for the removal of pharmaceuticals and artificial sweeteners: A pilot-scale approach. *Science of the Total Environment, 544*, 10–17.

Meffe, R., & de Bustamante, I. (2014). Emerging organic contaminants in surface water and groundwater: A first overview of the situation in Italy. *Science of the Total Environment, 481*, 280–295.

Modi, A., & Bellare, J. (2019). Efficient removal of dyes from water by high flux and superior antifouling polyethersulfone hollow fiber membranes modified with ZnO/cGO nanohybrid. *Journal of Water Process Engineering, 29*, 100783.

Mukherjee, S., Kumar, J. S., Nagar, A., & Pradeep, T. (2022). Concepts of sustainability in clean water technologies. In P. J. Boul (Ed.), *Energy transition: Climate action and circularity* (pp. 625–657). https://doi.org/10.1021/bk-2022-1412.ch016

Mukherjee, I., Singh, U. K., & Singh, R. P. (2021). An overview on heavy metal contamination of water system and sustainable approach for remediation. *Water pollution and management practices*, 255–277.

Mumtaz, S., Khan, R., Rana, J. N., Javed, R., Iqbal, M., Choi, E. H., & Han, I. (2023). Review on the biomedical and environmental applications of nonthermal plasma. *Catalysts, 13*(4), 685.

Munoz-Cupa, C., Hu, Y., Xu, C., & Bassi, A. (2021). An overview of microbial fuel cell usage in wastewater treatment, resource recovery and energy production. *Science of the Total Environment, 754*, 142429.

Murugesan, P., Moses, J. A., & Anandharamakrishnan, C. (2020). Water decontamination using non-thermal plasma: Concepts, applications, and prospects. *Journal of Environmental Chemical Engineering, 8*(5), 104377.

Nawaz, A., Hafeez, A., Abbas, S. Z., Haq, I. U., Mukhtar, H., & Rafatullah, M. (2020). A state of the art review on electron transfer mechanisms, characteristics, applications and recent advancements in microbial fuel cells technology. *Green Chemistry Letters and Reviews, 13*(4), 365–381. https://doi.org/10.1080/17518253.2020.1854871

Ngwenya, N., Ncube, E. J., & Parsons, J. (2012). Recent advances in drinking water disinfection: Successes and challenges. *Reviews of Environmental Contamination and Toxicology*, 111–170.

Novak Babič, M., Gunde-Cimerman, N., Vargha, M., Tischner, Z., Magyar, D., Veríssimo, C., Sabino, R., Viegas, C., Meye, W., & Brandão, J. (2017). Fungal contaminants in drinking water regulation? A tale of ecology, exposure, purification and clinical relevance. *International Journal of Environmental Research and Public Health, 14*(6), 636.

Nwadike, B. I., Falodun, O. I., & Ogunjobi, A. A. (2024). Bacterial and viral contaminants in drinking water: Why do they really matter to us. In *Environmental pollution and public health* (pp. 3–28). Elsevier.

Oberascher, M., Rauch, W., & Sitzenfrei, R. (2022). Towards a smart water city: A comprehensive review of applications, data requirements, and communication technologies for integrated management. *Sustainable Cities and Society, 76*, 103442.

Ojha, A. (2020). *Materials in electrochemical detection of water pollutants* (pp. 161–185). Sensors in Water Pollutants Monitoring.

Olaolu, D. T., Akpor, O. B., & Akor, C. O. (2014). Pollution indicators and pathogenic microorganisms in wastewater treatment: Implication on receiving water bodies. *International Journal of Environmental Protection and Policy, 2*(6), 205–212.

Omarova, A., Tussupova, K., Berndtsson, R., Kalishev, M., & Sharapatova, K. (2018). Protozoan parasites in drinking water: A system approach for improved water, sanitation and hygiene in developing countries. *International Journal of Environmental Research and Public Health, 15*(3), 495.

Oscar Omondi, D., & Caren Navalia, A. (2021). Constructed wetlands in wastewater treatment and challenges of emerging resistant genes filtration and reloading. *Inland Waters: Dynamics and Ecology*.

Pandian, A. M. K., Rajamehala, M., Singh, M. V. P., Sarojini, G., & Rajamohan, N. (2022). Potential risks and approaches to reduce the toxicity of disinfection by-product–a review. *Science of the Total Environment, 822*, 153323.

Parris, K. (2014). Impact of agriculture on water pollution in OECD countries: Recent trends and future prospects. *Water Quality Management*, 33–52.

Pearson, J. L., Michael, P. R., Ghaffour, N., & Missimer, T. M. (2021). Economics and energy consumption of brackish water reverse osmosis desalination: Innovations and impacts of Feedwater quality. *Membranes, 11*(8), 616.

Qi, Y., & He, K. (2025). Science and technology for water purification: Achievements and strategies. *Water, 17*(1), 91. https://doi.org/10.3390/w17010091

Rahman, M. E., Bin Halmi, M. I. E., Samad, M. Y. B. A., Uddin, M. K., Mahmud, K., Abd Shukor, M. Y., Abdullah, S. R. S., & Shamsuzzaman, S. M. (2020). Design, operation and optimization of constructed wetland for removal of pollutant. *International Journal of Environmental Research and Public Health, 17*(22), 8339. https://doi.org/10.3390/ijerph17228339

Rajkhowa, S., Sarma, J., & Das, A. R. (2021). Radiological contaminants in water: Pollution, health risk, and treatment. In *Contamination of water* (pp. 217–236). Academic.

Ralebitso-Senior, T. K., Senior, E., Di Felice, R., & Jarvis, K. (2012). Waste gas biofiltration: Advances and limitations of current approaches in microbiology. *Environmental Science & Technology, 46*(16), 8542–8573.

Rane, N. (2023). Integrating leading-edge artificial intelligence (AI), internet of things (IOT), and big data technologies for smart and sustainable architecture, engineering and construction (AEC) industry: Challenges and future directions. *Engineering and Construction (AEC) Industry: Challenges and Future Directions (September 24, 2023)*.

Rani, N., Sangwan, P., Joshi, M., Sagar, A., & Bala, K. (2019). Microbes. In Shah, M., Rodriguez-Couto, S. (eds), *Microbial Wastewater Treatment* (pp. 83–102). https://doi.org/10.1016/b978-0-12-816809-7.00005-1

Riley, S. M., Ahoor, D. C., & Cath, T. Y. (2018). Enhanced biofiltration of O&G produced water comparing granular activated carbon and nutrients. *Science of the Total Environment, 640*, 419–428.

Sadigov, R. (2022). Rapid growth of the world population and its socioeconomic results. *The Scientific World Journal, 2022*(1), 8110229.

Sankhla, M. S., Kumari, M., Nandan, M., Kumar, R., & Agrawal, P. (2016). Heavy metals contamination in water and their hazardous effect on human health-a review. *International Journal of Current Microbiology and Applied Sciences, 5*(10), 759–766.

Schroeder, E. D. (2002). Trends in application of gas-phase bioreactors. *Reviews in Environmental Science and Biotechnology, 1*, 65–74.

Sharma, L. K., & Naik, R. (2024). Wetland ecosystems. *Springer Nature*, 3–32.

Shivalkar, S., Singh, V., Sahoo, A. K., Samanta, S. K., & Gautam, P. K. (2020). Bioremediation: A potential ecological tool for waste management. In Kumar, V., Saxena, G., Shah, M.P. (eds), *Bioremediation for Environmental Sustainability*, Approaches to Tackle Pollution for Cleaner and Greener Society (pp. 1–21). https://doi.org/10.1016/b978-0-12-820318-7.00001-0

Siddha, S., & Sahu, P. (2022). Traditional and existing methods of urban water supply and their loopholes. *Current Directions in Water Scarcity Research, 6*, 245–271.

Singh, P. K., Kumar, U., Kumar, I., Dwivedi, A., Singh, P., Mishra, S., Seth, C. S., & Sharma, R. K. (2024). Critical review on toxic contaminants in surface water ecosystem: Sources, monitoring, and its impact on human health. *Environmental Science and Pollution Research, 31*, 56428–56462.

Sinha, S., Chugh, P., Abiha, U., & Singh, R. (2022). Biofiltration: An emerging and promising technology for the treatment of water and air pollutants. In M. Shah, S. Rodriguez-Couto, &

J. Biswas (Eds.), *An innovative role of biofiltration in wastewater treatment plants (WWTPs)* (pp. 459–476). https://doi.org/10.1016/b978-0-12-823946-9.00010-3
Somma, S., Reverchon, E., & Baldino, L. (2021). Water purification of classical and emerging organic pollutants: An extensive review. *ChemEngineering, 5*(3), 47.
Sosa-Hernández, D. B., Vigueras-Cortés, J. M., & Garzón-Zúñiga, M. A. (2016). Mesquite wood chips (Prosopis) as filter media in a biofilter system for municipal wastewater treatment. *Water Science and Technology, 73*(6), 1454–1462.
Stephens, G. L., Slingo, J. M., Rignot, E., Reager, J. T., Hakuba, M. Z., Durack, P. J., Worden, J., & Rocca, R. (2020). Earth's water reservoirs in a changing climate. *Proceedings of the Royal Society A, 476*(2236), 20190458.
Sun, A. Y., & Scanlon, B. R. (2019). How can big data and machine learning benefit environment and water management: A survey of methods, applications, and future directions. *Environmental Research Letters, 14*(7), 073001.
Suraj, M., & Anuradha, T. (2018). *Plasma based water purifier: Design and testing of prototype with different samples of water*. Preprints.
Tatinclaux, M., Gregoire, K., Leininger, A., Biffinger, J. C., Tender, L., Ramirez, M., Torrebts, A., & Kjellerup, B. V. (2018). Electricity generation from wastewater using a floating air cathode microbial fuel cell. *Water-Energy Nexus, 1*(2), 97–103.
Tharali, A. D., Sain, N., & Osborne, W. J. (2016). Microbial fuel cells in bioelectricity production. *Frontiers in Life Science, 9*(4), 252–266.
Truu, J., Truu, M., Espenberg, M., Nõlvak, H., & Juhanson, J. (2015). Phytoremediation and plant-assisted bioremediation in soil and treatment wetlands: a review. *The Open Biotechnology Journal, 9*(1).
U.S. Environmental Protection Agency. (2025). *Types of Drinking Water Contaminants*. https://www.epa.gov/ccl/types-drinking-water-contaminants. Accessed on June, 11, 2025.
U.S. Environmental Protection Agency. (n.d.). *What is water pollution?* U.S. Environmental Protection Agency. https://www.epa.gov/nutrient-policy-data/what-water-pollution. Accessed on June, 11, 2025.
Ustaoğlu, F., Tepe, Y., & Taş, B. (2020). Assessment of stream quality and health risk in a subtropical Turkey river system: A combined approach using statistical analysis and water quality index. *Ecological Indicators, 113*, 105815.
Vaghasiya, J. V., & Pumera, M. (2024). The rise of 3D/4D-printed water harvesting materials. *Materials Today, 78*, 46–74. https://doi.org/10.1016/j.mattod.2024.06.007
Van de Walle, A., Kim, M., Alam, M. K., Wang, X., Wu, D., Dash, S. R., Rabaey, K., & Kim, J. (2023). Greywater reuse as a key enabler for improving urban wastewater management. *Environmental Science and Ecotechnology, 16*, 100277.
Varsha, V. S., Boreda, T., Pailla, S. R., Kambhampati, Y., Gourav, T., Yadavalli, R., Laxmi, G. V., Nadimpalli, S., & Reddy, C. N. (2025). Nanotechnology and microbes: Revolutionizing water management. In Y. K. Mohanta, B. Mishra, & R. N. Pudake (Eds.), *Nano-microbiology for sustainable development* (pp. 293–329). Springer. https://doi.org/10.1007/978-3-031-78845-1_12
Verma, S. K. (2018). Sources of nuclear pollutants and their controls. In *Pollutants from energy sources: Characterization and control* (pp. 139–147). Singapore.
Villanueva, C. M., Cordier, S., Font-Ribera, L., Salas, L. A., & Levallois, P. (2015). Overview of disinfection by-products and associated health effects. *Current Environmental Health Reports, 2*, 107–115.
Vishwanathan, A. S. (2021). Microbial fuel cells: A comprehensive review for beginners. *3 Biotech, 11*(5), 248.
Vymazal, J. (2022). The historical development of constructed wetlands for wastewater treatment. *Land, 11*(2), 174.
Wei, X., Shi, Y., Fei, Y., Chen, J., Lv, B., Chen, Y., Zheng, H., Shen, J., & Zhu, L. (2016). Removal of trace phthalate esters from water by thin-film composite nanofiltration hollow fiber membranes. *Chemical Engineering Journal, 292*, 382–388.

Wen, L. L., Zhang, Y., Chen, J. X., Zhang, Z. X., Yi, Y. Y., Tang, Y., Rittmann, B. E., & Zhao, H. P. (2017). The dechlorination of TCE by a perchlorate reducing consortium. *Chemical Engineering Journal, 313*, 1215–1221.

World Health Organization. (2003). Nitrate and nitrite in drinking-water: Background document for development of WHO Guidelines for drinking-water quality. In Nitrate and nitrite in drinking-water: Background document for development of WHO guidelines for drinking-water quality.

Wu, H., Gai, Z., Guo, Y., Li, Y., Hao, Y., & Lu, Z. N. (2020). Does environmental pollution inhibit urbanization in China? A new perspective through residents' medical and health costs. *Environmental Research, 182*, 109128.

Yadu, A., Sahariah, B. P., & Anandkumar, J. (2021). Microbiological wastewater treatment. In: Shah, M., Rodriguez-Couto, S., Mehta, K. (eds), *The* Future of effluent treatment plants, (pp. 165–182). https://doi.org/10.1016/b978-0-12-822956-9.00010-6

Yahaya, T. O., Oladele, E. O., Fatodu, I. A., Abdulazeez, A., & Yeldu, Y. I. (2021). The concentration and health risk assessment of heavy metals and microorganisms in the groundwater of Lagos, Southwest Nigeria. ArXiv.org.

Zahmatkesh, S., Amesho, K. T. T., & Sillanpää, M. (2022). A critical review on diverse technologies for advanced wastewater treatment during SARS-CoV-2 pandemic: What do we know? *Journal of Hazardous Materials Advances, 7*, 100121.

Zeng, H., Yin, C., Zhang, J., & Li, D. (2019). Start-up of a biofilter in a full-scale groundwater treatment plant for iron and manganese removal. *International Journal of Environmental Research and Public Health, 16*(5), 698.

Zhang, M., Deng, Y. L., Liu, C., Lu, W. Q., & Zeng, Q. (2023). Impacts of disinfection byproduct exposures on male reproductive health: Current evidence, possible mechanisms and future needs. *Chemosphere, 331*, 138808.

Zhang, C., Wei, K., Zhang, W., Bai, Y., Sun, Y., & Gu, J. (2017). Graphene oxide quantum dots incorporated into a thin film nanocomposite membrane with high flux and antifouling properties for low-pressure nanofiltration. *ACS Applied Materials & Interfaces, 9*(12), 11082–11094.

Zhang, C., Wen, L., Wang, Y., Liu, C., Zhou, Y., & Lei, G. (2020). Can constructed wetlands be wildlife refuges? A review of their potential biodiversity conservation value. *Sustainability, 12*(4), 1442. https://doi.org/10.3390/su12041442

Zhong, S., Zhang, K., Bagheri, M., Burken, J. G., Gu, A., Li, B., Ma, X., Marrone, B. L., Ren, Z. J., Schrier, J., & Zhang, H. (2021). Machine learning: New ideas and tools in environmental science and engineering. *Environmental Science & Technology, 55*(19), 12741–12754.

Ziara, R. M., Dvorak, B. I., & Subbiah, J. (2018). Sustainable waste-to-energy technologies: Bioelectrochemical systems. In *Sustainable food waste-to-energy systems* (pp. 111–140). Academic.

Chapter 12
Advances and Roles of Microbial Technologies Operations in Minimizing Environmental Footprints: A Review of Bioreactors

Oluwafemi Oladipupo Kolajo and Bridget Tomilayo Tola-Winjobi

Abstract The application of microbial technologies has become prominent in the scientific world, spanning various fields and prompting multi-disciplinary collaborations. A prominent example of microbial technology which has found application in water purification, wastewater management, pharmaceutical, food processing, manufacturing of therapeutic products, energy and agriculture etc. are bioreactors. While the use of bioreactors can be regarded as an emerging field in some applications, it has been proven as a viable methods in the field of biotechnology. These engineered systems facilitate the growth of microorganisms or plant cells in a controlled environment for specific scientific roles. This review presented the types of bioreactors recently reported, their configuration, characteristics and applications in the various fields of environmental management. The review covers the continuously stirred tank bioreactors (CSTBRs), Fluidized bed bioreactors (FBRs), Photobioreactors (PBRs), Airlift Bioreactors (ABRs) and Membrane Bioreactors (MBRs). Bioreactors have significantly enhanced the efficiency and effectiveness of processing organic waste into valuable products such as biogas, fertilizers, and biofuels, while also advancing applications in water purification, antibiotic and enzyme production, and pharmaceutical innovation. The review presented current trends and prospects of bioreactor application and highlighted their potential to reduce environmental footprint and resource recovery.

Keywords Microbial technologies · Bioreactors applications · Organic waste · Environmental footprints

O. O. Kolajo (✉)
Department of Agricultural and Environmental Engineering, University of Ibadan, Ibadan, Nigeria
e-mail: oluwafemikolajo@gmail.com

B. T. Tola-Winjobi
Department of Agricultural Engineering, Abiola Ajimobi Technical University, Ibadan, Nigeria

M. I. Adeoba et al. (eds.), *Water Resources Management in Nigeria*, World Water Resources 10, https://doi.org/10.1007/978-3-032-22813-0_12

12.1 Introduction

Microbial technologies refer to the use of microorganisms like fungi, viruses and bacteria in applications where the metabolic capabilities and genetic properties of these microbes are used to produce useful products and solve environmental problems (Regonesi, 2023). The profiling, culturing, and utilization of these microorganisms take place in specialised vessels called bioreactors. Bioreactors are specialized devices/vessels designed to maintain and promote the growth and cultivation of biological products of plants and animal cells by providing a controlled biological environment (Gonzalez-Perez and Persson (2016); Spier et al., 2011; Wang et al., 1999). These concepts have found application in a vast range of disciplines, including but not limited to biotechnology, pharmaceutical, and environmental engineering. The aim of developing a bioreactor is to simulate an optimal condition suitable for enabling the conversion of substrates into desired products, some of which include vaccines, biofuels, and enzymes (Kaur & Sharma, 2021). Applicable materials for conducting bioreactors are usually carefully selected and must neither be corrosive nor toxic to the biological environment (Spier et al., 2011). They must also have the ability to tolerate high pH and pressure as may be required for the optimum growth of micro-organisms amongst other characteristics based on the intended use of the bioreactor (Malmir et al., 2022; Oliveira, 2023). Bioreactors can be classified as fed-batch, batch or continuous based on their design, functionality and expected output (Gargouri et al., 2011). The bioreactor class determines the modalities for nutrient supply, culture harvesting and the expected productivity from the system (Kaur and Sharma, 2021).

For optimum performance of the bioreactor, parameters such as pH, oxygen levels, culture concentration and the type and amount of mixing required are put into consideration. Another rather important parameter required for the success of the bioreactor is the choice of microorganism employed. Microorganisms can either be aerobic or anaerobic (Regonesi, 2023). The bottom line for the optimum application of bioreactors in solving problems lies in the understanding of the means to manipulate the microbial community for the purpose intended. To the authors' best knowledge; there is dearth of information on studies directing their foci on the recent advancements made in the bioreactors research space. This review therefore aimed to explain the roles of bioreactors in environmental resilience and sustainability. The current trends and prospects are bibliometrically analysed for journal output in the last 5 years from a reputable journal repository—Web of Science to broaden the scope to see the potentials embedded in the use of bioreactors and microbial technologies. This will enable the documentation of advances recently made in the construction, use and application of bioreactors, by highlighting some design parameters of interest, successes, drawbacks and challenges involved in the use and operations of bioreactors.

12.2 Review of Existing Bioreactors' Role and Advancements

This section of the review focuses on the details, types, operation parameters, uses and technological advances made on bioreactors in recent times. Contributions of each type to the reduction of environmental footprints were also reviewed and identified.

12.2.1 Continuous Stirred Tank Bioreactor (CSTB)

The continuous stirred tank bioreactors (CSTB) have found application in biotechnology and chemical engineering (Younesi et al., 2008; Mekuto et al., 2017). They have been successfully used for yeast fermentation, mammalian tissue cultivation and treatment of industrial and hydrocarbon-rich wastewater. CSTBs are operated to continuously allow for constant input of substrates and the concomitant removal of products for the optimization of production efficiency and stability (Marcos et al., 2004). For optimum output productivity, a steady state is sustained in the vessel to achieve and homogeneous concentration of reactant with the aid of a stirring device throughout the process. Such stirring encourages the hydro dynamism property which is critical in determining the reactor's performance (Kanaujiya et al., 2022). Residence time distribution (RTD) and Reynolds number are the two main parameters to measure the system's efficiency. Feedback loops and automated sensors are also incorporated into the system to maintain the parameters within the system.

Gargouri et al., (2011) applied an aerobic CSTB system to the bioremediation of hydrocarbon-rich industrial wastewater to optimize a feasible bioprocess. The experiment reported a highly and efficiently decontaminated wastewater. A chemical oxygen demand (COD) reduction of 95% and a total residue petroleum hydrocarbon reduction of 97.5% were also reported from the experiment. The resulting wastewater was found non-toxic based on a phytotoxicity test. The experiment had a total volume of 20 L and a working volume of 12 L for the continuous experiment. The temperature in the system was monitored and maintained at 30 °C using an installed thermostat jacket. Oxygen was injected into the system through a porous diffuser. Industrial wastewater, the only carbon source in the system was fed continuously to stimulate microbial growth within the system. With the aid of a masterflex peristaltic pump polytetrafluoroethylene (PTFE) tubing, the slurry was pumped into the reactor. Performance evaluation of the system was done by monitoring the pH, COD, Nitrogen content and biomass. The result showed that the optimised and specific hydro-dynamism of the CSTB which led to the development of an efficient microbial consortium are the bedrock for the encouraging results reported from this study.

The use of CSTB was equally adopted by Acharya et al. (2019). As shown in Fig. 12.1, ethanol production by syngas fermentation was studied, using a 3 L low-cost laboratory scale experimental design CSTB, made with transparent PVC-9002–86-20 pipe type and a plexiglass sheet VH-100 Acrylic resin type for the base and the lid. The working volume for this study was 2 L. A gas diffuser was adopted to supply the oxygen needed for the system. Temperature, pH, gas flow and membrane support were monitored and a 2 L anaerobic glass jar was used to monitor the continuous supply of the media as well as the extraction of the effluent as the experiment progressed. The result from the experiment is a continuous production of bioethanol (maximum at 3.75 g/L during syngas fermentation at 15 mL/min flow, 300 rpm, 0.25 mL/min liquid flow) with Acetic acid (maximum 14.97 g/L during syngas fermentation at 10 mL/min gas flow, 300 rpm, 0.25 mL/min liquid flow) which is made possible by the continuous addition of fresh broth media to the CSTB.

The CSTB also found application when the performance evaluation of toxicity removal of phthalate mixture by Gordonia sp. was studied (Kanaujiya et al., 2022). Phthalates which are also known as plasticisers are the foundation material in the food packaging industry. The experiment was carried out using a 5 L autoclave laboratory-scale glass fermenter with a calculated number of phthalates added for

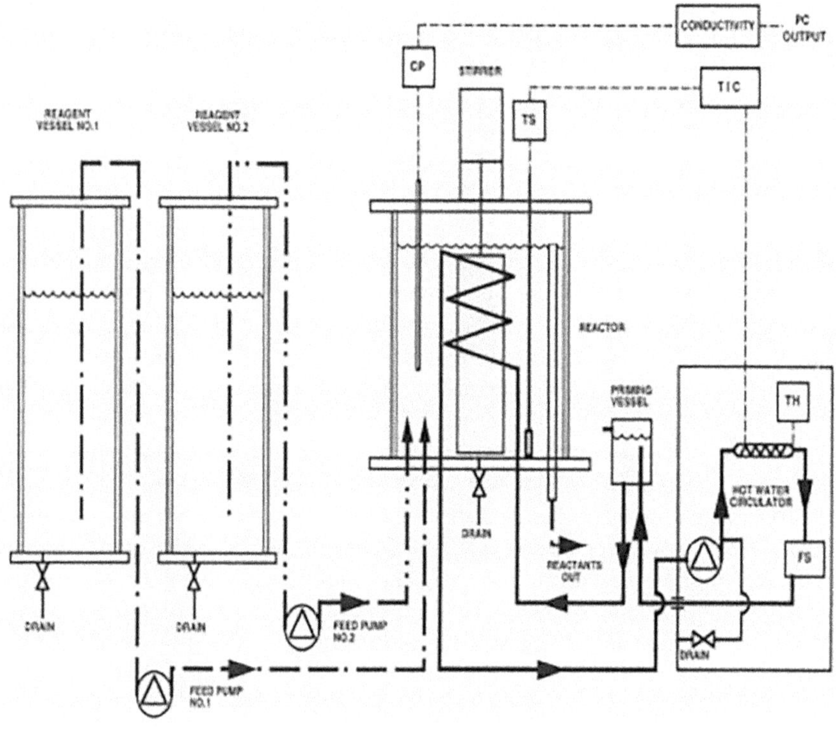

Fig. 12.1 Schematic representation of a CSTR experiment setup. *Source* Mataloni et al., (2025)

emulsification of the compound. Sensors were connected to the system to monitor the pH, dissolved oxygen (DO) and the agitation and foam levels while the experiment proceeded. Other operating parameters in the system were 28 °C temperature, 400 rpm for agitation, 1.0 vvm aeration rate, and 100% v/v of 48 h grown Gordonia sp. which was aseptically added into the CSTB as the inoculum. The result from the study showed a complete degradation of the phthalates up to 1250 mg/L of the total concentration when the experiment was conducted in the CSTB. When the result from the experiment was compared with a batch shake flasks experiment, the CSTB had an advantage for achieving high degradability of highly recalcitrant phthalates by Gordonia sp. due to controlled conditions of mixing, aeration and temperature which was achievable in the CSTR system.

The efficiency of CSTR in enhancing easier mixing ensures uniform reaction conditions and ease of control leading to consistent product formation and efficient pollutant treatment. The latest advances for CSTR include the inclusion of advanced sensors for real-time monitoring of critical parameters aimed at achieving precise process adjustments. These are the points of attraction for adopting CSTR for wide application to reduce environmental footprints.

12.2.2 *Fluidized Bed Bioreactor (FBBR)*

The specific design of this type of bioreactor is the suspension of the solid particles, support media and microorganisms in a fluid as shown in Fig. 12.2. This is done to facilitate mixing, mass and heat transfer which all contribute to the overall efficiency of the system (Bello et al., 2017; Özkaya et al., 2019; Ulson de Souza et al., 2008). Hydraulic fluidization where the rising flow of the fluid overcomes the gravitational force acting on the solid particle is the basic operational principle adopted in this type of bioreactor (Kuyukina et al., 2009; Mendonça da Silva et al., 2020). This has found application in the production of gasoline, wastewater treatment and industrial production of polymers. When a system requires high biomass concentration and large surface area for metabolic activities in the bioreactor, a fluidized bed gives the best advantage (Özkaya et al., 2019). A fluidized bed can accommodate hybrid microbial systems and can function as both aerobic and anaerobic in operation (Kim et al., 2011). This is a huge advantage as it allows researchers and industrialists to be able to optimize conditions specific to the desired bioproduct without compromising efficiency and sustainability.

Nelson et al. (2017) presented the circulating fluidized bed bioreactor (CFBBR) though still at the laboratory and pilot scale as developed in the University of Ontario (Fig. 12.3)a-c has shown great potential as an alternative to the conventional wastewater treatment system. The CFBBR has proven to present excellent mixing and mass transfer characteristics which makes the system rather effective in treating 90% of influent organic matter, and 80% of Nitrogen and producing less than one-third as much biological sludge as the activated sludge. In addition, the CFBBR waste handling and recovery from dynamic loading are properly handled with reduced

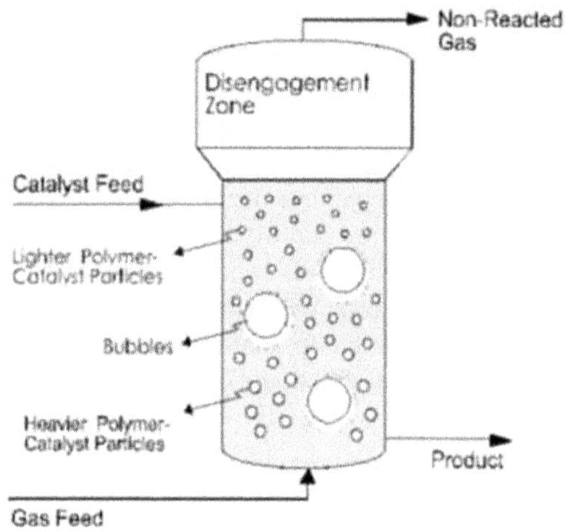

Fig. 12.2 Schematics of Fluidized bed bioreactor system. *Source* Nelson et al. (2017)

system size and footprint. Large volumes of wastewater have been treated with a smaller reactor volume at shorter retention times.

A 3.6 L batch-fed aerobic FBBR was used to study COD reduction in the effluents from the textile industry in Brazil with 95% concentrated nitric acid under a magnetic stirrer serving as the support for the system which lasted for 20 min at ambient temperature (Ulson de Souza et al., 2008). The study aimed to test the degradability of selected effluents from the textile industry using an aerobic process. The aerobic unit was comprised of a three-phase FBBR with a glass column having 140 cm height and an internal diameter of 5.8 cm. Air was introduced into the lower end through three porous diffusers positioned at 120°. A rotameter was used to control the air produced from the compressor as well as the flow rate into the system. Feeding into the system was achieved by gravity and controlled manually by a strangulation valve. Feedstock into the system was sampled from the stabilization tank, continuous washing machine and neutralization tanks. Parameters monitored to evaluate the performance of the system included DO, pH, COD and colour, following standard methods (1995) spectrophotometry, using area method consisting of spectral ranges of 300–700 nm. Results from the study indicated that the system's adherence to biomass tolerance was satisfactory since there was no sudden change in the COD. Regardless of the pH of the feed material, the pH remained at about 8.5 demonstrating a close lid system. Colour removal was however insignificant and this further confirmed that aerobic processes are inefficient when biodegradation of dyes of this class is involved.

Another study by Kim et al. (2011) made use of an Anaerobic Fluidized Bed Bioreactor (AFMBR) for wastewater treatment. Fouling was identified as the major drawback to this application, however, positioning the membrane directly in contact with the granulated activated charcoal (GAC) was a leeway to solving the fouling hurdle. The study employed a 120-day continuous fed evaluation using a two-stage anaerobic treatment system operated at 35 °C and fed with synthetic wastewater with

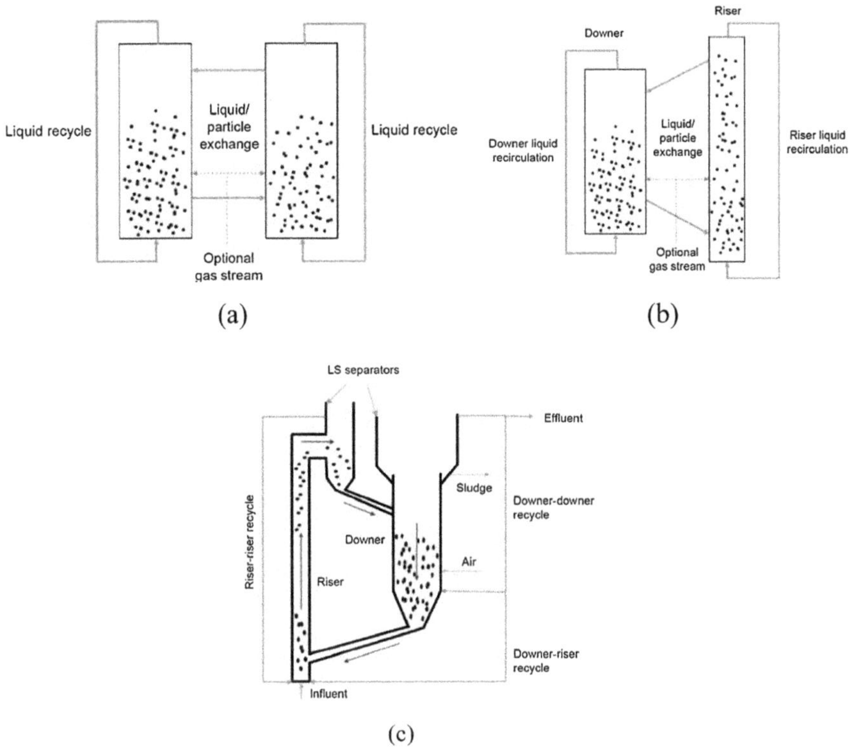

Fig. 12.3 **a** Conventional twin fluidized-bed system **b** Liquid fluidized bed with particle circulation **c** CFBBR pilot scale design. *Source* Nelson et al. (2017)

an average COD of 513 mg/L. The first stage has a hydraulic retention time of 2.0–2.8 h followed by the second stage with a retention time of 2.2 h, The COD removal from this study was reported at 88% and 87% in the first stage and second stage, respectively and 99% overall COD removal efficiency. The total energy required for the two-stage system was reported at 0.058 KWh/m^3 sourced from using only 30% of the gaseous methane energy produced from the system.

Tabiś et al., (2020) reported a novel biological water treatment with the use of a Fluidized-bed bioreactor. The system suggested steady-state characteristics determined with the aid of a developed mathematical model for a double-substrate aerobic microbiological process. The advantages of the system over conventional processes were noted to include its ability to run many aerobic microbiological processes simultaneously to not only increase the achieved conversion rate of carbonaceous substrate but also improve the process safety by eliminating the phenomenon of biomass leaching from the bioreactor. The improvement of efficiency, scalability and process control remained the focus of recent advances in the use of FBBR.

12.2.3 Photobioreactors (PBR)

The photobioreactor (PBR) design leverages the cultivation of photosynthetic organisms such as algae and cyanobacteria whose metabolism is light-dependent (Huang et al., 2017). The major parameters needed to be carefully selected in the design of this bioreactor type include light intensity, wavelength, temperature and nutrient availability within the system (Benner et al., 2022; Katsuda et al., 2000; Pagliolico et al., 2017; Pires et al., 2017; Posten, 2009). The system can be further classified as either an open photo or closed system bioreactor. Ponds are general examples of open photobioreactors. They are generally less expensive and are easier to maintain, although they face the problem of contamination. The closed system bioreactors can be further classified as tubular or flat-panel (whose better internal conditions can be better controlled for optimum productivity). The photobioreactor must have efficient light utilization, this has been enhanced by increasing innovations in LED technologies and light distribution techniques to ensure light exposure and enhance the photosynthetic efficiency of cultivated organisms (Kolajo et al., 2022; Sato et al., 2025). Photo bioreactors have garnered attention due to their ability to address environmental challenges. Algae, for instance, is applicable for use in this reactor as they can absorb carbon dioxide from the environment which is a major carbon sequestration breakthrough. Photo bioreactor has also found application in wastewater treatment technologies such as Microbial Fuel Cell Technology (MFCT) (Sato et al., 2021, 2023) whereby the algae supply the cathode with oxygen and the nutrients required for the Algae growth are synthesized from the MFCT system.

Recognizing algae's ability and their affinity for growth in a photo reactor has increased the interest in bioprocessing research. Mass cultivation of photoautotrophs (especially Algae) has been studied in open raceways because of the associated low cost, but these systems have many drawbacks including, temperature and water level control as well as high chances of contamination, among others (Katsuda et al., 2000). A photobioreactor operated in a closed structure and locked off from the outside environment with no possible exchange of gases or pollutants has the advantage of minimising toxicity and providing an improved environment (Quinn et al., 2012). Such close bioreactors give room for higher cell concentrations and complex biopharmaceutical processing (Yoo et al., 2013). They are considered the most suitable system for the cultivation of microalgae, thus representing a new type of algae culture device (Sato et al., 2021). Photo bioreactors have become an important key to resolving the efficient production of photoautotrophs and ensuring the quality of value-added products derived from them (Sato et al., 2023).

The need to produce microalgae for use as a shading system in sustainable buildings was the focus of the study by Pagliolico et al. (2017) and a photo-reactor system using plastic bags was explored. The algae growth rate and the light transmission of the experiment were monitored. Other factors necessitating the selection of PBRs are cost, volume, surface area to volume ratio, and thickness of the reactor. The experiment explored different configurations for the bioreactors and performance evaluation between and across the system was optimized. *Scenedesmus obliquus*

CCAP 276/38, a freshwater microalga was the microorganism utilized for the study. The microorganism was maintained at 4 °C on a protease-peptone solid medium (PP agar) containing 0.02 mg/L of $MgSO_4.7H_2O$, 0.02 g of K2HPO4, 0.2 g of KNO_3, 1 g of Protease-Peptone, and 20 g of agar per litre. The medium was inoculated and incubated for 15 days at room temperature. A standardized inoculum of 10% v/v was provided for each cultural condition tested. The study reported the use of disposable plastic bags with circular cubicles for the cultivation of microalgae that are used as screens for windows in sustainable buildings.

Photobioreactors were also employed when the cultivation of *Haematococcus pluvallis,* which was found to be a major source for the production of Astaxanthin, a viable food supplement and a major ingredient in the production of eye drops due to its powerful antioxidant properties (Yoo et al., 2012). The material used in the construction of the bioreactors was locally sourced and principally made of cast polypropylene film. The material is transparent and its durability in water is high. A large air/water interface is generated by injecting numerous large air bubbles into the aqueous broth column of diameter 10 cm and height of 90 cm as the experiment proceeded. The Algae growth was sustained by illuminating the bioreactor at 40 mol photon/m^2/s with cool white fluorescent lamps which was changed to 200 mol photon/m^2/s during the induction phase where the dark/light cycle was 8:16 and the supply of 2% CO_2-enriched air was done through the bottom of the medium with a micro sparger at the rate of 0.6 L/min. The study presented a successful culturing of the algae using a photo-bioreactor system.

12.2.4 Airlift Bio-reactors

The airlift bioreactor (ALR) has found application in cell culturing and enzyme reactions in different fields of biological and biochemical processes as it can provide efficient mass transfer and mixing while maintaining a gentle and stable environment for cellular growth (Giaveno et al., 2007). The fundamental principle enshrined in this method is the use of a gas usually air to achieve circulation within the reactor (Kong et al., 2014). The introduced air creates a density difference between the gas-laden liquid and the surrounding liquid leading to the formation of upward and downward flow patterns. This flow pattern ensured efficient nutrient, oxygen and microorganism transfer while simultaneously removing metabolic by-products which is crucial for sustaining cellular activity. The low shear force achieved during mixing is the major advantage of the airlift bioreactor. It is beneficial to sensitive cells and microorganisms that are prone to damage from intense agitations. The continuous flow also ensures process monitoring and control of basic operating parameters like pH and DO levels (Yen and Liu, 2014).

Al-Mashhadani et al. (2015) used Computational Fluid Dynamics (CFD) to show that microbubbles generated by fluidic oscillation can provide an effective, low-energy means of achieving high interfacial area for mass transfer and improved liquid oscillation for mixing in an Airlift bioreactor which will be a viable alternative to stirred tank bioreactors for bioprocessing which have gaseous reactant as their byproducts. While accepting the airlift bioreactor as the optimum choice, they presented different possible configurations to achieve results with the airlift bioreactor some of which included cylindrical design which is the conventional German design, egg-shaped design and rectangular cross-section designs which is noted for its limited use due to poorer mixing efficiency. The cylindrical bioreactor's diameter of the draught tube to the bioreactor ratio (d/D) as used in the study is 0.6, the total volume is 15 L with a working volume of 9 L. The remaining 6 L left in volume served as the head-pan volume needed to condense the vapour water and to return the vapour to the medium as shown in Fig. 12.4.

In a similar study, pneumatic (airlift) reactors have their distinguishing characteristics of being gas-liquid or gas-liquid-solid contacting devices that have found application in wastewater treatment (Cozma & Gavrilescu, 2012). The adsorption process is reported to be better in ALRs than in other types of bioreactors (Fluidized, moving bed or packet bed designs) and this is due to the configuration and hydraulic conditions of the ALRs which are important key parameters for enhancement of adsorption efficiency in wastewater treatment. As reported by (Cozma & Gavrilescu, (2012), *t*he control of the liquid circulation in the down corner, no zone of irregular flow at the top and the bottom, and efficient heat and mass transfer are major identifiable features that distinguish ALRs for processes where high mass and heat transfer

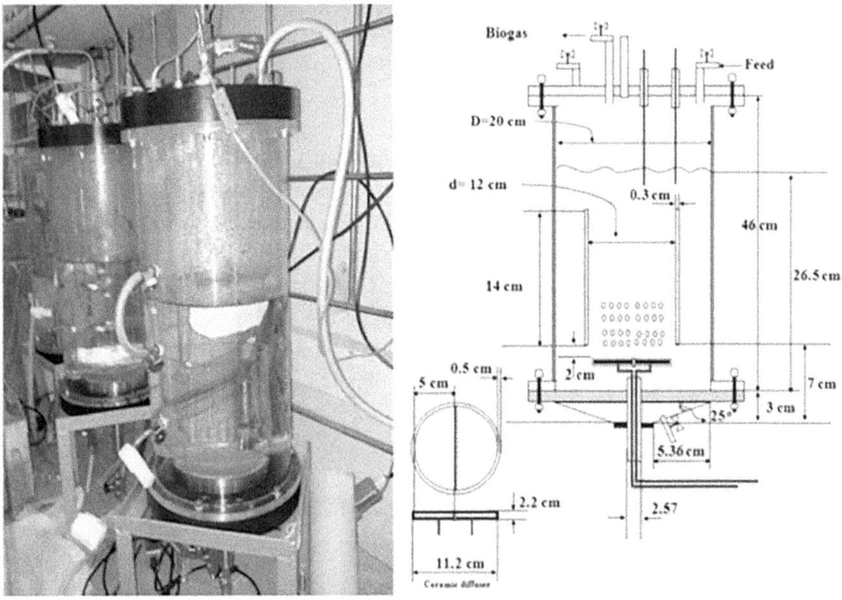

Fig. 12.4 Airlift bioreactor configuration and dimension. *Source* AL-Mashhadani et al. (2015)

efficiencies are required for the cultivation of cells when liquids with high velocity may result. Specifically, ALRs are usually applied when.

Conventional biological treatments are involved, but recent studies have also shown their applications in integrating processes of nitrification-denitrification in advanced wastewater treatment (AL-Mashhadani et al., 2015). While the basic features of ALRs have proven to meet the need for Activated sludge (conventional biological treatment), integrating ALRs with other alternatives such as sequencing bioreactors, membrane bioreactors and ultrasonic bioreactors, among others has been found necessary if ALRs are to be used for advanced wastewater treatment as microbial acclimatization is faster rather than in the conventional systems (AL-Mashhadani et al., 2015). Current studies relating to ALRs are at the bench scale, more studies are however being done to upscale the technology.

An airlift bioreactor was used for the study involving the feasibility of the production of adventitious root biomass and caffeic acid derivative of *Echinacea purpurea* (Jeong et al., 2009). A 5 L capacity bioreactor containing 4 L of strength medium was adopted and monitored for 4 weeks in the bioreactor. The result showed a tenfold increase in the biomass and secondary compound. A linear relationship was established between the concentration of biomass and the sucrose consumption rate in the experiment. It was concluded that the optimum inoculum density required in the system is 7 g/L while the aeration rate should be kept at 0.1 vvm for optimum biomass production.

12.2.5 Membrane Bioreactors (MBRs)

This system represents a significant advancement in wastewater treatment technologies in that it combines biological treatment (activated sludge) with a membrane filtration process to enhance the efficiency and efficacy of wastewater treatment (Gede Wenten et al., 2020; Melin et al., 2006; Mutamim et al., 2012; Wang et al., 2008). This approach presents a concurrent removal of contaminants while achieving high-quality effluent standards. At the core of Membrane bioreactors (MBRs) is the use of membranes, which selectively allow water and certain substances to pass while retaining suspended solids, pathogens and other contaminants(Liao et al., 2006; Mutamim et al., 2013). MBRs present an advantage over the conventional system of wastewater treatment in that the two steps of having a secondary clarifier installed in wastewater treatment to separate treated water from the biological sludge are compressed into an efficient unit operation. MBRs can display a high level of resistance to variation in inflow characteristics making them suitable for use for different types of influents which are required for a municipal wastewater treatment facility (Kaur & Sharma, 2021; Wang et al., 1999).

Melin et al., (2006) studied the use of MBRs for wastewater treatment and reuse for artificial groundwater recharge, indirect potable reuse as well as for industrial process water production in Belgium and the potential roles in promoting a more sustainable water use pattern. Although operated similarly to the conventional sludge system,

MBR does not require secondary clarifiers and tertiary steps like sand filtration in its operation, because the effluent is separated from activated sludge using low-pressure membrane filtration. Such low-pressure membrane filtration can be microfiltration (MF) or Ultrafiltration (UF) submerged or in an external circulation arrangement. Membranes used in the submerged systems are either hollow fibre membrane or plate membrane module design. Pre-treatment with a grid distance of a maximum of 3 mm is incorporated into the design of the system, putting into consideration specifics to prevent membrane clogging by fibres, hairs or other extreme contents. Usually, submerged MBRs were operated at low differential pressure below the critical permeate flux especially when fouling control is possible. Reported efficiency from MBRs is 99% TSS, 98.8–100% Turbidity, 89–98% COD, 97% BOD and 90% Nh3-N removal.

Membrane's ability to prevent bacteria and the use of Ultraviolet light to present viruses are the two corroborating attributes that make MBRs suitable for providing a considerable level of physical disinfection. Compared to conventional activated sludge systems, MBR has a smaller footprint and smaller reactor volumes, reduced sludge production, higher and more consistent effluent quality as a result of the filtration carried out by the membranes and lower sensitivity to contaminant peaks, among others (Regonesi, 2023). MBRs have also been found to be effective in the removal of microbiological parameters like total coliform, faecal coliform and even bacteriophage. Even though the phenomenal high efficiency of endocrine-disrupting trace pollutants which exhibit estrogenic activity has been reported in MBRs as compared to conventional activated sludge (Kim et al., 2011), the cause of this attribute cannot yet be explained. Arising from the efficiency and successes recorded in the pilot scale testing of MBR in Europe, many regions have upscaled the use of membrane bioreactors for wastewater treatment. Rodingen, Denmark upscaled to 2400 m^3/day, Markranstandt, Denmark with 3600 m^3/day, while Swanage, Lowestoft, Westbury, Cardigan, Buxton and Campbeltown in the UK upscaled with 12, 720, 14,160, 3536, 8640, 10,627 and 2678 m^3/day, respectively (Santos-Clotas et al., 2020). The study proved that MBRs have a significant potential to become a key element of wastewater reclamation and reuse schemes worldwide.

The application of MBRs in drinking water was studied by (Bodzek et al., 2011) where they emphasised the need for an intentional selection of membrane material which will depend on the size of the contaminant to be removed. The use of MBRs has also been explored for the treatment of high-strength industrial wastewater (Mutamim et al., 2012). Different operating conditions adopted during the study including sludge and membrane behaviours were reported. In the textile industry, a reactor volume of 500 L was adopted in an aerobic side-stream with a UF module and a membrane surface area of 0.28 m^2, pore size 0.025 μm and MLSS of 5000–15,000 mg/L. Factors influencing membrane fouling were identified and presented in Fig. 12.5. Membrane cleaning can be achieved by physical cleaning (Backwashing or membrane brushing) or chemical cleaning which serves not only to remove membrane fouling but to also disinfect the membrane.

MBRs have also been studied on their potential application in the treatment of Agricultural wastewater (Cicek, 2003) as MBRs have proven to be effective in

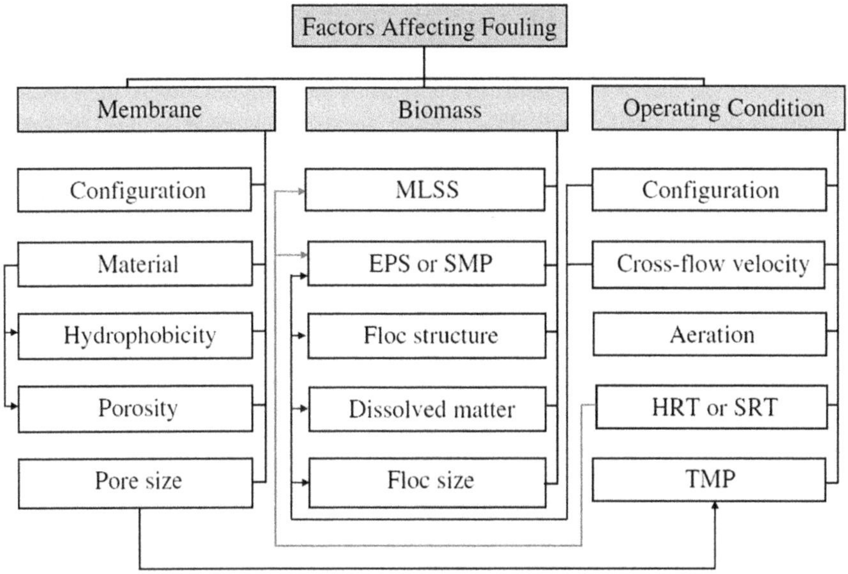

Fig. 12.5 Identified factors affecting fouling in MBRs. *Source* Nburu et al. (2020)

removing organic and inorganic contaminants as well as biological entities from wastewater. The types of membranes that have been explored in the treatment of domestic and municipal wastewater are ceramic and polymeric ultrafiltration (shaped into hollow fibre submerged, tubular external, cartridge disc external as well as plate and fibre submerged (Santos-Clotas et al., 2020). MBRs have also been successfully applied to the treatment of landfill leachate (Mutamim et al., 2013) which is usually characterized by high concentrations of organic and inorganic compounds including heavy metals. MBRs were also able to treat human excreta as present in domestic wastewater. The use of MBR provides a self-sustaining waste treatment for intensive livestock production as it has proven effective in handling growing environmental concerns of generated waste in terms of manure, wastewater, odour, ammonia and methane (Liao et al., 2006). MBR is a significant technological improvement because of the possibility of pretreatment and applied materials (nan-materials). This has made MBRs an environmentally friendly source of fresh water in many regions of the world necessitated by the need to meet scarce freshwater resources and the logic of adopting techniques that encourage water re-use rather than discharge as wastewater (Cicek, 2003; Mutamim et al., 2013).

12.3 Challenges with the Operations of Bioreactors

The gains of the use of bioreactors in the field of biotechnology were identified to include its use as a rapid and efficient plant micro propagation system, however, its use for commercial agricultural crops and forest trees are still under-explored (Ziv et al., 2003). Bioreactor culture system had been applied to herbaceous and woody plant species utilizing liquid media to overcome intensive manual handling which gives it a huge advantage over agar cultures with a better control of the culture conditions evident from the direct contact of the plant tissue with the medium, optimal nutrient and growth regulators supply, continuous aeration and circulation, filtration of the medium for controlling exudates and contamination and the production of clusters of buds, meristems for automated dispensing. However, the constraint with the application of this technology lies in a phenomenon called hyper hybridity, which is seen in morphogenic shoot and leaf malformation, due to the continuous immersion of the tissue in the medium. The malformations are manifested in glossy hyper hydrous leaves with distorted anatomy (Ziv et al., 2003).

In the same vein, Hansmann et al., (2013) described the constraints of the adaptation of the gains of bioreactors from a regulatory authority's point of view. Stating that every technology used for the production of a therapy should be proven through stringent regulatory guidelines. They reported that the possibility of an increased costs and increased risk of unsuccessful commercialization in inherent to the successful adoption of bioreactors in the field of Tissue Engineering, if regulations and guidelines are not developed to guide the operation and upscale of the use of emergence of bioreactors, it will perpertually be undermined in its use for clinical applications. Similarly, Fu et al., (2021) also reported that the adoption of bioreactors in tissue engineering will have to overcome a major challenge being that static cell culture on scaffolds cannot simulate the physiological environment required for the development of natural cartilage. However, Wendt et al. (2009) reported that the adoption of a 3D model to study various aspects of cell physiology and pathology, as well as to engineer implantable tissues may be the ice breaker to the identified limitation of the use of bioreactors.

12.4 Recent Advances and Prospects for Bioreactors

The use and application of bioreactors kept gaining prominence amongst researchers for water purification, wastewater management, and other applications. A search of journals using the inclusion and exclusion protocol from the Web of Science database between the years 2020 and 2024 showed 1020 open-access journal articles on the subject matter. A bibliometric analysis conducted on the generated results is presented in Fig. 12.6. The result of the analysis showed interconnected keywords attributed to researchers. Keywords with the most prominence included wastewater treatment, tissue engineering, anaerobic digester, denitrification, metabolic engineering, and mechanical simulation amongst others. In a study conducted by Nikita et al., (2023) in agreement with an earlier study by Sharma & Sharma (2009); Zhong, (2010), the importance of technologically conscious effort meant to improve the design of bioreactor for current realities including wastewater teatment was established. Specifically, Nikita et al. (2023) reported the inclusion of a robust and advanced control process control system that can ease process intensification, the adoption of advanced monitoring approaches, modelling techniques, soft sensors and process analytical technologies for either of the bioreactor designs can not be over-emphasised if they will continually be relevant in the advanced application like biotherapeutics.

As observed from Fig. 12.6, some other prospects for bioreactor application are for cartilage tissue engineering, gene expression and immortalization. In a study by Fu et al. (2021), prospects for articular cartilage regeneration are been investigated in the scientific field of Tissue Engineering to provide structural and functional substitutes to native tissues required for the development of natural cartilage all made possible through advanced technological application of bioreactors. Sharma et al. (2022) made it clear in their study that advancement in bioreactor technology is the backbone of the bioprocessing industry especially as it affects the enormous pressure on biopharma industries to meet the increasing need for treatment of prevailing and new diseases. These positions by these scholars are a build-up on the earlier submission by Martin

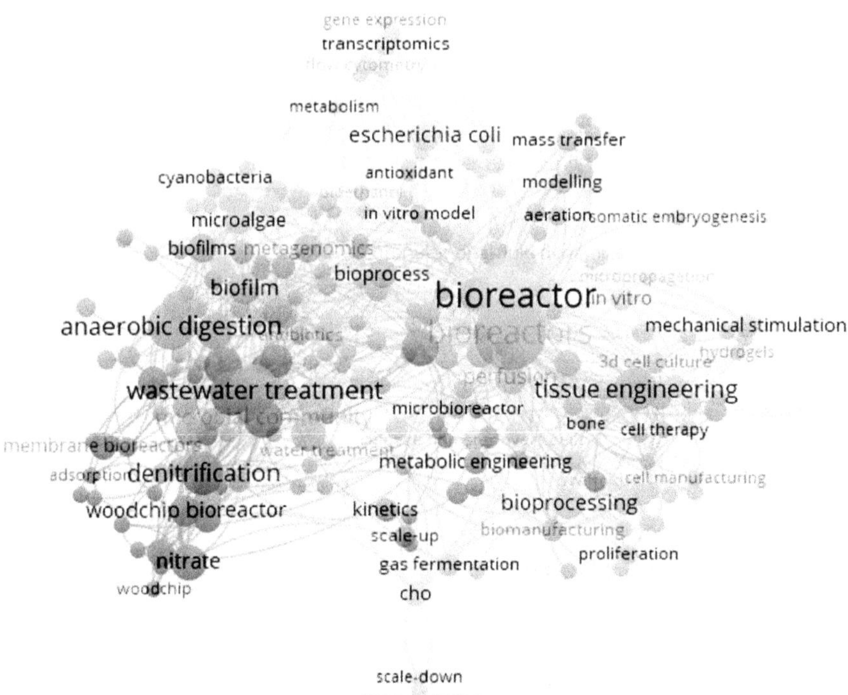

Fig. 12.6 Trends in research on bioreactors in the past 5 years (2020–2025)

and Vermette (2005) who have earlier reported a three-dimensional mammalian tissue growth in bioreactors and the importance of cell seeding. The indispensability of bioreactors in cell-based therapy for providing standardized, reproducible cell-based products for regenerative medicine and testing pharmacologic agents has also been reported (Zhong, 2010).

To support sustainable development, sewage treatment must not only reduce carbon emissions but also recover renewable resources. A major advancement in water purification processes is in the utilization of electrochemical anaerobic membrane bioreactors (Electro-AnMBRs). AnMBRs are promising low-carbon emission technologies, but their widespread use is limited by membrane fouling and poor nitrogen, and phosphorus recovery. Yuhan Zhang (2024) addressed these limitations by developing an electrochemical resource recovery anaerobic membrane bioreactor (eRAnMBR) using a magnesium plate and conductive membrane as dual nodes. This system achieved high recovery of carbon, nitrogen, and phosphorus along with 95% membrane anti-fouling efficiency. The eRAnMBR offers a simplified process, low maintenance, and reduced carbon footprint, representing innovation in sustainable wastewater treatment.

12.5 Conclusion

Microbial technologies as reported is a deal breaker in the field of biotechnology and water treatment. The use of bioreactors has been reported in various fields and the prospect is still enormous. The finetuning of processes, monitoring techniques, and material selection are the foci for recent advances in the adoption of microbial technologies for other applications. Understanding the design parameters and the potential efficiency of the application of bioreactors is attracting attention. The peculiarity of each of the reported bioreactors reported and the adoption of sensors for monitoring essential parameters is the driving force for the adoption of these technologies. Bioreactor's effectiveness and advantage over the conventional waste management system have been explored. Practical case studies leading to the adaptation of the technology have been explained in the course of the review as a modification over and above the current system of waste management for environmental sustainability. The burden of creating a sustainable system that is both cost-effective and environmentally sustaining will depend largely on fine-tuning the gains already reported from the use of bioreactors. Depending on the nature and characteristics of the waste to be treated, appropriate bio-reaction technologies can be selected from the pull presented in this review. Suitable modification in terms of configuration and integration can also be explored.

References

AL-Mashhadani, M. K. H., Wilkinson, S. J., & Zimmerman, W. B. (2015). Airlift bioreactor for biological applications with microbubble mediated transport processes. *Chemical Engineering Science, 137*, 243–253. https://doi.org/10.1016/j.ces.2015.06.032

Bello, M. M., Abdul Raman, A. A., & Purushothaman, M. (2017). Applications of fluidized bed reactors in wastewater treatment—A review of the major design and operational parameters. *Journal of Cleaner Production, 141*, 1492–1514. https://doi.org/10.1016/j.jclepro.2016.09.148

Benner, P., Meier, L., Pfeffer, A., Krüger, K., Oropeza Vargas, J. E., & Weuster-Botz, D. (2022). Lab-scale photobioreactor systems: Principles, applications, and scalability. *Bioprocess and Biosystems Engineering, 45*(5), 791–813. https://doi.org/10.1007/s00449-022-02711-1

Bodzek, M., Konieczny, K., & Kwiecińska, A. (2011). Application of membrane processes in drinking water treatment–state of art. *Desalination and Water Treatment, 35*(1–3), 164–184. https://doi.org/10.5004/dwt.2011.2435

Cicek, N. (2003). A review of membrane bioreactors and their potential application in the treatment of agricultural wastewater. *Canadian Biosystems Engineering / Le Genie Des Biosystems Au Canada, 45*, 37–49.

Cozma, P., & Gavrilescu, M. (2012). Airlift reactors: Applications in wastewater treatment. *Environmental Engineering and Management Journal, 11*(8), 1505–1515. https://doi.org/10.30638/eemj.2012.189

Fu, L., Li, P., Li, H., Gao, C., Yang, Z., Zhao, T., Chen, W., Liao, Z., Peng, Y., Cao, F., Sui, X., Liu, S., & Guo, Q. (2021). The application of bioreactors for cartilage tissue engineering: Advances, limitations, and future perspectives. *Stem Cells International, 2021*. https://doi.org/10.1155/2021/6621806

Gargouri, B., Karray, F., Mhiri, N., Aloui, F., & Sayadi, S. (2011). Application of a continuously stirred tank bioreactor (CSTR) for bioremediation of hydrocarbon-rich industrial wastewater effluents. *Journal of Hazardous Materials, 189*(1–2), 427–434. https://doi.org/10.1016/j.jhazmat.2011.02.057

Gede Wenten, I., Friatnasary, D. L., Khoiruddin, K., Setiadi, T., & Boopathy, R. (2020). Extractive membrane bioreactor (EMBR): Recent advances and applications. *Bioresource Technology, 297*, 122424. https://doi.org/10.1016/j.biortech.2019.122424

Giaveno, A., Lavalle, L., Patricia, C., & Donati, E. (2007). Airlift reactors: Characterization and applications in biohydrometallurgy. *Microbial Processing of Metal Sulfides, 169–191*. https://doi.org/10.1007/1-4020-5589-7_9

Gonzalez-Perez, A., & Persson, K. M. (2016). Bioinspired materials for water purification. *Materials, 9*(6). https://doi.org/10.3390/ma9060447

Hansmann, J., Groeber, F., Kahlig, A., Kleinhans, C., & Walles, H. (2013). Bioreactors in tissue engineering-principles, applications and commercial constraints. *Biotechnology Journal, 8*(3), 298–307. https://doi.org/10.1002/biot.201200162

Huang, Q., Jiang, F., Wang, L., & Yang, C. (2017). Design of Photobioreactors for mass cultivation of photosynthetic organisms. *Engineering, 3*(3), 318–329. https://doi.org/10.1016/J.ENG.2017.03.020

Jeong, J. A., Wu, C. H., Murthy, H. N., Hahn, E. J., & Paek, K. Y. (2009). Application of an airlift bioreactor system for the production of adventitious root biomass and caffeic acid derivatives of Echinacea purpurea. *Biotechnology and Bioprocess Engineering, 14*(1), 91–98. https://doi.org/10.1007/s12257-007-0142-5

Kanaujiya, D. K., Sivashanmugam, S., & Pakshirajan, K. (2022). Biodegradation and toxicity removal of phthalate mixture by Gordonia sp. in a continuous stirred tank bioreactor system. *Environmental Technology and Innovation, 26*, 102324. https://doi.org/10.1016/j.eti.2022.102324

Katsuda, T., Arimoto, T., Igarashi, K., Azuma, M., Kato, J., Takakuwa, S., & Ooshima, H. (2000). Light intensity distribution in the externally illuminated cylindrical photo-bioreactor and its application to hydrogen production by Rhodobacter capsulatus. *Biochemical Engineering Journal, 5*(2), 157–164. https://doi.org/10.1016/S1369-703X(00)00054-1

Kaur, I., & Sharma, A. D. (2021). Bioreactor: Design, functions and fermentation innovations. *Research & Reviews in Biotechnology & Biosciences Journal, 8*(2), 116–125.

Kim, J., Kim, K., Ye, H., Lee, E., Shin, C., McCarty, P. L., & Bae, J. (2011). Anaerobic fluidized bed membrane bioreactor for wastewater treatment. *Environmental Science and Technology, 45*(2), 576–581. https://doi.org/10.1021/es1027103

Kolajo, O. O., Pandit, C., Thapa, B. S., Pandit, S., Mathuriya, A. S., Gupta, P. K., Jadhav, D. A., Lahiri, D., Nag, M., & Upadhye, V. J. (2022). Impact of cathode biofouling in microbial fuel cells and mitigation techniques. *Biocatalysis and Agricultural Biotechnology, 43*, 102408. https://doi.org/10.1016/j.bcab.2022.102408

Kong, L., Holtz, C. T., Nairn, C. J., Houke, H., Powell, W. A., Baier, K., & Merkle, S. A. (2014). Application of airlift bioreactors to accelerate genetic transformation in American chestnut. *Plant Cell, Tissue and Organ Culture, 117*(1), 39–50. https://doi.org/10.1007/s11240-013-0418-8

Kuyukina, M. S., Ivshina, I. B., Serebrennikova, M. K., Krivorutchko, A. B., Podorozhko, E. A., Ivanov, R. V., & Lozinsky, V. I. (2009). Petroleum-contaminated water treatment in a fluidized-bed bioreactor with immobilized Rhodococcus cells. *International Biodeterioration and Biodegradation, 63*(4), 427–432. https://doi.org/10.1016/j.ibiod.2008.12.001

Liao, B. Q., Kraemer, J. T., & Bagley, D. M. (2006). Anaerobic membrane bioreactors: Applications and research directions. In. *Critical Reviews in Environmental Science and Technology, 36*(6). https://doi.org/10.1080/10643380600678146

Malmir, N., Fard, N. A., Aminzadeh, S., Moghaddassi-Jahromi, Z., & Mekuto, L. (2022). An overview of emerging cyanide bioremediation methods. *PRO, 10*(9). https://doi.org/10.3390/pr10091724

Marcos, N. I., Guay, M., Dochain, D., & Zhang, T. (2004). Adaptive extremum-seeking control of a continuous stirred tank bioreactor with Haldane's kinetics. *Journal of Process Control, 14*(3), 317–328. https://doi.org/10.1016/S0959-1524(03)00070-2

Martin, Y., & Vermette, P. (2005). Bioreactors for tissue mass culture: Design, characterization, and recent advances. *Biomaterials, 26*(35), 7481–7503. https://doi.org/10.1016/j.biomaterials.2005.05.057

Mataloni, S., Lo Conte, S., Della Pietra, M., Santoni, F., Zingone, A., & Verdone, N. (2025). Effect of operating conditions on molten carbonate electrolysis cell performance: An experimental study. *Journal of Power Sources, 652*. https://doi.org/10.1016/j.jpowsour.2025.237692

Mekuto, L., Ntwampe, S. K. O., Utomi, C. E., Mobo, M., Mudumbi, J. B., Ngongang, M. M., & Akinpelu, E. A. (2017). Performance of a continuously stirred tank bioreactor system connected in series for the biodegradation of thiocyanate and free cyanide. *Journal of Environmental Chemical Engineering, 5*(2), 1936–1945. https://doi.org/10.1016/j.jece.2017.03.038

Melin, T., Jefferson, B., Bixio, D., Thoeye, C., De Wilde, W., De Koning, J., van der Graaf, J., & Wintgens, T. (2006). Membrane bioreactor technology for wastewater treatment and reuse. *Desalination, 187*(1–3), 271–282. https://doi.org/10.1016/j.desal.2005.04.086

Mendonça da Silva, J., Erro, E., Awan, M., Chalmers, S. A., Fuller, B., & Selden, C. (2020). Small-scale fluidized bed bioreactor for long-term dynamic culture of 3D cell constructs and in vitro testing. *Frontiers in Bioengineering and Biotechnology, 8*, 1–16. https://doi.org/10.3389/fbioe.2020.00895

Mutamim, N. S. A., Noor, Z. Z., Hassan, M. A. A., & Olsson, G. (2012). Application of membrane bioreactor technology in treating high strength industrial wastewater: A performance review. *Desalination, 305*, 1–11. https://doi.org/10.1016/j.desal.2012.07.033

Mutamim, N. S. A., Noor, Z. Z., Hassan, M. A. A., Yuniarto, A., & Olsson, G. (2013). Membrane bioreactor: Applications and limitations in treating high strength industrial wastewater. *Chemical Engineering Journal, 225*, 109–119. https://doi.org/10.1016/j.cej.2013.02.131

Nburu, J., Gukeburger, E., Nwagi, P., Kinyua, R., & Hoinkis, J. (2020). Cost-benefit analysis for a containerrized MBR system for treatment of fish processing wastewater for industries in Kisumu,

Kenya and comparison to wastewater stabilization ponds and activated sludge process. *Journal of Water Technology and Treatment Methods, 3*(1).

Nelson, M. J., Nakhla, G., & Zhu, J. (2017). Fluidized-bed bioreactor applications for biological wastewater treatment: A review of research and developments. *Engineering, 3*(3), 330–342. https://doi.org/10.1016/J.ENG.2017.03.021

Nikita, S., Mishra, S., Gupta, K., Runkana, V., Gomes, J., & Rathore, A. S. (2023). Advances in bioreactor control for production of biotherapeutic products. *Biotechnology and Bioengineering, 120*(5), 1189–1214. https://doi.org/10.1002/bit.28346

Oliveira, V. B. (2023). Microbial fuel cells as a promising power supply for implantable medical devices. *Energies, 16*(6). https://doi.org/10.3390/en16062647

Özkaya, B., Kaksonen, A. H., Sahinkaya, E., & Puhakka, J. A. (2019). Fluidized bed bioreactor for multiple environmental engineering solutions. *Water Research, 150*, 452–465. https://doi.org/10.1016/j.watres.2018.11.061

Pagliolico, S. L., Lo Verso, V. R. M., Bosco, F., Mollea, C., & La Forgia, C. (2017). A novel photo-bioreactor application for microalgae production as a shading system in buildings. *Energy Procedia, 111*, 151–160. https://doi.org/10.1016/j.egypro.2017.03.017

Paucar, N. E. (2022). *Electricity Generation, Organic Removal, and Nutrient.*

Pires, J. C. M., Alvim-Ferraz, M. C. M., & Martins, F. G. (2017). Photobioreactor design for microalgae production through computational fluid dynamics: A review. *Renewable and Sustainable Energy Reviews, 79*, 248–254. https://doi.org/10.1016/j.rser.2017.05.064

Posten, C. (2009). Design principles of photo-bioreactors for cultivation of microalgae. *Engineering in Life Sciences, 9*(3), 165–177. https://doi.org/10.1002/elsc.200900003

Quinn, J. C., Yates, T., Douglas, N., Weyer, K., Butler, J., Bradley, T. H., & Lammers, P. J. (2012). Nannochloropsis production metrics in a scalable outdoor photobioreactor for commercial applications. *Bioresource Technology, 117*, 164–171. https://doi.org/10.1016/j.biortech.2012.04.073

Regonesi, G. (2023). Bioreactors: A Complete Review. Microalgae X. https://doi.org/10.13140/RG.2.2.11630.79685.

Santos-Clotas, E., Cabrera-Codony, A., Comas, J., & Martín, M. J. (2020). Biogas purification through membrane bioreactors: Experimental study on siloxane separation and biodegradation. *Separation and Purification Technology, 238*, 116440. https://doi.org/10.1016/j.seppur.2019.116440

Sato, C., Alikaram, G., Kolajo, O. O., Dudgeon, J., Hazard, R., Apollon, W., & Kamaraj, S. (2025). An algae cultivator coupled with a hybrid photosynthetic—air—cathode microbial fuel cell with ceramic membrane interface. 1–35.

Sato, C., Apollon, W., Luna-Maldonado, A. I., Paucar, N. E., Hibbert, M., & Dudgeon, J. (2023). Integrating microbial fuel cell and hydroponic technologies using a ceramic membrane separator to develop an energy–water–food supply system. *Membranes, 13*(9). https://doi.org/10.3390/membranes13090803

Sato, C., Paucar, N. E., Chiu, S., Mahmud, M. Z. I. M., & Dudgeon, J. (2021). Single-chamber microbial fuel cell with multiple plates of bamboo charcoal anode: Performance evaluation. *PRO, 9*(12). https://doi.org/10.3390/pr9122194

Sharma, R., Harrison, S. T. L., & Tai, S. L. (2022). Advances in bioreactor Systems for the Production of biologicals in mammalian cells. *ChemBioEng Reviews, 9*(1), 42–62. https://doi.org/10.1002/cben.202100022

Sharma, A. K., & Sharma, M. K. (2009). Plants as bioreactors: Recent developments and emerging opportunities. *Biotechnology Advances, 27*(6), 811–832. https://doi.org/10.1016/j.biotechadv.2009.06.004

Spier, M. R., Vandenberghe, L. P. D. S., Medeiros, A. B. P., & Soccol, C. R. (2011). Application of different types of bioreactors in bioprocesses. *Bioreactors: Design, Properties and Applications.*

Tabiś, B., Boroń, D., & Bizon, K. (2020). Biological water treatment by a hybrid fluidized-bed bioreactor: Theoretical study. *Chemical Engineering and Technology, 43*(5), 983–994. https://doi.org/10.1002/ceat.201900344

Ulson de Souza, A. A., Brandão, H. L., Zamporlini, I. M., Soares, H. M., Ulson, G., de Souza, S. M., & de A. (2008). Application of a fluidized bed bioreactor for cod reduction in textile industry effluents. *Resources, Conservation and Recycling, 52*(3), 511–521. https://doi.org/10.1016/j.resconrec.2007.06.004

Wang, H. H., Krstić, M., & Bastin, G. (1999). Optimizing bioreactors by extremum seeking. *International Journal of Adaptive Control and Signal Processing, 13*(8), 651–669. https://doi.org/10.1002/(SICI)1099-1115(199912)13:8<651::AID-ACS563>3.0.CO;2-8

Wang, Z., Wu, Z., Mai, S., Yang, C., Wang, X., An, Y., & Zhou, Z. (2008). Research and applications of membrane bioreactors in China: Progress and prospect. *Separation and Purification Technology, 62*(2), 249–263. https://doi.org/10.1016/j.seppur.2007.12.014

Wendt, D., Riboldi, S. A., Cioffi, M., & Martin, I. (2009). Potential and bottlenecks of bioreactors in 3D cell culture and tissue manufacturing. *Advanced Materials, 21*(32–33), 3352–3367. https://doi.org/10.1002/adma.200802748

Yen, H. W., & Liu, Y. X. (2014). Application of airlift bioreactor for the cultivation of aerobic oleaginous yeast Rhodotorula glutinis with different aeration rates. *Journal of Bioscience and Bioengineering, 118*(2), 195–198. https://doi.org/10.1016/j.jbiosc.2014.01.002

Yoo, J. J., Choi, S. P., Kim, J. Y. H., Chang, W. S., & Sim, S. J. (2013). Development of thin-film photo-bioreactor and its application to outdoor culture of microalgae. *Bioprocess and Biosystems Engineering, 36*(6), 729–736. https://doi.org/10.1007/s00449-013-0898-2

Yoo, J. J., Choi, S. P., Kim, B. W., & Sim, S. J. (2012). Optimal design of scalable photo-bioreactor for phototropic culturing of Haematococcus pluvialis. *Bioprocess and Biosystems Engineering, 35*(1–2), 309–315. https://doi.org/10.1007/s00449-011-0616-x

Younesi, H., Najafpour, G., Ku Ismail, K. S., Mohamed, A. R., & Kamaruddin, A. H. (2008). Biohydrogen production in a continuous stirred tank bioreactor from synthesis gas by anaerobic photosynthetic bacterium: Rhodopirillum rubrum. *Bioresource Technology, 99*(7), 2612–2619. https://doi.org/10.1016/j.biortech.2007.04.059

Zhong, J. J. (2010). Recent advances in bioreactor engineering. *Korean Journal of Chemical Engineering, 27*(4), 1035–1041. https://doi.org/10.1007/s11814-010-0277-5

Ziv, M., Chen, J., & Vishnevetsky, J. (2003). Propagation of plants in bioreactors: Prospects and limitations. *Acta Horticulturae, 616*, 85–93. https://doi.org/10.17660/ActaHortic.2003.616.6

Chapter 13
Innovative Water Remediation Technologies Using Clay Nanoparticles and Allied Nanomaterials

Olumuyiwa Idowu Ojo, Olumuyiwa Adewale Obijole, and Jeremiah Oludele Ojediran

Abstract Water pollution is an escalating global concern aggravated by rising industrialization, urbanization, agricultural run-off and natural causes. Conventional water treatment techniques often exhibit limitations in effectively and sustainably removing emerging contaminants such as heavy metals, pharmaceuticals, dyes, inorganic contaminants and pathogenic microorganisms. This review critically examines recent advancements in the use of clay nanoparticles and allied nanomaterials for water remediation. We discuss the structural and physicochemical properties of various clay minerals, synthesis and functionalization methods for clay-based nanocomposites, mechanisms underlying contaminant removal, performances in both laboratory and real-world applications, and highlight key challenges including scalability, stability, and environmental safety. Future research directions focus on green synthesis approaches, multifunctional composite materials, integration with smart monitoring technologies, and large-scale implementation strategies. Comprehensive tables summaries clay types, nanocomposite categories, and performance metrics for better understanding and enhanced clarity.

Keywords Clay nanoparticles · Nanocomposites · Water remediation · Adsorption · Ion exchange · Antimicrobial activity · Green synthesis

---

O. I. Ojo (✉)
Department of Agricultural Engineering, Ladoke Akintola University of Technology, Ogbomoso, Nigeria
e-mail: oiojo@lautech.edu.ng

O. A. Obijole
Department of Chemistry, Adeyemi Federal University of Education, Ondo, Nigeria
e-mail: obijoleoa@aceondo.edu.ng

J. O. Ojediran
VC Office, Bells University of Technology, Ota, Ogun State, Nigeria

M. I. Adeoba et al. (eds.), *Water Resources Management in Nigeria*, World Water Resources 10, https://doi.org/10.1007/978-3-032-22813-0_13

13.1 Introduction

Rapid urbanization, extensive industrial activity, and population growth and other natural causes Natural sources such asfluoride-rich rocks or fluorapatite which seeps into water aquifers through weathering, rainwater flow and leaching from fluoride rich rocks (Bretzler & Johnson, 2015; Kut et al., 2016) have contributed substantially to the contamination of freshwater resources worldwide (Naseer et al., 2025). World Health Organization (WHO, 2019), noted that there are about five million deaths per year because of drinking polluted water. Ganvir et al. (2002) reported that around 1.4 billion people don't have access to portable water. Common water pollutants include heavy metals (Pb^{2+}, Cd^{2+}, Hg^{2},$^{+}$ As^{2+}, Cr^{4+}), organic dyes, pharmaceuticals, pesticides, inorganic pollutants such as F^{-}, CN^{-}, SO_4^{-2}, $NH_3$and pathogenic microorganisms, all posing serious risks to human health and aquatic ecosystems (Gitari et al., 2022; Rezania et al., 2024).

Conventional water treatment processes, such as coagulation, electrocoagulation flotation/electrodialysis (Loganathan et al., 2013) flocculation, sedimentation, filtration, membrane filtration process (Otieno et al., 2012; Kumar et al., 2017), nanofiltration (Waghmare & Arfin, 2015) and chemical disinfection and precipitation (Aldaco et al., 2005) and adsorption (Ghorai & Pant, 2005; Obijole et al., 2019; 2021) exhibit limitations in removing emerging contaminants effectively, often incurring high costs, incomplete pollutant removal, and generating secondary waste (Amuda, et al., 2007; Mukhopadhyay et al., 2022).

Nanotechnology has emerged as a transformative approach, offering materials with enhanced surface reactivity, enlarged surface area, and tunable chemical properties that improve pollutant capture and degradation. Among these, clay nanoparticles are particularly promising due to their natural abundance, low cost, environmental compatibility, and superior ion exchange and adsorption capacities (Mukhopadhyay et al., 2022; Naseer et al., 2025). Moreover, their integration with polymers, carbon-based materials (e.g., graphene oxide, carbon nanotubes), and metal/metal oxide nanoparticles yields multifunctional nanocomposites with improved removal efficiency and antimicrobial properties (Deng et al., 2011). This review provides a comprehensive analysis of these recent advances, focusing on nanocomposite synthesis, pollutant removal mechanisms, applications, and prospects for large-scale sustainable water purification.

13.2 Clay Nanoparticles: Structure and Properties

Clay minerals are hydrous aluminosilicates characterized by layered structures that provide high surface area and cation exchange capacities (CEC) central to water remediation (Fayomi, et al., 2011; Ijagbemi et al., 2009; Naseer et al., 2025). Common clays include montmorillonite, bentonite, kaolinite, and sepiolite, which differ in their layer arrangements, surface chemistry, and adsorption properties.

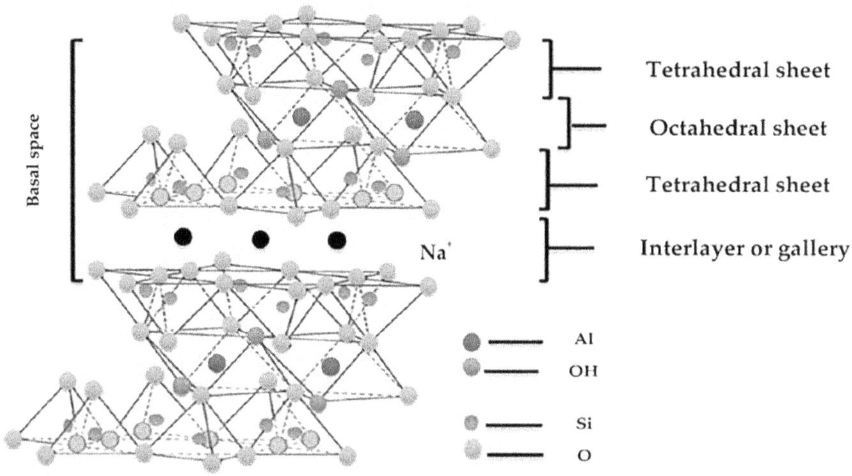

Fig. 13.1 Schematic structure of montmorillonite adapted from Nuruzzaman et al. (2016); Uddin (2017)

13.2.1 *Structural Features*

Montmorillonite and bentonite, classified in the smectite group, exhibit a 2:1 layer structure comprising two silica tetrahedral sheets sandwiching one alumina octahedral sheet, allowing intercalation of water molecules and exchangeable cations (Nuruzzaman et al., 2016; Uddin, 2017). Kaolinite possesses a simpler 1:1 structure with one tetrahedral and one octahedral sheet, offering greater structural stability but lower CEC. Sepiolite, a fibrous 2:1 clay, deposits channels conducive for organic pollutant adsorption (Rezania et al., 2024) (Fig. 13.1).

13.2.2 *Physicochemical Properties*

Specific surface area varies significantly from 10–30 m^2/g for kaolinite to 700–800 m^2/g for bentonite and influences adsorption capacity (Ijagbemi et al., 2009). Smectite clays' swelling capacity and negative surface charge promote adsorption of heavy metals and organic molecules; their surface chemistry is modulated by pH, ionic strength, and ion competition affecting selectivity and kinetics (Mukhopadhyay et al., 2022). Natural clay aggregation at the nanoscale may reduce active sites; hence, functionalization and nanocomposite formation enhance performance.

Table 13.1 Comparison of common clay minerals used in water remediation

Lay type	Structure	Specific surface area (m^2/g)	Cation exchange capacity (meq/g)	Stability	Typical applications
Montmorillonite	2:1	600–800	80–50	Moderate	Heavy metal adsorption, ion exchange
Bentonite	2:1	700–800	70–150	Moderate	Adsorption, catalysis support
Kaolinite	1:1	10–30	3–15	High	Stable adsorbent, filtration
Sepiolite	2:1	200–300	20–40	Moderate	Organic pollutant adsorption

Source Ijagbemi et al. (2009); Jiang et al. (2010); Mukhopadhyay et al. (2022)

13.2.3 Environmental Compatibility

Clays are non-toxic, biodegradable, and regenerable materials with demonstrated environmental safety, making them sustainable candidates for remediation. However, natural clays show limitations in selectivity and sorption rates, which nanostructuring helps overcome (Naseer et al., 2025) (Table 13.1).

13.3 Clay-Based Nanocomposites

Clay nanoparticles combined with polymers, carbonaceous materials, and metal/metal oxide nanoparticles form nanocomposites exhibiting synergistic effects such as enhanced adsorption capacity, selectivity, and antimicrobial activity (Mukhopadhyay et al., 2022).

13.3.1 Clay-Polymer Nanocomposites

Natural polymers like chitosan, alginate, and cellulose intercalated or grafted on clay surfaces enhance biocompatibility and introduce functional groups (amine, carboxyl) that interact selectively with contaminants via electrostatic attraction, chelation, or hydrogen bonding (Han et al., 2010; Hernández-Hernández et al., 2018). For example, chitosan/montmorillonite composites achieve up to 93% removal of Pb^{2+} and exhibit significant antimicrobial activity (Han et al., 2010; Deng et al., 2011).

Synthetic polymers such as polyacrylamide improve mechanical robustness and chemical stability.

13.3.2 Clay-Carbon Nanocomposites

Incorporation of activated carbon, graphene oxide (GO), and carbon nanotubes (CNTs) produces composites with large surface area, porosity, and hydrophobicity, tailored for organic pollutants and pharmaceuticals (Mukhopadhyay et al., 2022). GO/clay nanocomposites target aromatic compounds through π-π interactions, while CNTs enhance electrical conductivity enabling photocatalytic degradation (Rezania et al., 2024).

13.3.3 Clay-Metal and Metal Oxide Nanocomposites

Metal nanoparticles (silver, copper) and metal oxides (ZnO, TiO_2) immobilized on clay surfaces provide potent antimicrobial and photocatalytic functions (Deng et al., 2011; Gitari et al., 2020). These composites generate reactive oxygen species under light irradiation, disrupting microbial membranes and degrading organic contaminants, while clay matrices prevent nanoparticle aggregation and leaching, ensuring stability and recyclability (Table 13.2).

Table 13.2 Summary of clay-based nanocomposite types and key features

Nanocomposite type	Components	Key advantages	Typical contaminants targeted	References
Clay-polymer	Clay + chitosan, alginate, etc	Biocompatible, selective adsorption	Heavy metals, dyes, pathogens	Han et al. (2010); Wu and Chen (2013)
Clay-carbon	Clay + activated carbon, GO, CNTs	High surface area, photocatalytic activity	Organic pollutants, pharmaceuticals	Azizi et al. (2017); Shahadat and Isamil (2018)
Clay-metal/metal oxide	Clay + Ag, ZnO, TiO_2 nanoparticles	Antimicrobial, catalytic degradation	Microbes, organic pollutants	Deng et al. (2011); Undabeytia et al. (2014)

13.4 Mechanisms of Contaminant Removal

Removal efficacy of clay-based nanomaterials arises from multifaceted physico-chemical mechanisms:

i. Ion Exchange: Exchangeable cations within clay interlayers (e.g., Na^+, Ca^{2+}) are replaced by heavy metal ions (Pb^{2+}, Cd^{2+}, Cr^{6+}). Efficiency depends on ionic radius, charge, hydration energy, pH, and presence of competing ions (Ijagbemi et al., 2009; Naseer et al., 2025).

ii. Physical Adsorption: High surface area and porosity facilitate adsorption of organic pollutants and dyes via Van der Waals forces and pore filling, often modelled by Langmuir and Freundlich isotherms (Wu & Chen, 2013).

iii. Electrostatic Attraction: Functionalization with charged polymers imparts surface charges enhancing attraction of oppositely charged contaminants (Han et al., 2010).

iv. Catalytic and Antimicrobial Activity: Metal/metal oxide nanoparticles generate reactive oxygen species that degrade organics under UV/visible light and disrupt microbial cell membranes causing inactivation (Deng et al., 2011; Undabeytia et al., 2014).

Synergistic operation of these mechanisms in engineered nanocomposites allows for simultaneous removal of multiple contaminant classes.

13.5 Synthesis and Modification Techniques

Preparation of clay-based nanocomposites involves:

i. Surface Functionalization: Grafting functional groups (amine, carboxyl, sulfonic acid) enhances affinity towards target pollutants via chemical bonding or electrostatics (Hernández-Hernández et al., 2018).
ii. Pillaring and Intercalation: Inserting metal oxides (Al_2O_3, TiO_2, Fe_2O_3) or organic molecules between clay layers expands surface area and creates active sites, improving adsorption and mechanical strength (Wu & Chen, 2013).
iii. Hybridization with Allied Nanomaterials: Integration with nanocellulose, graphene oxide, or CNTs introduces multifunctionality (adsorption, photocatalysis, conductivity) (Mukhopadhyay et al., 2022).
iv. Green Synthesis Approaches: Use of plant extracts, biopolymers, and microwave-assisted synthesis reduces environmental footprints while improving nanoparticle morphology and stability (Naseer et al., 2025).

Synthesis methods such as sol-gel, coprecipitation, and hydrothermal techniques affect particle size, shape, and distribution, impacting performance and scalability. The processing of fine particles hydrothermally to nanocrystalline materials could be dated back to around 1870s, but the synthesized materials could not be confirmed due to limitations in the instruments available then, except some chemical techniques (Byrappa et al., 2006). The advantages of hydrothermal technology over other conventional processes are energy saving, simplicity, cost-effectiveness, better nucleation control, pollution-free (since the reaction is carried out in a closed system), higher dispersion, higher rate of reaction, better shape control, and lower temperature of operation in the presence of an appropriate solvent, etc. (Yoshimura & Byrappa, 2008). Hydrothermal technology is now emerging as a vital tool in nanocomposites preparation. Examples of nanocomposites are zeolites: CNT, TiO_2: activated carbon, ZnO: activated carbon, TiO_2: CNT, TiO_2: HAp, ferrites: polymer, organic–inorganic composites, polymer and semiconductor composites (Byrappa et al., 2006: Dayananda et al., 2007; Obijole et al., 2021).

13.6 Performance and Applications

Clay-based nanocomposites demonstrate exceptional removal efficiencies across laboratory to pilot scales:

i. Heavy Metals: Chitosan/montmorillonite composites remove 90–95% of Pb^{2+}, Cd^{2+}, and Hg^{2+}, outperforming natural clays (Han et al., 2010).
ii. Organic Pollutants: GO/clay and clay-CNT composites adsorb dyes (methylene blue, Congo red) and pharmaceuticals with efficiencies exceeding 80% (Shahadat & Isamil, 2018).
iii. Microbial Inactivation: Silver- and copper-modified clays achieve > 99% bacterial removal, supporting applications in point-of-use filtration (Deng et al., 2011).

Nanocomposites are applied as powders, pellets, membranes, filters, and integrated in packed columns or bioreactors. Regeneration methods (acid/base washing, advanced oxidation) help maintain performance and sustainability (Naseer et al., 2025) (Table 13.3).

Table 13.3 Representative performance metrics of clay-based nanocomposites

Material	Contaminant	Removal efficiency (%)	Application scale	References
Chitosan/ montmorillonite	Pb^{2+}	93	Laboratory	Han et al. (2010)
GO/clay nanocomposite	Methylene blue dye	85	Pilot-scale	Azizi et al. (2017)
Silver-modified clay	E. coli bacteria	> 99	Point-of-use filtration	Deng et al. (2011)
Clay-CNT composite	Carbamazepine	78	Laboratory	Shahadat and Isamil (2018)l

13.7 Challenges and Future Prospects

Challenges

i. Scalability and Cost: Large-scale synthesis with consistent quality remains difficult. Costly raw materials like graphene and CNTs limit affordability (Mukhopadhyay et al., 2022).
ii. Stability and Regeneration: Long-term stability, fouling resistance, and facile regeneration without material degradation are essential (Shahadat & Isamil, 2018).
iii. Environmental and Health Risks: Potential nanoparticle release raises toxicity concerns necessitating comprehensive risk assessments, lifecycle analyses, and safe disposal practices (Bel Hadjltaief et al., 2016).

Future Directions

i. Development of multifunctional, recyclable nanocomposites capable of removing diverse contaminants concurrently.
ii. Industrial scale-up of green synthesis pathways reducing energy and chemical consumption.
iii. Integration of smart sensing and real-time monitoring systems to dynamically adjust treatment processes.
iv. Enhanced interdisciplinary collaboration to address regulatory, economic, and social implementation barriers (Mukhopadhyay et al., 2022;).

13.8 Conclusion

Clay nanoparticles and their nanocomposites represent a sustainable, cost-effective frontier in water remediation. Through multifunctional design combining clay with polymers, carbonaceous materials, and metal nanoparticles, these materials achieve superior removal of heavy metals, organic pollutants, and pathogens. Continued innovation in synthesis, performance optimisation, environmental safety, and scale-up is vital to advance practical applications, ultimately contributing to global water security and public health.

References

Aldaco, R., Irabien, A., & Luis, P. (2005). Fluidized bed reactor for fluoride removal. *Chemical Engineering Journal, 107*(1–3), 113–117.

Amuda, O. S., Ojo, O. I., & Edewor, T. I. (2007). Biosorption of Lead from industrial wastewater using Chrysophyllum albidum seed Shell. *Bioremediation Journal, 11*(4), 183–194.

Azizi, S. N., Ahmadi, F., & Asl, A. H. (2017). Clay-based nanocomposite particles for pollutant removals: A critical review. *Water Science & Technology, 85*(10), 3002–3022. https://doi.org/10.2166/wst.2017.211

Bel Hadjltaief, H., Da Costa, P., Galvez, M. E., & Zina, M. B. (2016). Clay-based nanocomposite materials used in treatment of wastewater. *Desalination and Water Treatment, 57*(60), 29160–29177. https://doi.org/10.1080/19443994.2016.1173208

Bretzler, A., & Johnson, C. A. (2015). The geogenic contamination handbook: Addressing arsenic and fluoride in drinking water. *Applied Geochemistry, 63*, 642–646.

Byrappa, K., Subramani, A. K., Ananda, S., Rai, K. L., Sunitha, M. H., Basavalingu, B., & Soga, K. (2006). Impregnation of ZnO onto activated carbon under hydrothermal conditions and its photocatalytic properties. *Journal of Materials Science, 41*(5), 1355–1362.

Dayananda, A. S., Sajan, C. P., Basavalingu, B., Byrappa, K., Soga, K., & Yoshimura, M. (2007). Hydrothermal preparation of ZnO: CNT and TiO 2: CNT composites and their photocatalytic applications. *Journal of Materials Science, 42*, 231.

Deng, S., Ting, Y. P., & Liu, Y. (2011). Silver nanoparticle-modified clay composites for microbial removal. *Journal of Colloid and Interface Science, 360*(2), 548–554. https://doi.org/10.1016/j.jcis.2011.04.038

Fayomi, O. S. I., Ojo, O. I., & Popoola, A. P. I. (2011). Investigating (Ochadamu) silica sand, clay and local oils for foundry core. *International Journal of the Physical Sciences, 6*(8), 1894–1904.

Ganvir, V., Biswas, K., & Kokil, S. (2002). A novel water filter technology for rural areas. In *28th WEDC conference: Sustainable environmental sanitation and water services.*

Ghorai, S., & Pant, K. K. (2005). Equilibrium, kinetics and breakthrough studies for adsorption of fluoride on activated alumina. *Separation and Purification Technology., 42*, 265–271.

Gitari, W. M., Izuagie, A. A., & Gumbo, J. R. (2020). Synthesis, characterisation and batch assessment of groundwater fluoride removal capacity of trimetal Mg/Ce/Mn oxide-modified diatomaceous earth. *Arabian Journal of Chemistry, 13*(1), 1–16.

Gitari, M. W., Obijole, O. A., & Mudzielwana, R. (2022). Synthesis of porous-hydroxysodalite from aluminosilicate clay soils: Application towards fluorideand pathogen removal from groundwater. *Environmental Science and Pollution Research.* https://doi.org/10.1007/s11356-022-23575-8

Han, R., Zou, W., Yu, W., Cheng, S., Wang, Y., & Shi, J. (2010). Chitosan/montmorillonite nanocomposites for removal of E. Coli and S. Aureus. *Journal of Hazardous Materials, 181*(1–3), 341–347. https://doi.org/10.1016/j.jhazmat.2010.05.024

Hernández-Hernández, A., Martínez-Vargas, S., & Romero-Cano, L. A. (2018). Removal of phenols using polymer-reinforced clay nanocomposites. *Environmental Science and Pollution Research, 25*(6), 5277–5287. https://doi.org/10.1007/s11356-017-0709-2

Ijagbemi, C. O., Baek, M. H., & Kim, D. S. (2009). Montmorillonite surface properties and sorption characteristics for heavy metals removal from aqueous solutions. *Journal of Hazardous Materials, 166*(1), 538–546. https://doi.org/10.1016/j.jhazmat.2008.11.076

Jiang, M. Q., Jin, X. Y., Lu, X. Q., & Chen, Z. L. (2010). Adsorption of Pb(II), cd(II), Ni(II), and cu(II) onto natural kaolinite clay. *Desalination, 252*(1–3), 33–39. https://doi.org/10.1016/j.desal.2009.11.005

Kut, K. M. K., Sarswat, A., Srivastava, A., Pittman, C. U., Jr., & Mohan, D. (2016). A review of fluoride in African groundwater and local remediation methods. *Groundwater for Sustainable Development, 2*, 190–212.

Loganathan, P., Vigneswaran, S., Kandasamy, J., & Naidu, R. (2013). Defluoridation of drinking water using adsorption processes. *Journal of Hazardous Materials, 248*, 1–19.

Mukhopadhyay, S., et al. (2022). A critical review on clay-based nanocomposite particles for pollutant removals. *Water Science and Technology, 85*(10), 3002–3022. https://doi.org/10.2166/wst.2017.211

Naseer, A., Younas, F., Munir, R., Sardar, M. F., Noreen, S., & Cui, Z. (2025). Clay-based nanocomposite materials used in treatment of wastewater: Recent advances and challenges. *Journal of Environmental Nanotechnology, 12*(2), 121–135. https://doi.org/10.1080/15422119.2024.2369872

Nuruzzaman M., Rahman M.M., Liu Y., Naidu R. Nanoencapsulation, (2016). Nano-guard for pesticides: A new window for safe application. Journal of Agricultural and Food Chemistry 1447–1483. https://doi.org/10.1021/acs.jafc.5b05214.

Obijole, O. A., Gitari, M. W., Ndungu, P. G., & Samie, A. (2019). Mechanochemically Activated Aluminosilicate Clay Soils and their Application for Defluoridation and Pathogen Removal from Groundwater. *International Journal of Environmental Research and Public Health, 16*(4), 654.

Obijole, O. A., Gitari, M. W., Mudzielwana, R., Ndungu, P. G., Samie, A., & Babatunde, A. W. (2021). Hydrothermally treated aluminosilicate clay (HTAC) forremediation of fluoride and pathogens from water: Adsorbent characterization and adsorption modelling. *Water Resources and Industry*, 100144.

Otieno, F. A. O., Olumuyiwa, I. O., & Ochieng, G. M. (2012). Groundwater: Characteristics, qualities, pollutions and treatments: An overview. *International Journal of Water Resources and Environmental Engineering, 4*(6), 162–170.

Rezania, S., Darajeh, N., Rupani, P. F., Mojiri, A., Kamyab, H., & Taghavijeloudar, M. (2024). Recent advances in the adsorption of different pollutants from wastewater using carbon-based and metal-oxide nanoparticles. *Applied Sciences, 14.*

Shahadat, M., & Isamil, S. (2018). Reusability of clay/polymer nanocomposites for pollutant adsorption. *Journal of Environmental Management, 223*, 517–528. https://doi.org/10.1016/j.jenvman.2018.06.052

Uddin, F. (2017). Montmorillonite: An introduction to properties and utilization. *Current Topics in Materials Science, 7*, 113–126. https://doi.org/10.5772/67853

Undabeytia, T., Morillo, E., & Maqueda, C. (2014). Antimicrobial activity of cationic clay nanocomposites. *Applied Clay Science, 87*, 129–135. https://doi.org/10.1016/j.clay.2013.12.013

Yoshimura, M., & Byrappa, K. (2008). Hydrothermal processing of materials: Past, present and future. *Journal of Materials Science, 43*(7), 2085–2103.

Waghmare, S. S., & Arfin, T. (2015). Fluoride removal by clays, geomaterials, minerals, low cost materials and zeolites by adsorption: A review. *International Journal of Science, Engineering and Technology Research, 4*(11), 3663–3676.

World Health Organization. (2019). *Preventing disease through healthy environments: Inadequate or excess fluoride: A major public health concern* (Vol. No. WHO/CED/PHE/EPE/19.4, p. 5). World Health Organization.

Wu, F. C., & Chen, K. Y. (2013). Adsorption of tannic acid by protonated clay with chitosan resin. *Bioresource Technology, 144*, 324–332. https://doi.org/10.1016/j.biortech.2013.06.096

Chapter 14
Adsorptive Kaolin Clay-Based Desalination for Coastal Water Resource Management

Rebecca Adepate Olaoye, Alimot Olawumi Olatunji, Victor Adetunji Ebo, and Rasaq Rahman Owolabi

Abstract Freshwater scarcity remains a critical challenge in coastal regions, particularly in developing nations. This study investigates kaolin clay, a locally available and cost-effective material, as a sustainable adsorbent for coastal water desalination. Raw and activated kaolin clay was characterized using X-ray diffraction (XRD) and SEM. Fresh coastal water was collected and characterized for physicochemical parameters, including salinity, pH, turbidity, total dissolved solids (TDS), and major ionic species. Batch adsorption experiments were conducted to evaluate the effects of contact time, adsorbent dosage, temperature, and feed water concentration on salt removal efficiency. The adsorption data were analyzed using the Langmuir and the Freundlich isotherm models. Results showed that activated kaolin achieved 65% salinity removal within 30 min of contact time, with maximum removal efficiencies of 76.8% at 3 g/100 mL adsorbent dosage and 78.1% at 60 °C. The adsorption process followed the Langmuir and the Freundlich isotherm models with high correlation coefficients ($R^2 = 0.9961$). The results revealed that the Langmuir isotherm provided a better fit to the experimental data with correlation coefficients ($R > 0.85$) for activated kaolin clay. These findings demonstrate that kaolin clay is an effective, energy-efficient, and economically viable adsorbent for treating highly saline coastal waters, offering promise for resource-limited communities requiring affordable desalination solutions.

Keywords Freshwater scarcity · Kaolin clay · Adsorbent · Seawater · Coastal water · Desalination

---

R. A. Olaoye (✉) · A. O. Olatunji · V. A. Ebo
Department of Civil Engineering, Ladoke Akintola University of Technology, Ogbomoso, Nigeria
e-mail: raolaoye@lautech.edu.ng

A. O. Olatunji
e-mail: olatunjialimot2016@gmail.com

R. R. Owolabi
Department of Geological Sciences, Achievers University, Owo, Ondo State, Nigeria
e-mail: owolabi.rr@achievers.edu.ng

M. I. Adeoba et al. (eds.), *Water Resources Management in Nigeria*, World Water Resources 10, https://doi.org/10.1007/978-3-032-22813-0_14

14.1 Introduction

Water is fundamental to human survival, yet its distribution on Earth remains uneven. Although water covers approximately three-quarters of Earth's surface, approximately 97% exists as saline water in oceans, leaving only about 2.5–3% as freshwater in glaciers, ice caps, groundwater, and surface water bodies (Gleick, 1993). Freshwater scarcity affects billions of people worldwide, particularly in coastal regions, arid and semi-arid environments, and small island nations. Reports indicate that nearly two-thirds of the global population experience water shortages for at least one month annually, with projections showing water demand will increase by more than 50% by 2050 (Robert et al., 2024). This scarcity extends beyond drinking water access to encompass food security, industrial productivity, public health, and environmental sustainability (Olaoye et al., 2020).

The consequences of freshwater scarcity are particularly severe in developing regions. Contaminated water serves as a vector for waterborne diseases, including cholera, diarrhea, and dysentery, which remain leading causes of mortality in regions with limited access to medical facilities and potable water (Khawaji et al., 2008; Olaoye et al., 2024). Access to clean, reliable freshwater is therefore essential for human health and economic development.

Seawater desalination has emerged as an essential alternative for augmenting freshwater supplies in water-scarce regions. However, conventional desalination technologies, including reverse osmosis (RO), multi-stage flash distillation (MSF), membrane distillation, and electrodialysis, often suffer from high energy consumption, membrane fouling, and substantial operational costs, limiting their large-scale application in regions lacking adequate financial resources (Elimelech & Phillip, 2011; Muhammad et al., 2022).

Adsorption-based desalination has emerged as a promising alternative due to its low energy requirements, cost-effectiveness, and environmental sustainability compared to thermal or membrane-based methods. Natural adsorbents such as clay minerals, activated carbon, and zeolites have demonstrated significant potential for removing salt ions from seawater. Kaolin-clay, a naturally abundant aluminosilicate, has emerged as a versatile material for low-cost desalination and water purification owing to its unique combination of structural, chemical, and surface properties. The mineral's crystalline layers of silica and alumina offer high chemical stability and mechanical strength, which are essential for maintaining integrity under variable salinity and temperature conditions typical of desalination systems (Mustapha et al., 2019). Its moderately high surface area and presence of hydroxyl functional groups at the particle edges provide reactive sites for ion adsorption and exchange processes. These physicochemical characteristics make kaolin clay an effective sorbent for a wide range of dissolved species, including sodium, calcium, magnesium, and chloride ions, which are key contributors to salinity and hardness in saline waters (Khan et al., 2023).

Beyond direct ion adsorption, kaolin clay's adaptability allows it to be chemically modified to enhance its adsorption efficiency and hydrophilicity. Such modifications increase the density of active sites, thereby improving ion removal performance and compatibility with other treatment media (Khan et al., 2023). In membrane-based desalination systems, kaolin clay has also shown substantial promise as an inexpensive filler. Its integration into polymeric or ceramic membranes not only improves hydrophilicity and water permeability but also reduces fouling and enhances long-term operational stability (Aloulou et al., 2022; Fareed et al., 2022). These effects are attributed to kaolin clay's ability to form microchannels that facilitate water transport while maintaining selective rejection of ions and organic pollutants. Collectively, these attributes demonstrate that kaolin clay's role in desalination extends well beyond traditional adsorption. Its low cost, locally available, tunable surface chemistry, and dual functionality as both an adsorbent and membrane enhancer position it as a sustainable and scalable material for developing hybrid desalination systems. In regions where conventional desalination remains expensive, kaolin clay-based technologies can bridge the gap between affordability and performance, contributing to improved water security and resource management (Mustapha et al., 2019; Khan et al., 2023).

Despite the growing global interest in low-cost adsorbents for desalination, research on kaolin clay applications remains largely confined to wastewater and industrial effluent treatment, with limited emphasis on saline or coastal water systems. Previous studies have demonstrated that activated kaolin clay exhibits enhanced ion removal efficiency (Mierzwa et al., 2019; Umudi et al., 2022), making it a viable adsorbent for desalination. For instance, kaolin clay used in column configurations successfully reduced chloride concentrations in seawater collected from the axis of the Niger Delta (Umudi, 2021). Similarly, acid-activated kaolin clay has proven effective in removing reactive dyes and heavy metals from aqueous media (Faisle et al., 2023; Avornyo et al., 2025). However, these studies are limited in scope, particularly regarding ion-specific adsorption from natural coastal waters and the comparative performance of kaolin clay from different Nigerian deposits.

Most available Nigerian kaolin clay research focuses on removing organic dyes, heavy metals, or nutrients from wastewater (Abu et al., 2023; Faisle et al., 2023, Mustapha et al., 2019; Khan et al., 2023; Ogundipe et al., 2023a), leaving a paucity of data on its performance in treating high-salinity waters. The adsorption isotherms, adsorption capacities, and competitive ion behaviors under coastal water conditions remain understudied. Consequently, the applicability of kaolin clay for desalination or water treatment is still uncertain and limited. Only a few studies, such as Umudi (2021), have examined chloride removal from seawater using kaolin, yet there is insufficient data on multi-ion systems containing sodium, magnesium, calcium, sulphate and other ions. The absence of comprehensive adsorption isotherms and thermodynamic analyses for these ions limits understanding of kaolin clay's full desalination potential. Moreover, many laboratory experiments employ batch systems with synthetic saline solutions rather than real coastal water samples, thus failing to capture the complex ionic interactions that occur under natural environmental conditions.

Few among the natural clays that have been utilized for desalination are the raw and acid-activated local bentonite, which was investigated for the removal of only sodium from saline water in Ethiopia by Musie and Gonfa (2022).

The performance of activated kaolin clay, particularly after acid modification, has not been rigorously evaluated for desalination. Although activation is known to increase surface area and active site density, its impact on ion exchange and adsorption selectivity in saline environments is not well-documented. Likewise, few studies have adopted flow-through or column systems that simulate real-time water treatment. In addition, kaolin clay sources vary significantly in mineralogical composition and surface properties. The absence of techno-economic and field-scale feasibility studies also constrains adoption in resource-limited coastal regions. Addressing these gaps through systematic evaluation of activated kaolin clay for salt ion adsorption and coastal water desalination will not only expand scientific understanding but also provide sustainable, locally sourced solutions for water security in vulnerable coastal communities in Nigeria.

This study assesses activated kaolin clay derived from Nigerian deposits for its effectiveness in reducing salinity and other physicochemical constituents from coastal water samples, thereby contributing to practical solutions for water security in resource-limited coastal regions.

14.2 Materials and Methods

14.2.1 Materials and Sample Collection

Raw kaolin clay samples were sourced from a deposit located at coordinates 3° 38.6828′ E and 6° 56.4649′ N in Oyo town, Oyo State, Nigeria. Hydrochloric acid of analytical grade was used for clay activation. Fresh coastal water samples were collected at Oniru beach, Lekki, Lagos (3° 51′ 13.6188″ E, 6° 26′ 20.2452″ N) in sterile bottles pre-rinse with sample following standard sampling protocols.

14.2.2 Adsorbent Preparation and Activation

Collected kaolin clay samples were pretreated to remove debris and subsequently air-dried at ambient temperature for three days to eliminate moisture. The samples were crushed using a ceramic pestle and mortar, then sieved through a 75 μm mesh to obtain fine particles. Impurities were removed through physical separation followed by a wet-soaking method: 1000 g of sieved clay was soaked in 1000 cm^3 of distilled water for 24 h, then washed through a 100-μm sieve and allowed to settle. The resulting slurry was oven-dried at 110 °C for 24 h. Following drying, 500 g of the kaolinite product ($Al_2Si_2O_5(OH)_4$) was calcined in a furnace at 700 °C for 3 h to obtain meta kaolinite. The calcined clay was further treated using an acid solution prepared by diluting 22 cm^3 of concentrated hydrochloric acid (36.46 M) in 4000 cm^3 of distilled water, resulting in an approximate acid concentration of 0.20 M HCl. This solution was mixed with 200 g of calcined clay slurry, stirred using a magnetic stirrer, covered, and allowed to settle for 2 days. The water was decanted, and this treatment was repeated twice to oxidize organic matter. Subsequently, 10 cm^3 of 36.46 M HCl was added for bleaching, the mixture was stirred and allowed to settle for 2 days, then washed multiple times with distilled water, oven-dried, crushed, and stored for further analysis.

14.2.3 Adsorbent Characterization

The raw and activated adsorbents were characterized using X-ray Diffraction (XRD) to understand the crystalline structures of the kaolin clay and to identify the crystalline materials that exist in the solid specimen, while Scanning Electron Microscopy (SEM) analysis was used to reveal adsorbent texture, roughness, and particle shape, which directly relates to the availability of active sites and adsorption capacity.

14.2.4 Coastal Water Characterization

The collected raw coastal water was characterized for salinity, pH, turbidity, total dissolved solids (TDS), potassium (K^+), sulfate (SO_4^{2-}), calcium (Ca^{2+}), magnesium (Mg^{2+}), sodium (Na^+), chloride (Cl^-), copper (Cu), and iron (Fe) using methods described in the American Public Health Association (APHA, 2005) guidelines.

14.2.5 Batch Adsorption

Batch adsorption experiments were conducted using raw coastal seawater to evaluate the removal of major ions (Na^+, K^+, Ca^{2+}, Mg^{2+}, Cl^-, SO_4^{2-}), TDS, and turbidity by activated kaolin. All experiments were performed in 250 cm^3 Erlenmeyer flasks containing 100 mL of coastal water solution and agitated using an orbital shaker at 150 rpm to ensure uniform mixing. The use of raw seawater without pre-treatment reflects realistic field conditions for coastal water management applications.

The experimental setup was designed to study the influence of contact time, adsorbent dosage, temperature, and feed concentration on adsorption efficiency. All tests were performed in duplicate, and the mean values were used for analysis. After each experiment, samples were filtered through filter paper, and the residual concentrations of target ions were analyzed. All experiments were conducted at ambient lab temperature.

14.2.6 *Effect of Contact Time*

To determine the time required to achieve adsorption equilibrium, activated kaolin of 1.0 g was added to 100 mL of coastal water in a 250 cm^3 conical flask. The mixture was shaken at 150 rpm at predetermined time intervals (0, 5, 10, 15, 20, 25, 30 min) using an orbital shaker. After each interval, the mixture was filtered using filter paper, and the residual concentrations of ions were determined. The contact time corresponding to maximum removal efficiency was considered the optimum contact time.

14.2.7 *Effect of Adsorbent Dosage*

The experiment was repeated using varying kaolin masses (0.4, 0.6, 0.8, 1.0, 2.0, and 3.0 g) with 100 mL coastal water at the optimum contact time of 30 min, shaken at 150 rpm. The dosage that yielded the highest removal efficiency with minimal excess adsorbent was selected for subsequent tests.

14.2.8 *Effect of Temperature*

Adsorption was investigated at temperatures ranging from 20, 30, 40, 50, and 60 °C using 1.0 g of adsorbent in 100 mL coastal water, shaken at 150 rpm until equilibrium.

14.2.9 *Effect of Feed Concentration*

Coastal water was used at its original concentration (baseline) and diluted with distilled water to obtain solutions of 20, 40, 60, 80, and 100 mg/L ionic strength. Each batch experiment was carried out using 1.0 g of the activated kaolin clay in 100 mL of solution, agitated at 150 rpm for the specified contact times. All experiments were conducted in duplicate to ensure reproducibility isotherm analysis and efficiency calculations.

The removal efficiency (Re, %) and the amount of adsorbate absorbed at equilibrium q_e, mg/g) were calculated using Eqs. 14.1 and 14.2:

$$R_e(\%) = \frac{(C_o - C_e)}{C_o} \times 100 \tag{14.1}$$

$$q_e = \frac{(C_o - C_e)}{m} \times V \tag{14.2}$$

Where

q_e adsorbed amount per unit mass of kaolin at equilibrium (mg/g);
Re percentage removal at equilibrium (%);
C_0 initial ion concentration (mg/L);
Ce equilibrium ion concentration (mg/L);
V solution volume (L);
m kaolin mass (g).

The adsorption data were fitted to Langmuir and the Freundlich isotherm models. The Langmuir model assumes monolayer adsorption on homogeneous surfaces, while the Freundlich model describes adsorption on heterogeneous surfaces. Six forms of the Langmuir equation were evaluated to identify the best-fitting model for each pollutant. The correlation coefficient (R^2) was used to evaluate model fit quality.

14.2.10 Investigation of Suitable Adsorption Isotherm

An adsorption isotherm is important to describe how adsorbates interact with adsorbents. The correlation between the experimental data and the model-predicted values was evaluated using the correlation coefficient (R^2). Langmuir and the Freundlich isotherm models are the most used isotherms in presenting data of adsorption from solutions because they work well across various adsorbents, including activated clay and carbon-based materials. They also provide well-established models to describe the adsorption process, capturing different aspects of how adsorbate (ions and molecules) interacts with adsorbents like activated kaolin clay. Furthermore, a dimensionless constant called the Langmuir equilibrium parameter (R_L) was used to determine the nature of the adsorption process as shown in Eq. (14.3), (Ayawei et al., 2017; Al-Ghouti and Da'ana, 2020).

$$R_L = \frac{1}{K_L C_O + 1} \tag{14.3}$$

where C_o and K_L are the initial concentration of the pollutants (mgL^{-1}) and Langmuir adsorption constant (Lmg^{-1}), respectively.

For adsorption that takes place on heterogeneous surfaces, Freundlich isotherm is applicable. The surface heterogeneity and the exponential distribution of active sites and their energy are defined by the Freundlich isotherm. The Freundlich isotherm may be expressed linearly as shown in Eq. (14.4) (Emmanuel et al., 2019). It also shows the relative distribution of energy and the heterogeneity of the adsorbate sites;

$$\log q_e = \log K_f + \frac{1}{n} \log C_e \tag{14.4}$$

Table 14.1 Langmuir and Freundlich isotherm models

Model	Equation	Plot	Equation no.
Langmuir	$q_e = q_m - \frac{1}{K_L}\frac{q_m}{C_e}$	q_e *against* $\frac{1}{C_e}$	14.5
Freundlich	$\log q_e = \log K_F + \log C_e$	$\log q_e$ against $\log C_e$	14.6

where K_f is the adsorption capacity (Lmg^{-1}), and 1/n is the adsorption intensity.

The Langmuir isotherm assumes basically that adsorption takes place at specific homogeneous sites within the adsorbent. Literature usually presents linear forms of the Langmuir isotherm (Bolster and Hornberger, 2007). The two equations applied in this study are shown in Eqs. (14.5) and (14.6) for the Langmuir and Freundlich isotherms, respectively and the corresponding equations for obtaining the plots are presented in Table 14.1.

14.3 Results and Discussion

14.3.1 Characterization of Adsorbent

14.3.1.1 X-Ray Diffraction (XRD)

The (XRD) patterns of the raw (Fig. 14.1, Table 14.2) and activated kaolin samples (Fig. 14.2, Table 14.3) reveal significant structural and mineralogical transformations resulting from the activation process. In the raw kaolin, distinct peaks were observed at 2θ values between 11° and 74°, which correspond primarily to the crystalline planes of kaolinite, quartz, and anatase. The dominant reflections at 11.0°, 16.98°, and 26.72° 2θ indicate that kaolinite is the major mineral phase, while minor peaks at 32.57° and 50.00° 2θ correspond to quartz. The presence of anatase (TiO_2) was also detected around 29.78° and 45.00° 2θ, suggesting a minor titanium oxide component in the raw sample. These sharp and well-defined peaks reflect the high crystallinity and structural order characteristic of natural kaolin. The relative abundance of kaolinite (63.63%) and quartz (29.35%) confirms the predominance of aluminosilicate layers in the raw sample.

Following acid activation, notable changes in the diffraction pattern were observed. The characteristic kaolinite peaks diminished in intensity, indicating partial dissolution of the alumina layers and disruption of the octahedral sheets. New diffraction peaks emerged at 9.0°, 18.5°, and 21.68° 2θ, which correspond to quartz, microcline, quartz and the other secondary mineral phases such as illite, dolomite, and calcite. The appearance of a broad reflection around 9° 2θ, corresponding to an increased interlayer spacing (d = 9.824 Å), suggests a reduction in crystallinity and the development of amorphous regions. These structural modifications are typical indicators of enhanced surface porosity and higher cation exchange potential in acid-activated clays.

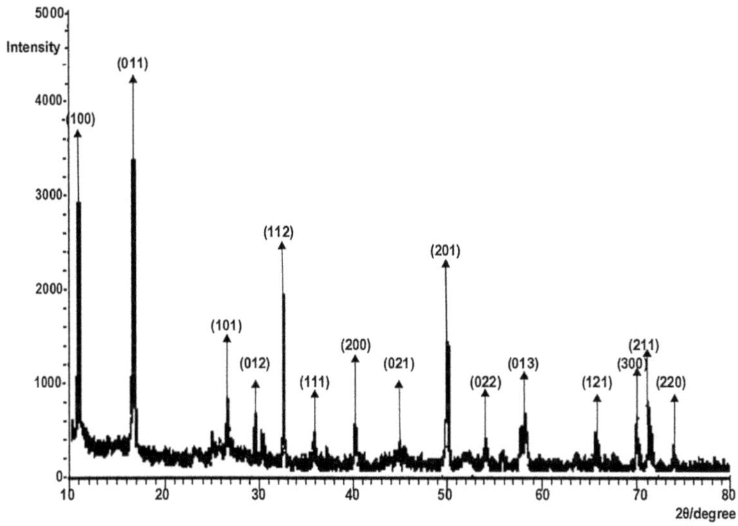

Fig. 14.1 XRD for raw kaolin clay

Table 14.2 The XRD diffraction pattern of raw kaolin

Peak	2θ/degree	Plane	Intensity	d-Valve (Å)	Minerals	% Composition
1	11.00	1 0 0	61.13	5.5361	Kaolinite	17.55
2	16.98	0 1 1	69.25	5.2211	Kaolinite	19.88
3	26.72	1 0 1	17.10	3.3365	Kaolinite	4.91
4	29.78	0 1 2	15.22	3.0002	Anatase	4.37
5	32.57	1 1 2	39.16	2.7493	Quartz	11.24
6	36.00	1 1 1	6.21	2.4948	Kaolinite	1.78
7	40.15	2 0 0	16.32	2.2459	Kaolinite	4.65
8	45.00	0 2 1	9.12	2.0144	Anatase	2.62
9	50.00	2 0 1	33.10	1.8241	Quartz	9.50
10	54.26	0 2 2	9.38	1.6906	Kaolinite	2.69
11	58.20	0 1 3	16.22	1.5851	Quartz	4.66
12	66.00	1 2 1	13.75	1.4146	Quartz	3.95
13	70.00	3 0 0	16.00	1.3439	Kaolinite	4.59
14	71.00	2 1 1	17.32	1.3275	Kaolinite	4.97
15	74.02	2 2 0	9.10	1.2792	Kaolinite	2.61

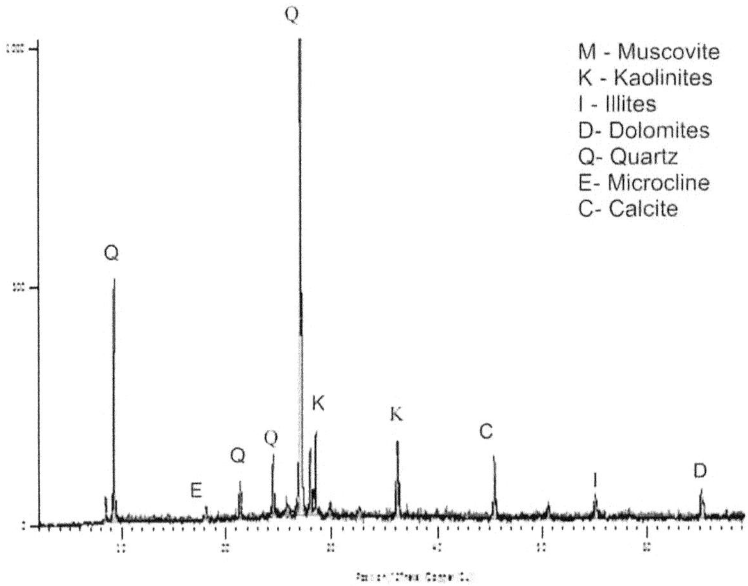

Fig. 14.2 XRD for activated kaolin clay

Table 14.3 The XRD diffraction pattern of activated kaolin

Peak	2θ (degrees)	Plane	Intensity	d-Value (Å)	Mineral	% Composition
1	9.0	1 0 0	25.1	9.824	Quartz	10.5
2	18.5	0 1 1	30.25	4.795	Microcline	8.2
3	21.68	1 0 1	65.2	4.097	Quartz	12.4
4	24.36	0 1 2	40.8	3.655	Quartz	9.8
5	27.28	1 1 2	50.65	3.27	Quartz	11.7
6	28.31	1 1 1	20.1	3.147	Kaolinite	6.5
7	36.26	2 0 0	45.32	2.476	Kaolinite	9.0
8	45.44	0 2 1	15.67	1.995	Calcite	4.6
9	55.7	2 0 1	35.48	1.649	Illite	8.3
10	65.83	0 2 2	18.55	1.417	Dolomite	6.0

The reduction of kaolinite from approximately 64% in the raw sample to about 15.5% in the activated form, along with the formation of new mineral species, implies a more heterogeneous and reactive surface. The increase in quartz intensity and the presence of exchangeable cations (Ca^{2+}, Mg^{2+}, and K^{+}) in dolomite, calcite, and illite further enhance the material's ability to interact with adsorbates.

Consequently, the activated kaolin is expected to demonstrate higher adsorption efficiency than the raw material due to its enlarged surface area, improved pore accessibility, and greater abundance of hydroxyl functional sites thereby improving cation exchange and ion adsorption capacity. These findings align with previous studies that reported improved adsorptive and ion-exchange behavior following acid modification of natural clays (Mustapha et al., 2019; Avornyo et al., 2025; Aloulou et al., 2022).

14.3.1.2 Scanning Electron Microscopy (SEM)

The surface morphology of the raw and activated kaolin clays was examined using Scanning Electron Microscopy (SEM) at magnifications of 8000× and 9000× , as shown in Fig. 14.3a–d. The SEM micrographs obtained at high magnifications provide valuable visual insight into how acid activation modifies the surface morphology of kaolin clay, thereby affecting its adsorption behavior. In the raw kaolin images (Fig. 14.3a–b), the particles appear largely compact, smooth, and layered, characteristic of intact kaolinite sheets stacked in a well-ordered, crystalline fashion. Such morphology limits the exposure of active sites and restricts the diffusion of ions into interior surfaces.

By contrast, the activated kaolin (Fig. 14.3c–d) displays markedly different features: rough surfaces, fissures, microcracks, and exfoliated layers are evident Musie and Gonfa (2022). These structural disruptions suggest that acid treatment has partially leached out alumina and destabilized the octahedral layers, leading to increased porosity, particle loosening, and generation of edge defects. The combination of micro-voids and crevices enhances the catalyst for ion access and the availability of adsorption sites. These morphological changes are beneficial for adsorption processes because they (1) increase the surface area accessible to solute species, (2) expose previously buried reactive hydroxyl groups, and (3) create channels and pores that reduce diffusion distance for ions (Na^+, Mg^{2+}, Ca^{2+}, Cl^-) into functional surfaces. Thus, the activated clay is expected to outperform the raw clay in ion removal efficiency.

In addition, the integrity of the clay particles seems maintained; there is no evidence of gross collapse or disintegration, implying that acid activation was moderate and did not degrade the structural backbone excessively. The observed microstructural modifications are consistent with improved ion exchange capacity and enhanced kinetics of adsorption. Scanning electron imaging thus confirms that acid activation converts kaolin from a dense, low-surface morphology to a more porous, defect-rich structure well suited for enhanced adsorption of ionic species in saline waters.

(a) SEM Image for Raw Kaolin Clay at Mag 8,000x

(b) SEM Image for Raw Kaolin Clay at Mag 9,000x

(c) SEM Image for Activated Kaolin Clay at Mag 8,000x

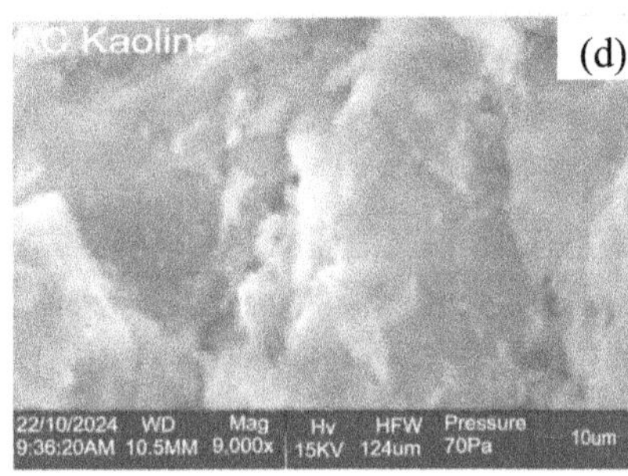

(d) SEM Image for Activated Kaolin Clay at Mag 9,000x

Fig. 14.3 SEM micrographs for **a–b** raw and **c–d** activated kaolin clay

14.3.2 Characterization of Coastal Water

Table 14.4 presents the physicochemical parameters of the collected coastal water compared with the WHO and U.S. EPA regulatory standards. The salinity measured 3500 mg/L, substantially exceeding the WHO guideline of 1000 mg/L. This high salt concentration necessitates desalination treatment for potability. Elevated potassium (1422.78 mg/L) far surpassed WHO recommendations (1–2 mg/L), potentially posing health risks to individuals with kidney disease. Sulphate concentration of 3400.9 mg/L exceeded WHO limits (250 mg/L) and can cause gastrointestinal effects. Calcium (1621 mg/L) and magnesium (1238 mg/L) concentrations were significantly elevated, contributing to water hardness and scaling potential. Sodium concentration of 1316 mg/L exceeded WHO standards (200 mg/L) and may elevate cardiovascular risks with chronic consumption. Chloride concentration of 1210 mg/L substantially exceeded the WHO limit of 250 mg/L, contributing significantly to the water's high salinity and unpalatability. The Oyeyemi et al. (2023) and Omajene et al. (2024) studies show elevated salinity, conductivity, sodium, chloride and other ions in coastal/estuarine aquifers in Nigeria, supporting the results of the present study.

Table 14.4 The physicochemical parameters of coastal water

Parameter	Measured value	WHO Guideline (2017)	U.S. EPA Secondary Standard (2018)	Compliance status	Implications
Salinity (mg/L)	3500	≤ 1000	≤ 1000	✗ Exceeds	Indicates high ionic load, typical of saline or coastal water
Potassium (mg/L)	1422.78	1–2	–	✗ Exceeds	Far above typical freshwater levels may reflect seawater intrusion or mineral dissolution
Sulphate (SO_4^{2-}) (mg/L)	3400.9	≤ 250	≤ 250	✗ Exceeds	May impart bitter taste and laxative effect, non-potable
Phosphate (PO_4^{3-}) (mg/L)	1.8	≤ 5	–	✓ Within limit	Within acceptable range; unlikely to affect potability
Calcium (Ca^{2+}) (mg/L)	1621	≤ 500	≤ 200 (hardness as $CaCO_3$)	✗ Exceeds	Contributes to excessive hardness and scaling
Magnesium (mg^{2+}) (mg/L)	1238	10–50	30–50	✗ Exceeds	Strongly exceeds safe limits; adds to water hardness
Sodium (Na^{+}) (mg/L)	1316	≤ 200	≤ 200	✗ Exceeds	Beyond palatability and health threshold, it may pose risk for hypertensive individuals
Chloride (cl^{-}) (mg/L)	1210	≤ 250	≤ 250	✗ Exceeds	Causes salty taste and corrosion, typical of seawater intrusion
pH	7.1	6.5–8.5	6.5–8.5	✓ Within limit	Acceptable range for drinking water quality
Electrical conductivity (EC) (mS/cm)	9.94	1.5 (recommended upper limit for drinking water)	0.7–1.5	✗ Exceeds	May cause scaling, corrosion, and poor taste.
Turbidity (NTU)	47.3	1–5	≤ 5	✗ Exceeds	High turbidity interferes with disinfection and aesthetic quality
Temperature (°C)	28.9	~ 25 (desirable)	–	✗ Slightly exceeds	Slightly above the ideal temperature for drinking water

(continued)

Table 14.4 (continued)

Parameter	Measured value	WHO Guideline (2017)	U.S. EPA Secondary Standard (2018)	Compliance status	Implications
Total dissolved solids (TDS) (mg/L)	2000	≤ 1000	≤ 500	✗ Exceeds	Indicates high salinity, unsuitable without desalination
Total suspended solids (TSS) (mg/L)	10.60	≤ 5	≤ 5	✗ Exceeds	Suggests particulate contamination or runoff influence
Copper (mg/L)	0.069	2.0	1.0	✓ Within limit	No expected health effects or aesthetic issues.
Iron (mg/L)	0.057	0.3	0.3	✓ Within limit	No metallic taste or staining tendency.

Total dissolved solids (2000 mg/L) and electrical conductivity (9.94 mS/cm) both far exceeded WHO guidelines, confirming high dissolved salt content. Turbidity (47.3 NTU) and total suspended solids (10.60 mg/L) significantly exceeded recommended limits (1–5 NTU and 5 mg/L, respectively), indicating substantial suspended matter including clay, silt, and organic materials. The pH (7.1) and temperature (28.9 °C) were within or near acceptable ranges. Copper and iron concentrations indicate minimal heavy metal contamination. Collectively, the raw water is clearly unsuitable for direct consumption, these results demonstrate that the coastal water exhibits high salinity and contains multiple ions (high ionic load, especially of chloride, sulphate, sodium, magnesium, and calcium) at concentrations unsuitable for direct consumption, necessitating treatment to meet potable water limits recommended by WHO and U.S. EPA regulatory standards for potable water.

Previous studies on Nigerian kaolin clay from Kutigi, Niger State, Nigeria showed its effectiveness for removing nutrient ions—Nitrate, Phosphate, Ammonium (N, P, NH_4) from domestic wastewater (Ogundipe et al., 2023b); likewise, kaolin has demonstrated excellent removal of heavy metals from wastewater (Ogundipe et al., 2023a; Faisle et al., 2023) while Abu et al. (2023) reported that Kaolin from Illela (Sokoto State) was efficient for copper II Sulphate adsorption.

14.4 Batch Adsorption Experiment

14.4.1 Effect of Contact Time

The effect of contact time on the reduction of salt, including Potassium K^+, Calcium Ca^{2+}, Magnesium Mg^{2+}, and Sulphate SO_4^{2-} in the coastal water onto the activated kaolin was determined as shown in Fig. 14.4. The equilibrium percentage adsorption on coastal water (65%) at 30 min. For the heavy metals' removal, the adsorption of K^+, Ca^{2+}, Mg^{2+}, and SO_4^{2-} attained equilibria of 45.82%, 57.68%, 44.76%, and 62.81% at 30 min each, respectively. From this result, it can be deduced that the optimum time for the heavy metals removal was uniform at 30 min contact time. Further decrease in contact time shows an increase in the percentage of metal removal. The slow uptake at the beginning is probably due to an attachment-controlled process caused by fewer available active sites for adsorption, while the slight decrease in percentage removal of metal ions may be due to the saturation of the surface of the adsorbent.

These results were obtained at pH = 7.1 with mixtures of 100 mL of coastal water with 1 g of adsorbent. The percentage adsorption of salt from coastal water onto the adsorbent decreases with an increase in time. Baskaran and Saravanane (2020) found that montmorillonite clay achieved equilibrium at 40 min, slightly longer than the present study. However, studies on heavy metal adsorption using natural kaolinite frequently reported faster equilibration. It was reported that Pb (II) and Cd (II) reached equilibrium within 20–25 min on activated kaolin (Ikhaji et al., 2019). The rapid establishment of equilibrium demonstrated here suggests that kaolin-based batch treatment systems could operate effectively with short detention times, making the process economically attractive for water-limited communities.

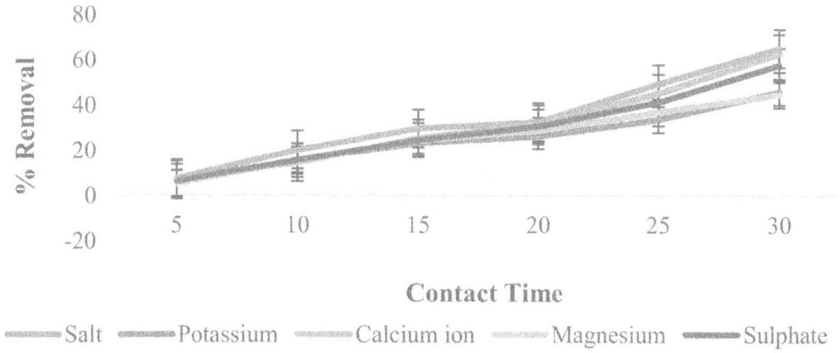

Fig. 14.4 The percentage removal of salinity and metal ions against contact time

14.4.2 *Effect of Adsorbent Dosage*

The effect of adsorbent dose on the percentage removal of some parameters like salt, k^+, Ca^{2+}, Mg^{2+} and SO_4^{2-} from coastal water is shown in Fig. 14.5. The effects of dosage on the adsorption of all the parameters absorbed onto the activated kaolin were investigated from 0.4 to 3 g. It is noteworthy that there were increases in the percentage of all the parameters with an increase in the adsorbent dosages used. An increase in adsorption of salt from 66.98 to 86.56% was obtained. A similar trend was observed for the adsorption of the heavy metal ions onto kaolin as well as for K^+ (56.78–71.65%), Ca^{2+} (42.36–68.54%), Mg^{2+} (45.63–73.21%), SO_4^{2-} (43.57–70.81%).

Al-Degs et al. (2020) demonstrated that enhanced salt ion adsorption occurred with higher kaolin modification and dosage up to 40 g/L for seawater treatment. However, the present study employed substantially lower adsorbent concentrations (0.4–3.0 g per 100 mL, or 4–30 g/L) yet achieved comparable removal percentages to some higher-dosage systems. This suggests that Nigerian kaolin, when properly activated through acid treatment at 700 °C, exhibits adsorption efficiency competitive with or superior to other clay materials under comparable operating conditions. Baskaran and Saravanane (2020) analyzed pH, Total Dissolved Solids (TDS), Chloride (Cl) and Sodium (Na) and found that the optimum clay of 60 g at a detention time of 40 min gives us a better result of removing 89.83% TDS, 96.62% of Chloride, and 98% of Sodium using montmorillonite clay. Liu et al. (2020) noted diminishing returns in removal efficiency at dosages exceeding 30 g/L, recommending optimal dosage selection based on cost-benefit analysis for full-scale systems.

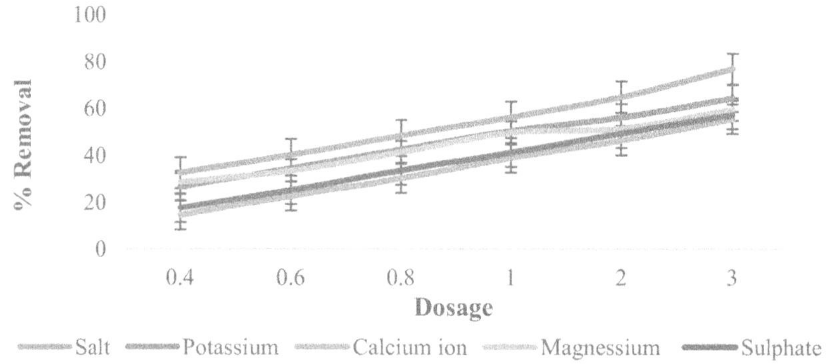

Fig. 14.5 The percentage removal of salinity and metal against dosage

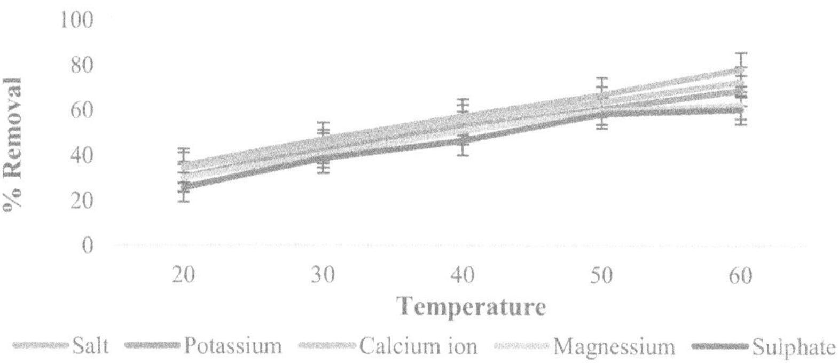

Fig. 14.6 The percentage removal of salinity and metal against temperature

14.4.3 Effect of Temperature

The effect of temperature on the adsorption was investigated in the temperature range 20–60 °C. Temperature is a highly important parameter in the metals onto kaolin, as shown in Fig. 14.6. As can be seen from the figures, the percentage adsorption of all parameters in the coastal water bent increased with an increase in temperature. With an increase in temperature, there was an increase in the adsorption of salt from (62.34–73.86%). A similar trend was observed for the heavy metals 79%, 96%, 75%, and 100% for Fe, Cu, Cr, and Zn respectively and other ions; k^+ (52.62–64.78%), Ca^{2+} (67.34–77.56%), Mg^{2+}(55.67–68.76%), SO_4^{2-}(52.87–67.43%) Faisle et al. (2023). The increase in adsorption may be due to the increase in kinetic energy at higher temperatures. This leads to an increase in the adsorption rates.

Umudi et al. (2022) reported that kaolin-based desalination showed temperature-dependent enhancement for chloride removal, with optimal performance around 50–60 °C. However, Ali et al. (2021) noted that while temperature increases adsorption rates, the relationship is not indefinitely linear; they found diminishing returns above 60 °C in their activated clay system. The current study's use of 60 °C as the maximum temperature aligns with practical considerations for resource-limited settings, as excessive heating would increase operational costs.

14.4.4 Effect of Concentration

The effect of concentration is shown in Fig. 14.7. The results show that decreasing the concentration of coastal water led to a high adsorption capacity. For example, at an initial salt concentration of 20 mg per liter, the adsorption capacity reached 27.5% whereas at 100 mg per liter, the adsorption capacity decreased to 1.5%. This trend was consistent across potassium, calcium, and sulphate ions. Zhang et al. (2020) observed similar trends in activated kaolin systems for salt removal, where

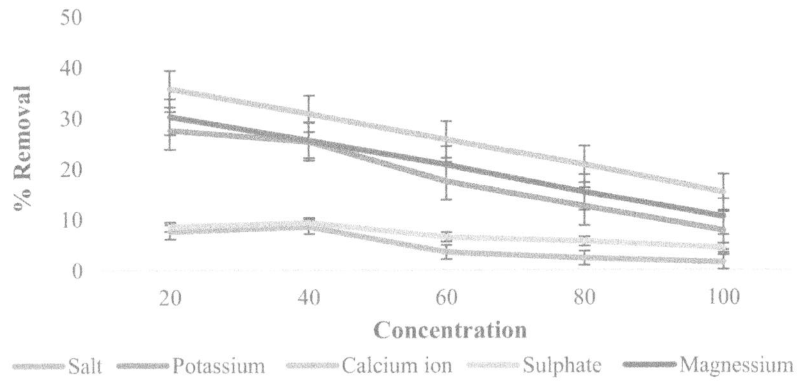

Fig. 14.7 The percentage removal of salinity and metal against concentration

lower initial salt concentrations (10–50 mg/L) yielded higher percentage removals compared to saturated seawater (> 30,000 mg/L). Baskaran and Saravanane (2020) reported consistent patterns in montmorillonite clay treatment of seawater at varying dilutions, confirming that this is a fundamental principle of batch adsorption systems rather than an artifact of specific materials or conditions. The present study showed only 1.5% removal at 100 mg/L but 27.5% at 20 mg/L, suggest that pretreatment through initial dilution or reverse osmosis concentration of the brine stream could optimize kaolin-based treatment for drinking water applications, though this would require assessment of energy and cost trade-offs for full-scale systems.

14.4.5 Isotherm Analysis

Langmuir and Freundlich isotherms were applied to experimental data obtained from the adsorptive kaolin clay for desalination to obtain the plots in shown in Figs. 14.8 and 14.9. The data obtained were screened and filtered for only the valid (positive) adsorption values from both the Langmuir and the Freundlich model, positive values for q_m, K_L, K_F; $0 < 1/n < 1$ (for the Freundlich model) and $R^2 \geq 0.7$) to obtain the refined data as shown in Tables 14.5 and 14.6. A previous study by Ogundipe et al. (2023) investigated the effect of Beneficiated Kaolin Clay (BKC) from Kutigi, Niger State, Nigeria, in removing nitrate, phosphate, and ammonium from domestic wastewater using both Langmuir and the Freundlich models.

Evaluation of Langmuir linear models revealed that Langmuir exhibited the highest R^2 values across all pollutants. Calcium and magnesium ion exhibits R^2 values of 0.9231 and 0.891, respectively, while potassium displays a larger q_m (10.48 mg/g) with R^2 value of 0.8631; these imply that cations fit better due to kaolin's negative charge (Mao et al. 2022). For salinity, models yielded $R^2 =$ 0.9961, indicating an excellent fit. Sulphate ion also demonstrated strong correlation

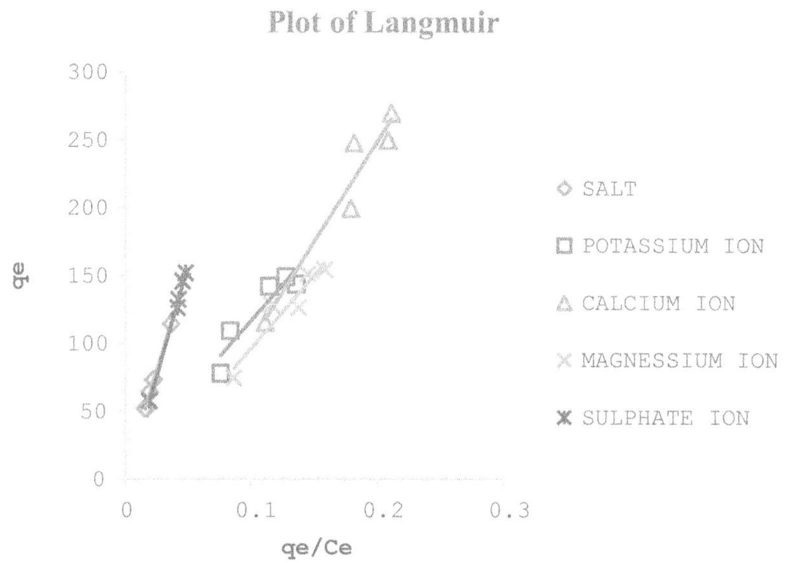

Fig. 14.8 The plot of Langmuir isotherm

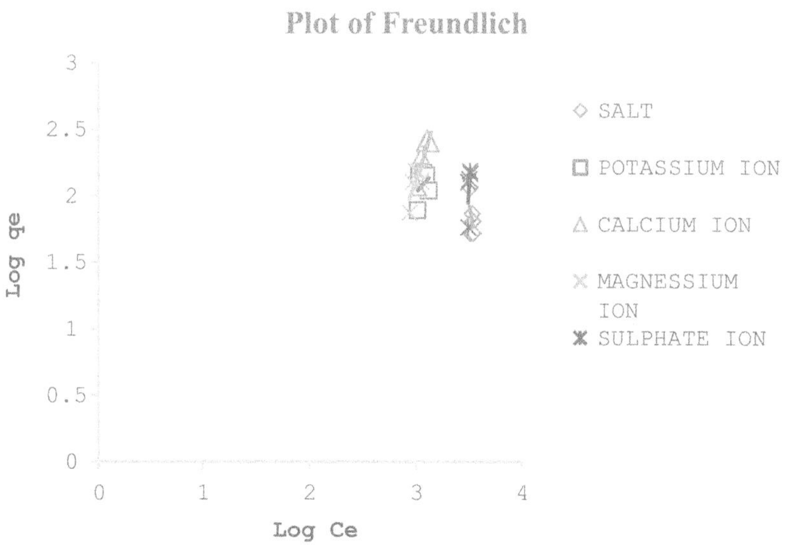

Fig. 14.9 The plot of Freundlich isotherm

Table 14.5 Langmuir isotherm parameters

Pollutant	q_m (mg/g)	K_L (L/mg)	R^2	Remarks
Salinity	4.98	3×10^{-4}	0.9961	Excellent fit (Langmuir)
Potassium ion	10.48	9×10^{-4}	0.8631	Good fit (Langmuir)
Calcium ion	0.052	1.75×10^{3}	0.9231	Very strong fit (Langmuir)
Magnesium ion	4.98	9×10^{-4}	0.8906	Strong fit (Langmuir)
Sulphate ion	4.00	3×10^{-4}	0.9956	Excellent fit (Langmuir)

Table 14.6 Freundlich isotherm parameters

Pollutant	K_L (L/mg)	1/n	R^2	Remarks
Potassium ion	0.5161	0.896	0.112	Poor fit, not retained for modeling
Calcium ion	0.0024	0.371	0.7196	Acceptable, modified to within 0–1 range
Magnesium ion	0.0184	0.492	0.4738	Moderate fit, may improve with transformation
Sulphate ion	0.0079	0.524	0.3384	Low fit, acceptable for comparison

($R^2 = 0.9956$), as did magnesium ($R^2 = 0.8906$). The high correlation coefficient ($R^2 > 0.85$) observed for all ions suggests that the adsorption process is well described by the Langmuir isotherm model, supporting the assumption of monolayer adsorption on homogeneous surfaces. The refined isotherm parameters show how activated kaolin clay interacts with major salt ions in coastal water. The available surface sites on the activated kaolin clay can be approximated as energetically similar and adsorption tends toward site saturation for these species (Al-Ghouti & Da'ana, 2020). The practical implication is that the adsorbent shows defined capacities for removing bulk salt and divalent anions, which supports the concept of kaolin-based treatment for salinity reduction. Musie and Gonfa (2022) also revealed that Langmuir isotherm provided a better fit to their experimental data with correlation coefficients of 0.967 and 0.979 for raw and activated bentonite, respectively, similar to that of kaolin clay in the present study.

The Freundlich isotherm model was applied to further describe the adsorption behavior of the different ions onto the activated kaolin clay. The Freundlich model assumes a heterogeneous surface with adsorption sites of varying energies and affinities, allowing for multilayer adsorption of adsorbate molecules (Foo & Hameed, 2010). The linear form of the Freundlich equation, expressed as $\log q_e = \frac{1}{n} \log C_e + \log K_f$, as described in eq. 14.3, was used to generate the plots shown in Fig. 14.9. In this study, only calcium and magnesium ions showed moderately good linearity ($R^2 = 0.7196$ and 0.4738, respectively), suggesting that their adsorption behavior partially conforms to the Freundlich model, indicating the possibility of multilayer formation or surface heterogeneity for these ions. The Freundlich fit for potassium is very poor ($R^2 = 0.112$), with 1/n ≈ 0.896; this combination suggests that while the material can take up a sizable amount of potassium, K^+, the adsorption behavior is not strongly heterogeneous (Freundlich). In high-ionic strength matrices such as coastal water, smaller monovalent cations like K^+ compete strongly for surface sites and are more affected by electrostatic screening; this can lead to apparent high q_m but poorer predictive certainty across concentrations (Avornyo et al., 2025).

Calcium is noteworthy because the Freundlich result ($R^2 \approx 0.720$) implies heterogeneous adsorption behavior. A very large K_l is typically symptomatic of unfitness or site specificity at low concentration ranges; however, the Freundlich $1/n < 1$ indicates favorable adsorption and surface heterogeneity, consistent with studies showing that divalent cations (Ca^{2+}) interact via stronger ion-exchange and inner-sphere complexation on modified clays (Rouibah et al., 2024). For coastal water applications, calcium's strong affinity can both help hardness removal and compete with other target ions, affecting overall selectivity.

The patterns in these tables align with reported properties of activated kaolin, acid, or thermal activation increases surface area, expose silanol/aluminol groups (Si–OH, Al–OH), and raise cation exchange capacity (Wang et al., 2024; Avornyo et al., 2025). Divalent cations (Mg^{2+}, Ca^{2+}) typically form stronger electrostatic and inner-sphere bonds with such sites than monovalent ions, explaining higher R^2 and consistency for Mg^{2+} and Ca^{2+} but mixed potassium, K^+ behavior. Sulphate ($SO_4{}^{2-}$) showing good Langmuir behavior suggests specific adsorption or precipitation processes at the surface under the experiment, but sulphate interactions can be complex in high Ca^{2+} or high ionic strength matrices and deserve further thermodynamic probing.

The divergence between Langmuir and Freundlich fits across ions underscores two practical points. First, single two-parameter models cannot fully describe adsorption across diverse ion types, and concentration ranges present in coastal water; hybrid or three-parameter models (Sips, Redlich–Peterson) often produce more realistic fits when heterogeneity and site interactions coexist (Foo & Hameed, 2010). Second, competitive adsorption and ionic strength effects not captured by single-ion isotherms will substantially affect field performance. Studies on modified clays have documented that background Na^+, K^+, and hardness levels suppress uptake of target ions and alter apparent isotherm constants; hence, column and multi-component experiments are the logical next step (Avornyo et al., 2025; Rouibah et al., 2024).

From a practical perspective for coastal water resource management, the observed q_m values (single-ion capacities in the order of 4–10 mg/g) indicate that activated kaolin clay is a useful, low-cost pretreatment adsorbent for salinity reduction and hardness control in resource-limited contexts. However, the actual efficacy at scale will depend on (a) contact time, (b) adsorbent regeneration and lifecycle costs, and (c) treatment of complex multi-ion seawater matrices. Recent work emphasizes that coupling kaolin adsorption with membrane polishing or electrodialysis can reduce energy needs while achieving potable targets (Avornyo et al., 2025; Wang et al., 2024).

The fit of Langmuir and Freundlich models indicates that kaolin clay surface adsorption is both described as monolayer uptake on homogeneous sites and/or multilayer adsorption on heterogeneous surfaces. However, the varying adsorption mechanisms for different ions highlight that the process is ion specific.

The poor fit of the Freundlich model for most ions in the present study contrasts with results from studies on heavy metal and organic contaminant adsorption on kaolin, which frequently yielded higher Freundlich R^2 values.

14.5 Conclusion and Recommendations

This study demonstrated the effectiveness of activated kaolin clay as a promising low-cost adsorbent for desalination of coastal and seawater, contributing to the search for alternative freshwater sources. The adsorption performance was strongly influenced by operational parameters such as contact time, adsorbent dosage, and temperature, which significantly reduced salinity and the concentrations of dominant ions (K^+, Ca^{2+}, Mg^{2+}, SO_4^{2-}). These findings confirm that locally sourced kaolin can serve as an accessible and environmentally sustainable material for the removal of dissolved salts from saline waters.

Beyond laboratory application, the implications of this study are broad. The abundance and affordability of kaolin clay deposits make them suitable for deployment in small or large -scale, decentralized desalination systems for coastal communities that lack grid-based treatment infrastructure. However, energy and chemical costs associated with acid activation represent a trade-off that must be optimized for real-world feasibility. Comparative energy analysis and life-cycle assessment are recommended to quantify the sustainability of kaolin activation relative to freshwater production gains.

Future work should focus on adsorbent regeneration and reuse efficiency, pilot-scale testing under field conditions, and integration with hybrid systems (e.g., membrane-assisted adsorption). Pretreatment of highly turbid coastal water like this (47.3 NTU) should be included to enhance adsorption efficiency and prolong the lifespan of the adsorbent. Overall, this research lays the groundwork for scaling kaolin-based adsorption desalination as a sustainable, community-adaptable water management solution for coastal Nigeria and similar environments worldwide.

Acknowledgement The authors sincerely appreciate the support of the Bioenvironmental Water and Engineering Research Laboratory, Ladoke Akintola University of Technology for providing the necessary facilities to conduct this research.

References

Al-Degs, Y. S., Al-Muhtaseb, A. H., & El-Gohary, F. (2020). Enhanced adsorption of salt ions by modified kaolin for seawater desalination. *Separation and Purification Technology, 238*, 116356.

Abu, T. O., Adegoke, H. I., & Odebunmi, E. O. (2023). The optimization of adsorption potentials of Nigerian kaolinite mineral through acid activation. *Fountain Journal of Natural and Applied Sciences, 12*(2) https://doi.org/10.53704/fujnas.v12i2.438

Al-Ghouti, M. A., & Da'ana, D. A. (2020). Guidelines for the use and interpretation of adsorption isotherm models: A review. *Journal of Hazardous Materials, 393*, 122383. https://doi.org/10.1016/j.jhazmat.2020.122383

Ali, E. S., Askalany, A. A., Harby, K., Diab, M. R., Hussein, B. R. M., & Alsaman, A. S. (2021). Experimental adsorption water desalination system utilizing activated clay for low grade heat source applications. *J. Energy Storage, 43*, 103219. https://doi.org/10.1016/j.est.2021.103219

Aloulou, H., Gassara, S., Ali, M. B. S., & Ben Salah, A. (2022). Development of ultrafiltration kaolin membranes for the removal of organic pollutants from wastewater. *Materials, 15*(3), 987. https://doi.org/10.3390/ma15030987

APHA—American Public Health Association. (2005). *Standard methods for the examination of water and wastewater* (21st edn.).

Avornyo, A. K., Katzourakis, V. E., Hasan, S. W., Banat, F., & Chrysikopoulos, C. V. (2025). NaCl as a simple yet unique activator of kaolin: Surface engineering for enhanced heavy metal adsorption. *Water, 17*(14), 2120. https://doi.org/10.3390/w17142120

Ayawei, N., Ebelegi, A. N., & Wankasi, D. (2017). Modelling and interpretation of adsorption isotherms. *Journal of Chemistry, 2017*, 3039817. https://doi.org/10.1155/2017/3039817

Baskaran, V., & Saravanane, R. (2020). Effect of montmorillonite clay in partial treatment of sea water, environmental nanotechnology. *Monitoring and Management, S2215-1532*(20), 30211–30217. https://doi.org/10.1016/j.enmm.2020.100326

Bolster, C. H., & Hornberger, G. M. (2007). On the use of linearized Langmuir equations. *Soil Science Society of America Journal, 71*(6), 1796–1806. https://doi.org/10.2136/sssaj2006.0304

Elimelech, M., & Phillip, W. A. (2011). The future of seawater desalination: Energy, technology, and the environment. *Science, 333*(6043), 712–717.

Emmanuel, U. I., Martins, O. O., & Nurudeen, O. A. (2019). Modeling in adsorption: fundamentals and applications. In *Composite Nanoadsorbents* (pp. 85–118). Elsevier. https://doi.org/10.1016/B978-0-12-814132-8.00005-8

Faisle, M., El- Rabiei, M., Farghali, A., Abd El-Hafeez, G., & Shehata, N. (2023). Kaolin and its application in removing heavy metal ions from real municipal wastewater. *Labyrinth: Fayoum Journal of Science and Interdisciplinary Studies, 1*(3), 17–22. https://doi.org/10.21608/ifjsis.2023.230083.1039

Fareed, H., Khan, A. L., Shukla, A. K., & Altaee, A. (2022). Brine desalination via pervaporation using kaolin-modified membranes. *Desalination, 541*, 116017. https://doi.org/10.1016/j.desal.2022.116017

Foo, K. Y., & Hameed, B. H. (2010). Insights into the modeling of adsorption isotherm systems. *Chemical Engineering Journal, 156*(1), 2–10. https://doi.org/10.1016/j.cej.2009.09.013

Gleick, P. H. (1993). *Water in crisis: A guide to the world's fresh water resources*. Oxford University Press.

Ikhaji, D., Ebhodaghe, S. O., & Okoro, U. C. (2019). Structural changes in kaolinite during thermal activation. *Applied Clay Science, 168*, 21–28.

Khan, S., Farid, M. U., & Ngah, W. S. W. (2023). Clay-based materials for enhanced water treatment: A review. *Environmental Science and Pollution Research, 30*, 12345–12367. https://doi.org/10.1007/s11356-023-26219-1

Khawaji, A. D., Kutubkhanah, I. K., & Wie, J. M. (2008). Advances in saltwater desalination technologies. *Desalination, 221*(1–3), 47–69. https://doi.org/10.1016/j.desal.2007.01.067

Liu, X., Chen, X., & Zhang, Y. (2020). Sustainable desalination using natural and modified adsorbents: Kaolin clay as a case study. *Journal of Environmental Management, 270*, 110878.

Mao X, Liu H, Chu Z, Chen T., Zou X, Chen D., Zhang X., Hu J. Adsorption of lead by kaolinite, montmorillonite, goethite, and ferrihydrite: Performance and mechanisms based on quantitative analysis. Clay Minerals 57(3–4):230–240. (2022). https://doi.org/10.1180/clm.2022.41.

Mierzwa, J. C., Rodrigues, C. C., Teixeira, A. C. S. C., & Boaventura, R. A. R. (2019). Adsorption processes for water treatment and desalination: Recent advances and applications. *Journal of Environmental Chemical Engineering, 7*(5), 103280.

Muhammad, N. N., Asad, A. Z., Hamdullah, K., Sagar, K., Muhammad, T. B. O., Yasmin, A. W., Kingshuk, D., Juhana, J., Muhammad, U., Mohd, R. J., & Irfan, A. B. (2022). Desalination technology for energy-efficient and low-cost water production: A bibliometric analysis. *Green Processing and Synthesis, 11*, 306–315. https://doi.org/10.1515/gps-2022-0027

Musie, W., & Gonfa, G. (2022). Adsorption of sodium from saline water with natural and acid activated Ethiopian bentonite. *Results in Engineering, 14*(2022), 100440. https://doi.org/10.1016/j.rineng.2022.100440

Mustapha, S., Ndamitso, M. M., Abdulkareem, A. S., Tijani, J. O., Shuaib, D. T., Mohammed, A. K., & Sumaila, A. (2019). Potential of using kaolin as a natural adsorbent for pollutant removal from wastewater. *Heliyon, 5*(4), e01568. https://doi.org/10.1016/j.heliyon.2019.e01568

Ogundipe, F. O., Saidu, M., Abdulkareem, A. S., & Busari, A. O. (2023a). Removal of heavy metals from domestic wastewater using beneficiated kaolin clay, silver oxide, and zinc oxide nanocomposite. *Nigerian Journal of Technological Development, 20*(3) https://doi.org/10.63746/njtd.v20i3.1651

Ogundipe, F. O., Saidu, M., Abdulkareem, A. S., & Busari, A. O. (2023b). Isotherm and thermodynamic adsorption studies of nitrate, phosphate, and ammonium onto beneficiated kaolin clay from Kutigi, Niger state. *Nigeria Journal of Inventive Engineering and Technology*. Retrieved from https://www.jiengtech.com/index.php/INDEX/article/view/56

Olaoye, R. A., Afolayan, O. D., Adeyemi, K. A., Ajisope, L. O., & Adekunle, O. S. (2020). Adsorption of selected metals from cassava processing wastewater using cow-bone ash. *Scientific African, 10*(e00653), 1–9. https://doi.org/10.1016/j.sciaf.2020.e00653

Olaoye, R. A., Olawale, O. A., Adejumobi, M. A., Ayanlere, S. A., Fagbenro, O. K., & Adeyemo, A. A. (2024). Evaluating water quality dynamics: A strategic tool for optimizing water treatment operations. *LAUTECH Journal of Engineering and Technology, 18*(3), 21–30. https://doi.org/10.36108/laujet/4202.81.0330

Omajene, A., Egbai, J. C., & Okolie, E. C. (2024). Investigation of saltwater intrusion into freshwater aquifers in some estuary environments in Niger Delta. *Journal of Water Resources and Ocean Science, 13*(4), 94–104. https://doi.org/10.11648/j.wros.20241304.11

Oyeyemi, K. D., Abuka-Joshua, J., Rotimi, O. J., Dieppois, B., Gomo, M., Olaojo, A. A., Falae, P. O., & Metwaly, M. (2023). Geoelectrical characterization of coastal aquifers in Agbado-Ijaye, Lagos, southwestern Nigeria; implications for groundwater resources sustainability. *Sustainability, 15*(4), 3538. https://doi.org/10.3390/su15043538

Ohlund, R. J., Dahdah, B. H., Guillen, G. R., & Childress, A. E. (2024). Augmenting Ocean water desalination with potable reuse: Concept feasibility in terms of cost and environmental impacts. *Desalination, 569*, 116941. ISSN 0011-9164. https://doi.org/10.1016/j.desal.2023.116941

Rouibah, K., Ferkous, H., Abdessalam-Hassan, M., Mossab, B. L., Boublia, A., Pierlot, C., Abdennouri, A., Avramova, I., Alam, M., Benguerba, Y., & Erto, A. (2024). Exploring the efficiency of Algerian kaolinite clay in the adsorption of Cr (III) from aqueous solutions: *Experimental and computational insights. Molecules, 29*(9), 2135. https://doi.org/10.3390/molecules29092135

Umudi E. Q. (2021). Porous clays as low-cost adsorbent to reduce chloride in seawater Journal of Research in Environmental Science and Toxicology, 10(5): 97–100. ISSN:2315-5698, 10.14303/jrest.2021.10.

Umudi, E. Q., Adaikpo, E., Obukohwo, I. E., Igborgbor, J., & Egheneji, A. (2022). Evaluation of kaolin as adsorbent for chlorine reduction in seawater from Forcados/Ogulagha Axis of Southwest Delta state, Nigeria. *Journal of Applied Sciences and Environmental Management, 26*(2), 237–242. https://doi.org/10.4314/jasem.v26i2.10

United States Environmental Protection Agency (U.S. EPA). (2018). National secondary drinking water regulations. EPA 816-F-08-013. Retrieved from https://www.epa.gov/sdwa/drinking-water-regulations-and-contaminants

Wang, P., Shen, X., Qiu, S., Zhang, L., Ma, Y., & Liang, J. (2024). Clay-based materials for heavy metals adsorption: Mechanisms, advancements, and future prospects in environmental remediation. *Crystals, 14*(12), 1046. https://doi.org/10.3390/cryst14121046

WHO (World Health Organization). (2017). Guidelines for drinking-water quality: Fourth edition incorporating the first addendum. WHO Press. Retrieved from https://apps.who.int/iris/handle/10665/254637

Zhang, L., Li, Y., & Liu, H. (2020). Application of activated kaolin in seawater desalination and heavy metal removal. *Desalination, 480*, 114312.

Part V
Analytical and Integrated Approaches

Chapter 15
Integrated Water Resources Management (IWRM) in Nigeria: Principles, Practice, and Policy Pathways

Mariam Iyabo Adeoba, Ayokunle Isaac Ilesanmi, Stephen Olubusoye Ajagbe, Sunday Emmanuel Olusola, Gabriel O. Oladipo, and Olushola Ahmed Olaleye

Abstract Integrated Water Resources Management (IWRM) represents a significant change from the older, separate ways of managing water resources. It encourages the coordinated planning and management of water, land and other related resources to improve social and economic conditions while also maintaining a healthy and sustainable environment. Nigeria has many water bodies and large river systems, but it still faces challenges in ensuring water security. These challenges include uneven distribution of water, seasonal changes, pollution, overuse and the effect of climate change. The objective of this chapter is to critically examine the principles, evolution, and implementation of Integrated Water Resources Management (IWRM) in Nigeria, assess its effectiveness in addressing water-related challenges, and propose strategic policy and research pathways for sustainable water governance aligned with national

M. I. Adeoba (✉)
UNISA Biomedical Research Group, Department of Mechanical, Bioresources and Biomedical Engineering, College of Science Engineering and Technology (CSET), University of South Africa, Florida, South Africa
e-mail: mariamsalami@yahoo.co.uk

A. I. Ilesanmi
Department of Fisheries and Aquaculture, University of Agriculture and Environmental Sciences, Umuagwo, Imo State, Nigeria

S. O. Ajagbe
Department of Wildlife and Ecotourism, Forestry Research Institute of Nigeria, Ibadan, Nigeria

S. E. Olusola
Depatment of Fisheries and Aquaculture Technology, School of Agriculture, Food and Natural Resources, Olusegun Agagu University of Science and Technology, Okitipupa, Nigeria

G. O. Oladipo
Department of Science Laboratory Technology, D.S Adegbenro ICT Polytechnic, Itori-Ewekoro, Ogun State, Nigeria

O. A. Olaleye
Department of Zoology, Faculty of Science, University of Johannesburg, Johannesburg, South Africa

M. I. Adeoba et al. (eds.), *Water Resources Management in Nigeria*, World Water Resources 10, https://doi.org/10.1007/978-3-032-22813-0_15

development goals and SDG 6. This chapter provides a comprehensive analysis of IWRM in Nigeria, exploring its conceptual foundations, relevance in addressing persistent water challenges such as uneven distribution, seasonal variability, pollution, and climate change impacts. It reviews national frameworks including the Water Resources Act (2004), National Water Policy, and the role of River Basin Development Authorities (RBDAs), highlighting gaps in coordination, stakeholder participation, and technical capacity. Using a desk-based synthesis of literature, policy documents, and institutional reports, it examines hydrological patterns, governance structures, and case studies from regions such as Hadejia-Jama'are, Benue River Basin, Lagos, and Cross River. Findings reveal that while IWRM offers opportunities for climate resilience, conflict mitigation, agricultural productivity, and urban water security, progress remains constrained by institutional fragmentation, inadequate funding, weak enforcement, and limited community engagement. Further studies should focus on (i) developing real-time water data systems and predictive models for climate variability; (ii) assessing socio-economic impacts of IWRM interventions; (iii) exploring innovative financing mechanisms and public-private partnerships; (iv) integrating indigenous knowledge into water governance; and (v) addressing emerging threats such as antimicrobial resistance in wastewater within IWRM frameworks. The chapter concludes with strategic recommendations for legal reforms, capacity building, innovative financing, and inclusive governance to strengthen IWRM implementation and achieve sustainable water management aligned with national development goals and SDG 6.

Keyword Water policy · IWRM · Water management · Water sustainability · Water integration

15.1 Introduction

Integrated Water Resources Management (IWRM) represents a paradigm shift from traditional, sectoral approaches to water governance toward a holistic framework that recognizes the interconnections between water, land, and related resources. Unlike earlier fragmented systems, IWRM promotes coordinated planning and management to optimize social and economic benefits while safeguarding environmental sustainability (Biswas, 2004; Savenije & Van der Zaag, 2008).

Globally, IWRM has emerged as a key strategy for addressing complex water challenges in the twenty-first century, including climate variability, population growth, and urbanization (Grigg, 2008; Rahaman & Varis, 2005). It emphasizes that water systems are deeply linked to ecosystems, economies, and societies, requiring integrated solutions rather than isolated interventions (Pahl-Wostl, 2007).

In Nigeria, despite abundant water resources such as major rivers like Niger and Benue, extensive lakes, and aquifers, water insecurity persists due to uneven distribution, seasonal variability, pollution, and weak governance structures (Akpabio & Subramanian, 2021; Ishaku et al., 2011). Rainfall patterns vary significantly, from over 3000 mm annually in the south to less than 500 mm in the arid north, creating stark regional disparities (Nicholson, 2013). Climate change further exacerbates these challenges, increasing the frequency of floods and droughts and threatening agricultural productivity and urban water security (Cervigni et al., 2015).

Water governance in Nigeria is shared among federal, state, and local authorities, with the Federal Ministry of Water Resources and River Basin Development Authorities (RBDAs) playing central roles. However, institutional fragmentation, poor coordination, and limited stakeholder engagement undermine effective implementation of policies such as the National Water Policy and Water Resources Act (Akpabio, 2012). These gaps have led to conflicts over water access, particularly between farmers and pastoralists, and have heightened socio-economic vulnerabilities (Ngene et al., 2021).

This chapter explores the conceptual foundations of IWRM, its evolution in Nigeria, and the status of implementation. It reviews policy frameworks, governance structures, and case studies from key regions, highlighting successes, limitations, and opportunities for improvement. Finally, it proposes strategic recommendations and future research directions to strengthen IWRM as a tool for sustainable water management and climate resilience.

15.2 Research Methodology

This chapter adopts a desk-based qualitative research design aimed at synthesizing existing knowledge on Integrated Water Resources Management (IWRM) in Nigeria. The approach emphasizes conceptual analysis, policy review, and case study evaluation to understand the evolution, implementation status, and challenges of IWRM within the Nigerian context.

15.2.1 Data Sources

The study relies on secondary data drawn from:

- Peer-reviewed journal articles are indexed in major databases such as *Web of Science*, *Scopus*, and *Google Scholar*.
- Policy documents and legal frameworks, including the National Water Policy (NWP), Water Resources Act (WRA), and operational reports from the Federal Ministry of Water Resources (FMWR) and River Basin Development Authorities (RBDAs).
- Institutional and international reports from organizations such as the African Ministers' Council on Water, UN-Water, Global Water Partnership, and the World Bank.

15.2.2 Search Strategy

A systematic literature search was conducted using keywords such as:

- *Nigeria water resources*, *water policy in Nigeria*, *river basin management*, *IWRM implementation*, and *water governance*. The review focused on publications between 1988 and 2025, ensuring coverage of historical governance evolution, policy developments, and contemporary implementation experiences.

15.2.3 Inclusion and Exclusion Criteria

- Inclusion: Studies addressing water governance, policy frameworks, IWRM principles, and case studies in Nigeria or comparable developing countries.
- Exclusion: Articles lacking empirical or policy relevance, opinion pieces without supporting evidence, and studies outside the African context.

15.2.4 Analytical Framework

The analysis employed thematic synthesis to identify patterns, trends, and gaps across sources. Key themes included:

- Policy and institutional frameworks for IWRM.
- Hydrological and climatic variability.
- Stakeholder participation and governance challenges.
- Case studies illustrating practical implementation.

Tables and figures were used to summarize rainfall patterns, governance structures, and implementation status for clarity and comparability.

15.2.5 Limitations

The methodology is constrained by:

- Dependence on secondary data, which may lack real-time insights.
- Limited stakeholder perspectives, as the study did not include primary interviews or surveys. Future research should incorporate field-based assessments, community-level surveys, and real-time hydrological data to complement this desk-based approach.

15.3 Thematic Findings

The thematic findings of this chapter synthesize insights from policy documents, institutional reports, and scholarly literature to reveal patterns and gaps in Nigeria's approach to Integrated Water Resources Management (IWRM). This section organizes the results into key themes, including hydrological variability, governance structures, stakeholder participation, and practical implementation across different regions. By examining rainfall patterns, institutional frameworks, and case studies, the analysis highlights both opportunities and persistent challenges such as institutional fragmentation, limited technical capacity, and weak community engagement that shape the effectiveness of IWRM in Nigeria. These findings provide the foundation for subsequent discussions on policy recommendations and future research directions.

To provide a visual representation of the conceptual underpinnings of Integrated Water Resources Management (IWRM). Figure 15.1 illustrates the framework that integrates water, land, and related resources within a holistic planning and management approach. This model emphasizes the interconnectedness of ecological, social, and economic systems, highlighting how coordinated strategies can optimize resource use while maintaining environmental sustainability. It serves as a foundation for understanding how IWRM principles are applied globally and adapted to Nigeria's context.

In order to contextualize Nigeria's water management challenges, Table 15.1 presents the major hydrological zones of the country alongside their annual rainfall

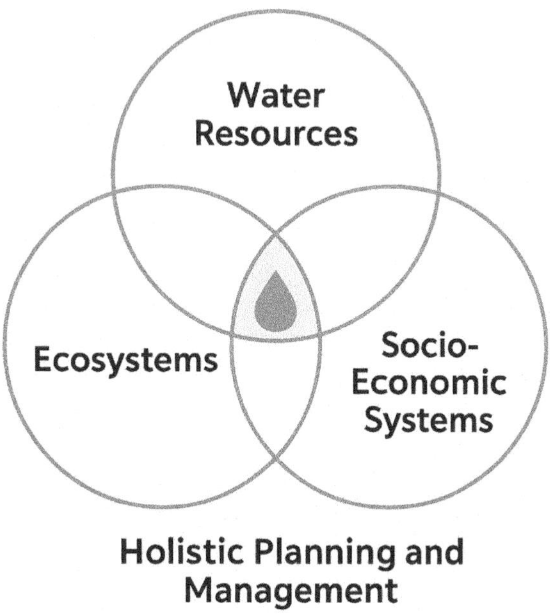

Fig. 15.1 Conceptual framework of Integrated Water Resources Management (IWRM), illustrating the intersection of water resources, ecosystems, and socio-economic systems within a holistic planning and management approach (adapted from Loucks & Van Beek, 2005; Young, 2005)

Table 15.1 IWRM principles and implementation status in Nigeria, summarizing policy commitments, current implementation levels, and key challenges (adapted from Akpabio & Subramanian, 2021)

Hydrological zone	Annual rainfall (mm)	Major rivers	Notable features
South-south	> 3000	Cross, Imo	Dense rainfall, mangroves
North-east	< 500	Yobe, Gongola	Arid, prone to desertification
North-central	800–1200	Benue, Kaduna	Agriculture and hydro dams
South-west	1200–2000	Ogun, Osun	Urban and peri-urban pressure
South-east	1500–2500	Anambra, Imo	High population density
North-west	600–1000	Sokoto-Rima, Zamfara	Irrigation agriculture
Middle Belt	1000–1500	Niger, Benue	Major confluence zone
Chad Basin	300–600	Komadugu-Yobe	Transboundary waters

patterns, dominant river systems, and notable ecological or socio-economic features. This classification highlights the stark regional disparities in water availability from the humid, high-rainfall South-South region to the arid North-East and underscores the need for tailored IWRM strategies that account for climatic variability and resource distribution. Understanding these patterns is critical for designing adaptive water governance frameworks that address both abundance and scarcity across different zones.

15.3.1 Evolution of Water Governance in Nigeria

RBDAs was created in 1970 with the major aim of managing water at different river basins. This led to early usage of IWRM, it was initiated in 1973 with 11 bodies which was later expanded to 12. As shown in Fig. 15.2, they focus mainly on developing water resources within various river basins, ensuring proper monitoring of farming, irrigation, fish stocks improvement, flood management and planning (Ayoade, 1988). Nigeria's means of managing water has gradually improved compared to colonial era, it has been integrated from a centralized controlled system to distribution of controls and how water is managed (Ukpai, 2022).

Nigeria National Water Policy (NNWP) was inaugurated in 2004 which overseas how Nigeria handles water resources. All core parties were involved and how water can be utilized efficiently and how the environment can be protected. Water serves as both social and economic necessity, which includes equal access, careful utilization and comprehensive way of linking water with other aspects of developing the

Fig. 15.2 Timeline of Water Governance Evolution in Nigeria, showing major milestones from fragmented approaches to integrated frameworks: pre-1970s fragmented management, 1976–1988 creation of River Basin Development Authorities, 2004 inauguration of the National Water Policy, and 2021 drafting of updated IWRM framework (adapted from Ayoade, 1988; Ukpai, 2022)

nation. Other importance of NNWP includes improving water organization teamwork, developing a reliable water data system, introducing communities' decisions and ensuring there is equality in water usage and management. Hence it aligns with IWRM and shows Nigeria's strivings to ensure it partake in international discussions on issues related to water sustainability and management (Ngene et al., 2021).

15.3.2 Principles of IWRM and Their Application in Nigeria

Table 15.2 shows the Important principles and policies that guided IWRM are made, set up and strategies which are put into action. A lot of years of experience are considered when setting up the policies guiding the world on how to handle water resources and how water systems are inter-connected. It is important to locate how these ideas work in Nigeria homes and communities (Akpabio, 2007; Ngene et al., 2021). These policies and principles are shown in Nigeria via RBDAs, they are set up to ensure proper management of water in accordance with natural river boundaries instead of

Table 15.2 IWRM principles and implementation status in Nigeria, summarizing policy commitments, current implementation levels, and key challenges (adapted from Akpabio & Subramanian, 2021)

IWRM principle	Policy commitment	Implementation status	Key challenges
Coordinated management	Strong—National Water Policy	Weak—limited inter-agency coordination	Institutional fragmentation, unclear mandates
Stakeholder participation	Moderate—some mechanisms exist	Weak—limited meaningful engagement	Capacity constraints, cultural barriers
Equitable access	Strong—constitutional right to water	Weak—significant disparities persist	Infrastructure deficits, funding limitations
Integrated objectives	Strong—triple bottom line approach	Moderate—some integration in planning	Sectoral silos, trade-off management

political borders or ethnic differences. It is done to ensure this policy aligns with the way water flows naturally. Basin approach shows the significance and impact of actions performed upstream affect downstream also which led to effective communication to reduce havoc in teamwork. Weak system planning, decision making, poor teamwork, bad communication and ethnic conflicting has become a major challenge in water management (Akpabio, 2007; Ngene et al., 2021).

Participant involvement has become a major significant to consider in decision-making aspect of management, thereby constituting to affected areas in water management decisions. Putting together the opinions, past experiences, knowledge and rights of people involved in water management creates a long-term commitment to water management and which will lead to community trust and individual commitment to ensuring water is properly managed. With real participants policies will directly meet the needs of each situation surrounding the water management issues and make direct impact on it (Akpabio, 2007; Ngene et al., 2021).

15.3.3 IWRM Framework Implementation in Nigeria

Integrated Water Resources Management Plan (IWRMP) was developed as a framework in Nigeria that guides the implementation of IWRM principles in the nation. It sets out goals for water improvement, development, management, and conservation, and defines the roles of organizations involved. It also aids the river basin level of involvement with the public and ensures the management sector adapts instantly to new rise of changes (Ngene et al., 2021). Moreover, RBDAs are involved in water utilization, laws and regulation, and protection of regional distribution, thus aligning with IWRM (Ngene et al., 2021).

Each state in Nigeria plays a vital role in implementing IWRM through water organizations, agencies, utility services and basin level participants. They manage water resources within their respective states, ensuring rural and urban water supply, water sanitation and liaising with federal agencies to resolve challenges due to shared borders with other states. Uneven commitment in state level has become a challenge in Nigeria. For instance, Lagos state developed urban water management which has completely solved water supply issues, sanitation and flood prevention. This agency is called Lagos state water regulatory commission (Ngene et al., 2021; Soladoye, 2025; Rukayat, 2019). Urban water management in Nigeria faces complex challenges beyond flooding and sanitation, particularly in densely populated cities like Lagos. While integrated planning and public-private partnerships have improved water supply and drainage systems, emerging threats such as antimicrobial resistance in wastewater add a new dimension to these challenges. Hospital effluents, often discharged without adequate treatment, introduce multidrug-resistant organisms into municipal water systems, increasing public health risks and complicating water quality management. Recent evidence from Nigerian hospitals highlights how untreated hospital wastewater serves as a reservoir for multidrug resistance, underscoring the need for stricter wastewater treatment protocols and monitoring within IWRM frameworks (Okpanachi et al., 2025). Addressing these issues requires coordinated action between health institutions, water agencies, and regulatory bodies to safeguard urban water security.

IWRM is implemented at local levels by LGAs and traditional rulers due to least interaction of federal agencies in local areas. Community-based committees are provided to oversea the water management, sanity and provision of water to the rural areas. Moreover, community duelers take full responsibility in maintaining water facilities and protecting watersheds thereby resolving conflicts among community members. Hence, this aligns with IWRM practices and local way of controlling water and making water decisions (Ngene et al., 2021; Rukayat, 2019; Soladoye, 2025).

Internationally, IWRM has implemented its policies by providing financial assistance, technicality and capacity building. NEWMAP which was founded by WB is among the largest efforts focused on watershed management in Nigeria. GEF also focused on managing water resources across country borders, especially Lake Chad and Niger River regions. EU-funded water resource management and policy have provided technical support and development in improving institutions and launching training programs (Eze et al., 2024; Ngene et al., 2021). Some private sectors have involved in management in Nigeria, especially in the provision of water supply and sanitation service in urban areas. Public-Private Partnership (PPPs) were introduced in various state in Nigeria which utilize the funding from private partners in water management and control. For instance, Lagos State Water Partnership shows how various private companies have been involved in delivering water services at a reasonable cost and ensure full access to it. Due to weak regulation and lack of funding affect these services (Dada et al., 2024; Obani, 2020).

15.3.4 Opportunities and Benefits of IWRM for Nigeria

Climate changes and variations in Nigeria have created a pressing need for adaptation of management strategies that will handle how water condition changes, severe weather condition and long-term climate changes. IWRM introduced a comprehensive framework that improves and encourages the coordination and planning that addresses this challenge (Oseke et al., 2021). Water resources are centralized to how climate changes and thus it impacts agriculture, energy, health, and urban development. IWRM achieve better and more efficient outcomes compared to single sector ideology which also aids the adaptation of various strategies during trade-offs (Oseke et al., 2021).

IWRM also involve drought control and build resilience in northern Nigeria. They face unpredictable rainfall and reoccurring droughts, which has put farmers livelihood to stake and economic risk. This management system combines weather tracking, water demand management, water supply increase via groundwater and transporting of water between basins. Thereby reducing water shortages (Ngene et al., 2021). The environment and communities benefit from good water managements and this includes increases in crop production, create business opportunities, improve public health, reduction in medical emergencies, boost tourism and other aspects of the economy as shown in Table 15.3. The indirect benefits outshined the direct cost spend on water projects and thus lead to broader economic reward as implemented by IWRM (Inabo & Arshed, 2019).

15.3.5 Challenges and Barriers to Effective IWRM

Integrating and implement water management policies in Nigeria is complicated due to various groups, government levels, other sub-sectors and frequent shift of powers as shown in Table 15.4. Various overlapping responsibilities, improper team-lead and poor coordination of the sectors or levels of government. Hence, these leads to frustration, confusion, nonchalant behaviours, conflicts between members of the organization. Various agencies responsible for the management and control of water across Nigeria operate separately with little or no communication link, which result to duplicate efforts and inconsistent policies, low solution to water problems across the nation (Ngene et al., 2021).

Federal and state working together in improving water management policies has been a major challenge in Nigeria due to low coordination and overlaps in area of responsibilities. Some state gives a constitutional rules or law on water management which keeps federal control out of being the major water resource manager, thus leading to confusion and unstable policies in regulating, controlling and managing water in Nigeria (Ngene et al., 2021).

Table 15.3 Benefits of IWRM implementation in Nigeria, outlining categories such as climate resilience, conflict mitigation, agricultural productivity, and economic development, along with potential impacts and implementation requirements (adapted from UN, 2015)

Benefit category	Description	Potential impact	Implementation requirements
Climate resilience	Enhanced adaptation to droughts and floods	Reduced vulnerability for 50+ million people	Early warning systems, adaptive capacity
Conflict mitigation	Reduced water-related disputes	Improved security in 200+ LGAs	Stakeholder platforms, resource sharing agreements
Agricultural productivity	Improved irrigation efficiency and yields	20–30% yield increases possible	Technical support, financing mechanisms
Urban water security	Reliable water services for growing cities	Service coverage improvements	Infrastructure investment, institutional strengthening
Ecosystem conservation	Protected wetlands and watersheds	Maintained ecosystem services	Community engagement, financing for conservation
Economic development	Job creation and income generation	Thousands of direct and indirect jobs	Skills development, market access

Table 15.4 Key hindrances to IWRM in Nigeria, detailing institutional, financial, technical, and policy-related barriers, their impacts on implementation, and potential solutions (adapted from Akpabio & Subramanian, 2021; Ishaku et al., 2011)

Barrier category	Specific challenges	Impact on implementation	Potential solutions
Institutional fragmentation	Overlapping mandates, weak coordination	Duplicated efforts, conflicts	Clear mandates, coordination protocols
Data and information	Limited monitoring, weak information systems	Poor planning, limited adaptive management	Investment in monitoring, information systems
Community engagement	Limited participation, weak feedback	Reduced legitimacy, missed local knowledge	Capacity building, inclusive processes
Financial constraints	Inadequate funding, limited cost recovery	Limited scope and sustainability	Innovative financing, improved cost recovery
Technical capacity	Skills gaps, limited training opportunities	Implementation failures, maintenance problems	Education and training programs
Legal and policy	Regulatory gaps, weak enforcement	Uncertain authority, limited compliance	Legal reform, strengthened enforcement

15.3.6 Case Studies of IWRM Practice in Nigeria

This section presents how Integrated Water Resources Management (IWRM) principles have been applied in practice in selected case studies from different regions of Nigeria. These examples highlight the diversity of implementation contexts, approaches, and outcomes, offering insights into both successes and persistent challenges. The cases include wetland management in the Hadejia-Jama'are Basin, multipurpose water use in the Benue River Basin, watershed restoration in Cross River, and integrated urban water management in Lagos. Each case demonstrates how local conditions, stakeholder engagement, and institutional arrangements influence the effectiveness of IWRM strategies. Table 15.5 summarizes the primary focus, key stakeholders, implementation approaches, major outcomes, and lessons learned from these initiatives.

Northern Nigeria basin Hadejia-Jama'are is a distinct instance of area that shows how integrated water management worked. They balance the need to protect, develop and maintain their environment with the support and trust allegiance of the communities. Wetlands that introduced vital services like flood reduction, clean water, underwater life protection and numerous provisions through agricultural processes such fishing, farming and aquatic rearing (Ayeni et al., 2019; Odunuga et al., 2011). It is supported by international partners and uses local members to manage the wetland. Community management committees were set to settle disagreement and utilization

Table 15.5 Case studies of IWRM Implementation in Nigeria, summarizing basin locations, primary focus areas, key stakeholders, implementation approaches, major outcomes, and lessons learned (adapted from Akpabio, 2012; Ishaku et al., 2011)

Basin/location	Primary focus	Key stakeholders	Implementation approach	Major outcomes	Lessons learned
Hadejia-Jama'are	Wetland management and water allocation	Communities, RBDA, NGOs	Participatory planning, community-based management	Improved biodiversity, enhanced livelihoods	Importance of inclusive engagement, long-term commitment
Benue	Irrigation and hydropower coordination	RBDA, power companies, farmers	Technical analysis, multi-purpose optimization	Revealed coordination gaps	Need for clear institutional arrangements
Cross River	Watershed restoration, forest conservation	Communities, state government, NGOs	Community-based natural resource management	Reduced erosion, improved water quality	Economic incentives essential for conservation
Lagos	Integrated urban water management	State agencies, private sector, communities	Master planning, public-private partnerships	Enhanced flood control, improved services	Coordination mechanisms critical for urban integration

of resources in a more sustainable and substantial way. Other ideas include preparation for flood control, moving water to upstream area while balancing the wetland downstream, thus it creates employment opportunity in the community and reduce pressure of water shortage or outage (Ayeni et al., 2019; Odunuga et al., 2011).

The second largest river system in Nigeria Benue River basin, also support important activities and usage for farming, fishing, electricity generation and water provision for the cities. This basin includes several dams such as Kainji, Jebba and Shiroro which serve as hydroelectricity generation for the nation (Youdeowei et al., 2019). Clear guidance for managing multi-purpose reservoirs are created. The aim of these guidelines is to balance various needs and goals while ensuring high benefits for all involvers. Detailed studies of water flowing patterns, future consumption of water and different utilization of water categorically when there's weather changes (Youdeowei et al., 2019).

Complicated water challenges are faced in Lagos State such as flooding, improper sanitation, etc. Hence IWRM was implemented to reduce the nemesis they face. Moreover, a unified planning committee has been created to implement and ensure residence abides by the rules and policies of IWRM. Hence, LSWSMP has provided a long-term means of account for water increase, usage and demand in the area (Awuah & Bijimi, 2024; Ngene et al., 2021; Olabode & Comte, 2024). Different water sources such as ground, surface and reused water are utilized for the integration of water management. It reduces being dependent on one type of water and makes the system more reliable and reduces water shortage. Lagos Lagoon and Ogun River produced large amounts of water to other distribution sections in the city (Awuah & Bijimi, 2024; Ngene et al., 2021; Olabode & Comte, 2024).

Controlling private and public wells, protecting and managing underground water sources and enhancing public wells helps to manage groundwater in Nigeria. In managing groundwater, it involves a challenge way due to no or not enough resources and funding to properly manage or enforce rules on it (Awuah & Bijimi, 2024; Ngene et al., 2021; Olabode & Comte, 2024).

15.3.7 Policy and Strategies Recommendation

This section offers an in-depth recommendation based on the challenges and opportunities IWRM has introduced to Nigeria water management. It aimed in analyzing issues found and how other international countries best practice policies that are suitable for Nigeria unique situation (Ngene et al., 2021). For IWRM in Nigeria, legal changes must be introduced while defining those who are responsible and accountable for managing and controlling water resources. Comprehensive water management laws should be set, defining the rules and how each one works. Also reorganizing sectors and agencies and clearly define the function of each of them, clarifying overlapping duties and effect of good communication amidst the agencies.

Finally, differentiate the aims and functions of the federal, state and local government agencies in charge of the basin and other water resources with guidelines and accountability system (Ngene et al., 2021).

Nigeria IWRM needs a clear mission, vision, enough funding and power to work in all agencies and all level of government. Helping agencies and keeping track of water usage should be the main job dealing with conflicts between various groups and organizations. Mixtures of different ways in raising funds should be introduced such as government, private investors, companies and international alliances. This should focus on management, control, building, use of budget and running of the water management sector (Ngene et al., 2021).

15.4 Future Directions

While Integrated Water Resources Management (IWRM) in Nigeria has made notable progress, several areas require further exploration to strengthen its implementation and sustainability. Future research and policy efforts should focus on:

1. Real-Time Water Data Systems and Predictive Modeling

Develop advanced hydrological monitoring systems and climate prediction models to improve adaptive water management and early warning capabilities.

2. Socio-Economic Impact Assessment

Conduct comprehensive evaluations of IWRM interventions at community and basin levels to measure their effects on livelihoods, health, and economic development.

3. Innovative Financing Mechanisms

Explore blended financing models, including public-private partnerships (PPPs), green bonds, and international climate funds, to address funding gaps in water infrastructure.

4. Integration of Indigenous Knowledge

Investigate how traditional water governance practices and cultural norms can complement modern IWRM strategies for more inclusive and context-sensitive solutions.

5. Emerging Threats in Water Systems

Examine issues such as antimicrobial resistance in wastewater and industrial effluents and incorporate mitigation strategies within IWRM frameworks.

6. Capacity Building and Institutional Strengthening

Design targeted training programs for technical staff, policymakers, and community stakeholders to enhance governance and operational efficiency.

By addressing these areas, Nigeria can advance toward a resilient, equitable, and sustainable water management system aligned with national development priorities and global commitments such as SDG 6.

15.5 Recommendations

To strengthen Integrated Water Resources Management (IWRM) in Nigeria, it is essential to enhance policy coherence and institutional frameworks that promote cross-sectoral collaboration among water, agriculture, energy, and environmental agencies. Stakeholder engagement should be prioritized by creating inclusive platforms that allow communities, private sector actors, and civil society to participate in decision-making processes. Capacity building programs for water managers and local authorities are critical to improve technical expertise and governance. Furthermore, investment in innovative technologies such as remote sensing, GIS, and digital monitoring systems will enable efficient water allocation and resource planning. Financial sustainability must be addressed through public-private partnerships and incentives for water conservation initiatives. Climate adaptation strategies should be integrated into water management plans to mitigate the impacts of droughts and floods, which are increasingly prevalent due to climate variability. Finally, research and knowledge-sharing networks should be strengthened to provide evidence-based solutions and foster collaboration between academia, industry, and policymakers for sustainable water resource management in Nigeria.

15.6 Conclusion

Nigeria's progress in implementing Integrated Water Resources Management shows both notable successes and ongoing difficulties. These challenges come from the complexity of institutions, missing policies, and uneven abilities across different levels of government. Although national plans like the Water Policy and IWRM Strategy have laid a solid base for coordinated planning, putting these plans into action is still hard because of poor teamwork, not enough money, and weak enforcement. To make water management work long-term, it's important to improve how institutions work together, involve more people in decision-making, and use flexible management methods. With clear changes, continued support, and fair leadership, Nigeria can improve how it manages water, build better resilience, use resources more fairly, and help the country grow economically, even as environmental and climate issues get worse.

References

Akpabio, E. M. (2007). Nigeria's water law: How is it translated in the Cross River basin? *International journal of regulation and governance, 7*(2), 157–184.

Akpabio, E. M. (2012). Water meanings, sanitation practices and hygiene behaviours in the cultural mirror: A perspective from Nigeria. *Journal of Water, Sanitation and Hygiene for Development, 2*(3), 168–181.

Akpabio, E. M., & Subramanian, S. V. (2021). Nigeria's water crisis: Drivers, impacts, and policy implications. *Sustainability, 13*(15), 8420.

Awuah, K. G. B., & Bijimi, C. K. (2024). Discovering the core stakeholders in the Nigerian urban water supply system. *Sustainable Water Resources Management, 10*(1), 8.

Ayeni, A. O., Ogunsesan, A. A., & Adekola, O. A. (2019). Provisioning ecosystem services provided by the Hadejia Nguru wetlands, Nigeria–current status and future priorities. *Scientific African, 5*, e00124.

Ayoade, J. O. (1988). Water resources and their development in Nigeria. *Hydrological Sciences Journal, 33*(4), 405–422.

Biswas, A. K. (2004). Integrated water resources management: A reassessment. *Water International, 29*(2), 248–256.

Cervigni, R., Liden, R., Neumann, J. E., & Strzepek, K. M. (2015). *Enhancing the climate resilience of Africa's infrastructure: The water and power sectors*. World Bank Publications.

Dada, M. O., Ogunlana, S. O., & Anakor, C. A. (2024). Perceptions of public and private sector stakeholders on the benefits and challenges of public-private-partnerships in Nigeria. *African Journal of Housing and Sustainable Development, 5*(1).

Eze, N. L., Obasi, F., Eze, H. O., & Otibeh, F. (2024). The role of communication in Enugu state Nigeria erosion and watershed management project (Ens-Newmap) sustainable delivery. *Caritas Journal of Management, Social Sciences and Humanities, 3*(1).

Grigg, N. S. (2008). Integrated water resources management: Balancing views and improving practice. *Water International, 33*(3), 279–292.

Inabo, O. A., & Arshed, N. (2019). Impact of health, water and sanitation as key drivers of economic progress in Nigeria. *African Journal of Science, Technology, Innovation and Development, 11*(2), 235–242.

Ishaku, H. T., Majid, M. R., & Johar, F. (2011). Water supply dilemmas in Nigeria: Looking towards integrated water resources management. *Journal of Environmental and Earth Sciences, 3*(3), 85–92.

Loucks, D. P., & Van Beek, E. (2005). *Water resources systems planning and management: An introduction to methods, models and applications*. UNESCO Publishing.

Ngene, B. U., Nwafor, C. O., Bamigboye, G. O., Ogbiye, A. S., Ogundare, J. O., & Akpan, V. E. (2021). Assessment of water resources development and exploitation in Nigeria: A review of integrated water resources management approach. *Heliyon, 7*(1).

Nicholson, S. E. (2013). The west African Sahel: A review of recent studies on the rainfall regime and its interannual variability. *International Scholarly Research Notices, 2013*(1), 453521.

Obani, P. (2020). Localizing the human right to water in Lagos state Nigeria. *Utrecht Law Review, 16*(2).

Odunuga, S., Okeke, I., Omojola, A., & Oyebande, L. E. K. A. N. (2011). *Hydro-climatic variability of the Hadejia-Jama'are river systems in north-Central Nigeria* (Vol. 344, pp. 163–168). IAHS-AISH Publication.

Olabode, O. F., & Comte, J. C. (2024). Water scarcity in the fast-growing megacity of Lagos, Nigeria and opportunities for managed aquifer recharge. *Wiley Interdisciplinary Reviews: Water, 11*(5), e1733.

Okpanachi, V., Jonah, J., Olubunmi, M., Abidemi, M., Rabiu, A., Amoh-Asante, N. A., Msemakweli, J., Obochi, V., & Amaechi, C. (2025). Hospital wastewater as a reservoir of multidrug resistance: Evidence from good shepherd male and female hospital wards, Anyigba Nigeria. *Microbes and*

Infectious Diseases. Advance online publication. https://doi.org/10.21608/mid.2025.419444. 3165.

Oseke, I. E. F., Anornu, G. K., Adjei, K. A., & Eduvie, M. O. (2021). Integrated water resources management approach in mitigating the potential impacts of climate change on hydrology in Gurara reservoir catchment, Northwest Nigeria. *Proceedings of the International Association of Hydrological Sciences, 384*, 355–361.

Pahl-Wostl, C. (2007). Transitions towards adaptive management of water facing climate and global change. *Water Resources Management, 21*(1), 49–62.

Rahaman, M. M., & Varis, O. (2005). Integrated water resources management: Evolution, prospectpahl and future challenges. *Sustainability: Science, Practice and Policy, 1*(1), 15–21.

Rukayat, B. O. (2019). Water supply regulation in Nigeria: Problems, challenges, solutions and benefits. Вестник Российского университета дружбы народов. *Серия: Экология и безопасность жизнедеятельности, 27*(1), 65–81.

Savenije, H. H., & Van der Zaag, P. (2008). Integrated water resources management: Concepts and issues. *Physics and Chemistry of the Earth, Parts A/B/C, 33*(5), 290–297.

Soladoye, O. (2025). Performance assessment of water infrastructure in Lagos state, Nigeria: Implications for sustainable development. *International Journal of Geography & Environmental Management, 11*(3), 161–179. www.iiardjournals.org

Ukpai, S. N. (2022). Water policy reform in the Nigeria water governance system: Assessment of water resources management based on OECD principles on water governance. *Water Policy, 24*(10), 1704–1722.

United Nations. (2015). Transforming our world: the 2030 Agenda for Sustainable Development. Resolution adopted by the General Assembly on 25 September 2015.

Youdeowei, P. O., Nwankwoala, H. O., & Desai, D. D. (2019). Dam structures and types in Nigeria: Sustainability and effectiveness. *Water Conservation and Management, 3*(1), 20–26.

Young, R. A. (2005). Determining the economic value of water, concepts and methods. *Washington, resources for the future*. Discussion paper, 101, 27–35.

Chapter 16
Multivariate of Analysis Water Quality Parameters: Insights from Correlation, Clustering, and Factor Analysis

Akinyinka Akinnusotu, Samuel Dare Oluwagbayide, Francis Olawale Abulude, and Oyebola Adebola Elemide

Abstract Assessing water quality is essential for comprehending the hydrochemical properties of water sources. This investigation uses Principal Component Analysis (PCA) and Factor Analysis (FA) to evaluate the interactions between major ions (Ca^{2+}, Mg^{2+}, Na^{+}, K^{+}), Total Dissolved Solids (TDS), and Total Hardness (TH) in a collection of 34 samples. The PCA outcomes show that Factor 1 accounts for 40.5% of the variance and is primarily influenced by Ca^{2+}, Na^{+}, K^{+}, and TH, implying that there is a predominant geogenic or human-caused origin; Factor 2, which is responsible for 23.5% of the variability, is strongly correlated with TDS, indicating another source controlling process, perhaps from dissolved mineral content or contamination places; and factor loadings show strong positive correlations between Na^{+}, K^{+}, and Ca^{2+}, indicating their common origin, most likely from rock-weathering or human activities. On the other hand, Mg^{2+} exhibits moderate influence across both factors, indicating distinct geochemical behavior. These relationships are visually reinforced by the biplot and loading plot, where Ca^{2+}, Na^{+}, and K^{+} align closely along the first factor while TDS dominates the second. The time series analysis of Ca^{2+} and Mg^{2+} depicts varying patterns, indicating potential temporal or spatial deviations in water chemistry, while spikes in Ca^{2+} level indicate regular impacts, like pollution incidents or changes in water locations. These results highlight the importance of ongoing water quality evaluation, especially in environmentally vulnerable areas.

A. Akinnusotu
Department of Science Laboratory Technology, University of Medical Sciences (UNIMED), Ondo, Nigeria

S. D. Oluwagbayide
Department of Agricultural and Bio-Environmental Engineering, The Federal Polytechnic, Ilaro, Ogun State, Nigeria

F. O. Abulude (✉)
Environmental and Sustainable Research Group, Science and Education Development Institute, Akure, Nigeria

O. A. Elemide
Department of Agricultural and Bio-Environmental Engineering, Federal College of Agriculture, Akure, Nigeria

M. I. Adeoba et al. (eds.), *Water Resources Management in Nigeria*, World Water Resources 10, https://doi.org/10.1007/978-3-032-22813-0_16

The findings of the investigation offer valuable information for managing water resources, highlighting the part that Ca^{2+} and Na^{+} play in water hardness and the unique impact that TDS has on total water purity. Gaining an awareness of these elements will assist with create environmentally friendly resource usage and water management strategies that work.

Keywords Water · Hydrochemical properties · PCA · TH · Human-caused origin · Ca^{2+} level

16.1 Introduction

Water is an essential and fundamental element that sustains life and the environment, is a key component of industry, is used by both humans and animals, and is a source of pollution in both homes and businesses. The three fundamental necessities that have the biggest effects on socioeconomic growth and the quality of life are access to enough water for irrigation, sanitation, and residential use. Water resources and ecosystems are generally under more stress due to population increase, urbanization, industry, agriculture, tourism, and changes in the climate and way of life (Ingrao et al., 2023). On the worldwide environmental sustainability agenda, the issue of water has drawn a lot of focus and visibility. This is a result of mounting pressure from elements like public health, population increase, climate change, and economic development models (Mehmood, 2019; Levallois & Villanueva, 2019; Muniz & Oliveira-Filho, 2023). The United Nations 2030 Agenda for Sustainable Development's Sustainable Development Goal number 6 (SDG6) is solely focused on water and addresses the problems of universal access to clean water and water in adequate amounts and quality. It also raises concerns about the management of water resources (UN-Water, 2016).

There are now water emergencies (Bogardi et al., 2021). With all of its ramifications for economies, cultures, and the environment, the drama of too little, too much, and dirty water is playing out right before our eyes. There is no such thing as an isolated crisis, as we are all aware.

Everyone might be fooling one another if we did not recognize the systemic difficulties encounter in a world where natural systems are highly interdependent, our economies are structured around international supply chains, and the seismic impacts of emergencies on societies rarely stop at national borders. According to SIWI (Stockholm International Water Institute, 2020), water is everyone's obligation, and the distribution and sustainable usage necessitate the active and informed involvement of all individuals in the management and development of water resources. The management of water resources benefits greatly from the analysis, evaluation, and monitoring of water quality since it offers a thorough grasp of the condition of the water (Moriarty et al., 2007; Cosgrove & Loucks, 2015). There has been a trend for the compilation of water quality data to increase through studies that assess water quality, even if there is still a dearth of data at the global level, primarily because less developed countries do not monitor their water quality (Mehmood, 2019; Connor & Coates, 2021).

According to Meybeck et al. (1996), water quality can be defined as a measure of the suitability of water in relation to natural quality, pollution effects, or specific use based on physical, chemical, and biological attributes. This measure offers objective evidence that is necessary for decision making in the management of water resources, for the use of water quality monitoring programs (Behmel et al., 2016), and for warning people about problems that are ongoing or emerging (such as eutrophication, chemical and microbial contamination, eutrophication, emerging contaminants, and climate change issues, among others), in assessing environmental state, assessing the effects on aquatic ecosystems (Dallas & Day, 2004), assessing compliance with regulatory criteria, preserving the beneficial applications of water, and evaluating temporal patterns of water quality (Haener et al., 2018).

Nigeria's abundance of rivers, streams, lagoons, and oceans, particularly in the southwest, greatly contributes to the country's overall economic growth (Abulude et al., 2023a). The Atlantic Ocean (Lagos State), the Osun River (Osun State), the Ala River, and the Arinran River (Ondo State) are noteworthy, to name a few. The region's economy, culture, and environment all greatly depend on these waterways (Gebre & Gebremedhin, 2019). Both urban and semi-urban residents use them for a variety of reasons (Anh et al., 2023). It contributes significantly to tourism, investment, and biodiversity conservation and is regarded as the foundation of all businesses established along the shore. However, anthropogenic activities like urbanization, intensive agriculture, rapid industrialization, population growth, urban runoff, municipal waste, overfishing, grass cutting along the shores, car washing on the lakeshore, and horticultural farming on the lakeshore, as well as natural activities like erosion and heavy rainfall, pose a serious threat to the waters' cleanliness (Oluwagbayide et al., 2025).

Water quality has been the subject of numerous studies by different researchers (Khoshnam et al., 2017; Mohammed, 2017; Sulaiman et al., 2022; in Southwest, Nigeria (Olajire & Imeokparia, 2001; Wahab et al., 2012; Irenosen et al., 2012; Oginni & Ojoawo, 2014; Odebunmi et al., 2014; Oluwafemi et al., 2015; Olowe et al., 2016; Olayiwola et al., 2019; Awomeso et al., 2020; Adesina et al., 2020). These areas with data are out of date. This means that ongoing validation of the water quality is crucial. The study's novelty likely stems from its specific focus on evaluating water quality in three different Nigerian states, providing regional viewpoints on water appropriateness for farming using multivariate analysis. A number of methods are used to evaluate groundwater, including statistical (including machine learning), graphical, geographical (using GIS), and indexical (numerical) methods (Unigwe et al., 2022; Egbueri & Agbasi 2022; Gautam et al., 2023; Singh et al., 2024). The most frequently employed indexical methods are sodium adsorption ratio (SAR), electrical conductivity (EC), residual sodium carbonate (RSC), Kelly's ratio (KR), percentage sodium (Na %), and permeability index (PI). Special achievements from this case study might involve raising awareness of local water quality problems and circumstances, providing new information on Ca^{2+}, Mg^{2+}, Na^{+}, K^{+}, TDS, and

TH, and assisting in the formulation of agricultural policies. The rivers, wells, boreholes, streams, and dam water used in this study are located in Southwest Nigeria's Ondo, Osun, and Ekiti States. The rivers are located in urban and suburban areas that are well-known for their cash and arable crops. Between 24 and 35 °C are the typical daily temperatures. Oluwagbayide and Abulude (2022) claim that the samples travel through confined woods, agricultural areas, and development, providing water resources and local fish. In addition to serving as water sources for the local population, some of the rivers and streams have been classified as recreational areas. These locations are the main ecotourism destinations for locals. Numerous studies have found that recreational rivers are vulnerable to contamination (Saana et al., 2016; Popoola et al., 2019; Ojekunle et al., 2020).

It should be noted that a number of studies have been conducted on the water in the southwestern region; data are produced, but they are outdated. Constant monitoring of the water and its surroundings is vital to guarantee the safety of the water for human and animal consumption as well as for agricultural purposes. Starting this study is not inappropriate. Therefore, the study's objective was to use multivariate statistical analysis (Correlation, Clustering, and Factor Analysis) to evaluate the current status (water quality metrics) of water samples from three Nigerian states: Ondo, Ekiti, and Osun.

16.2 Materials and Methods

16.2.1 Sample Collection and Analysis

The research region is situated in southwest Nigeria and is comprised of the states of Osun, Ekiti, and Ondo (Fig. 16.1). The climate of the research area often exhibits similar patterns (Abulude et al., 2023b). Housing, industry, transportation, population, and agriculture are all developing fast throughout the states. For this experiment, 35 water samples were chosen in November 2023 from boreholes (7), hand-dug wells (14), rivers (10), and rain (2). Using distilled water, nitric acid treatment, and membrane filtering with 0.45 micron pores, the samples were thoroughly cleaned after being collected in plastic bottles. The coordinates of the sampling sites were determined using a Garmin global positioning system. A portable multi-parameter meter known as the Temp/pH/TDS/EC meter (model MI 1399) was used to test the EC and TDS in situ as soon as possible after sample collection in the field (Oluwagbayide and Abulude, 2022). The water samples were treated with nitric acid to prevent the precipitation of trace components. The Limgis (2001) methods were used to quantify Na^+, Ca^{2+}, Mg^{2+}, TH, and K^+ in order to determine whether the water samples were suitable for agricultural and drinking purposes (Fig. 16.2).

The fundamental description of the 34 water samples' parameters is shown in Table 16.1. Ca^{2+} ranges from 0.31 to 1.81 mg/L, with a mean of 0.96 ± 0.38 mg/L. The standard deviation indicates that the calcium contents of the samples varied only

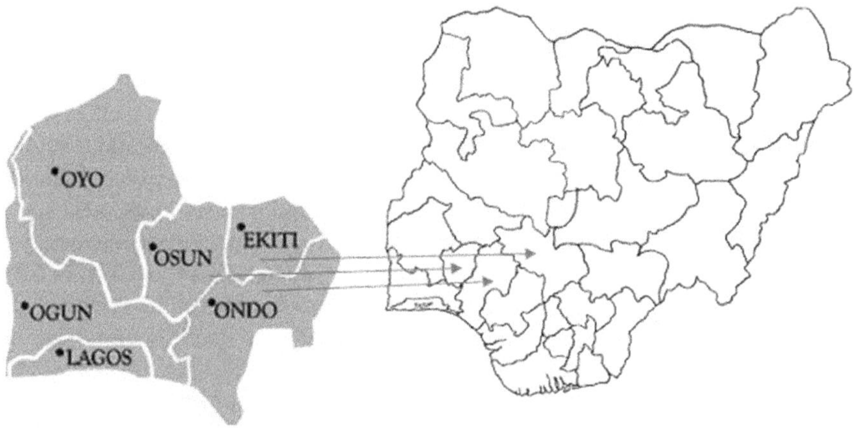

Fig. 16.1 The map of the selected states used for the water analysis

Fig. 16.2 Flow chart of the water samples analysis

slightly. It demonstrates that the variation was mild, which is normal for groundwater. With a deviation of 0.06 mg/L, the Mg^{2+} levels (0.15 to 0.40 mg/L) in this investigation have the lowest concentrations and the least fluctuation. A component of overall hardness is magnesium. Na^{+} readings range from 0.32 to 2.66 mg/L, with respective means of 1.01, 0.89, 0.51, 0.67, 0.89, and 1.22, as well as SE mean, StDev, Q1, median, and Q3. The large variable range could imply human impact or rock-water interaction. The highest mean parameter in the study is K^+ (5.97 ± 1.48 mg/L). The high average could likely be from fertilizer runoff or agricultural effect of the neighboring locations. TDS has a mean of 1.64 mg/L, a range of 0.34 and 6.39 mg/L, and a StDev of 1.12. The wide range points to several anthropogenic pollution sources for solutes. TH has Q1 (1.29) and Q3 (1.97), and it varies moderately.

The results of this study are compared with those of other studies conducted in various parts of Africa in Table 16.2. The study's Ca^{2+} findings ranged from 0.31 to 1.81. The findings were in good agreement with those published by Ishaku et al., (2016) for Dashen, Nigeria, and Eyankware et al., (2021) for Port Harcourt, Nigeria. When compared to this study, the findings for Lagos, Egypt, Burkina Faso, Ethiopia, and Tunisia exhibit significant variation. Despite being low, the Mg^{2+} values from this study are consistent with those for Lagos, Dashen, Manga (Burkina Faso), and Sierra Leone reported by Balogun et al., (2012), Ishaku et al., (2016), Millogo et al., (2024), and Kallon (2024), respectively. Findings for Port-Harcourt (Eyankware et al., 2021), Egypt (Alzahrani et al., 2025), and Tunisia (Gharbi et al., 2019). The study's Na^+ and K^+ values for Burkina Faso and Sierra Leone nearly match, indicating that the mineral concentrations of the water samples vary similarly in both countries. While the TDS levels recorded for Egypt (137.0–3942 mg/L), Ethiopia (368.0–1289 mg/L), Tunisia (1400–6500 mg/L), Egypt (416–6906.46 mg/L), and Port-Harcourt, Nigeria (109–723 mg/L) are significantly higher than the WHO standard guideline for drinking water, our analysis shows very low TDS. They will need to be treated before being used because this suggests that they may be salinized or contaminated by industrial or agricultural processes. Since the water's TH values are low, it can be classified as soft water that is acceptable for drinking and other household uses. High levels of TDS and Na^+ in water samples pose a health danger, particularly to those who have kidney issues and hypertension. Although the study's parameters appear to be modest, it is still a good idea to regularly check the samples.

The time series was shown in Fig. 16.3 as line graphs that showed variations in the several spikes. The y-axis shows the concentrations of each sample, while the x-axis (index) shows the sample locations. The range of Na^+ is between 3.5 and 10.5. This graph exhibits great variability and frequent oscillations of the peak, which is

Table 16.1 Basic description of the results of the samples

Variable	N	Mean	SE mean	StDev	Minimum	Q1	Median	Q3	Maximum
Ca2+	34	0.9567	0.0644	0.3758	0.3075	0.6628	0.8512	1.1597	1.8112
Mg2+	34	0.24803	0.00974	0.05677	0.14950	0.20250	0.24250	0.29237	0.40300
Na+	34	1.0124	0.0875	0.5102	0.3200	0.6725	0.8850	1.2163	2.6600
K+	34	5.972	0.255	1.484	3.917	4.635	5.750	7.229	9.792
TDS	34	1.635	0.192	1.120	0.338	0.859	1.512	2.053	6.388
TH	34	1.6787	0.0908	0.5295	0.7450	1.2942	1.6142	1.9684	3.0162

Table 16.2 Comparison of the water samples results with other research works

Country	Ca^{2+}	Mg^{2+}	Na^{+}	K^{+}	TDS	TH	References
Lagos, Nigeria	2.4–43.77	0.0–5.4	4.5–145	0.0–16.63	21–610	2.9–39.16	Balogun et al. (2012)
Dashen and environs, Nigeria	1.25–3.11	0.89–2.08	0.01–0.23	0.027–0.205	91–580	21.44–53.28	Ishaku et al. (2016)
West El-Minya region, Egypt	7.1–321.7	3.1–188.6	24–1957	1.6–29	416–6906.46	–	Alzahrani et al. (2025)
Port Harcourt, Nigeria	0.2–1.5	15.9–114.2	1.0–15.8	4.2–16.5	109–723	–	Eyankware et al. (2021)
Manga, Burkina Faso	44.0–324.0	18.44–96.19	0.0–57.78	2.30–11.96	0.00–2.10	158–1081.00	Millogo et al. (2024)
Kono, Sierra Leone	2.00–53.0	2.00–9.00	2.40–14.0	1.10–14.0	17.10–317.0	5.00–175.0	Kallon (2024)
Sidi Bouzid, Tunisia	96.0–455.0	58.0–326.0	132.0–1041.0	6.30–30.0	1400–6500	–	Gharbi et al. (2019)
Lake Hawassa, Ethiopia	–	–	–	–	368.0–1289	52.0–74.0	Zigde and Tsegaye (2019)
Egypt	14.0–128.0	8.5–121.5	6.0–1755	3.0–285	137.0–3942	46.5–800	Alshahrani et al. (2025)
Southwest, Nigeria	0.31–1.81	0.29–0.40	1.22–2.66	7.23–9.79	2.05–6.39	1.97–3.02	This study

largest at index 25. Urban discharge, natural brine intrusion, and agricultural runoff are the causes of the increases. TDS fluctuates between 1 and 3, with the Oba-Ile borehole water having the largest jump (index 27). The TDS levels in the water samples are often low and consistent, which indicates both good water quality and sporadic pollution. The K^+ ranges from 4 to 10.5 and exhibits a variety of patterns. The largest peak was seen near the Ala River, Oba-Ile housing development (index 25). This could be because of soil leaching, human activity, or even shifts in pollution levels. Ca^{2+} fluctuated moderately, ranging from 0.6 to 1.8. The maxima are located on indices 5, 14, 17, and 32 (Iju (stream); Owena (well); Igbara-odo (well); and the borehole used for the FECA students' project. The mineral contents in the areas have an effect on the Ca^{2+} contents. These time series' outcomes are consistent with the conclusions drawn from the boxplot. Mg^{2+} is comparatively stable and has a small range of change (0.15 and 0.4 mg/L). The Owena River in Ondo State has the greatest value, although there are some moderate variations in the uneven spikes. The concentrations' uniformity points to a constant hydrological or geological effect, which could be caused by underlying rocks or steady source inputs. Although the water samples are generally healthy, manmade activity in the water's immediate surrounds can sometimes cause problems. It is advised to monitor continuously. TH like all other parameters showed variations. The highest peaks are exhibited by water from samples numbers 13 (Arinran River, Osun State), 15 (Owena River, Ondo State), and 32 (Rain water harvested in FECA, Ondo State).

Previous studies have investigated contamination and ecotoxicological impact on the Lower Passaic River (Soetan et al., 2022). To gain a further understanding of potential source and effect relationships between the toxic metals and water quality indicators evaluated in this study, metal and water quality data were analyzed using the Pearson correlation method. Figure 16.4 shows the existing associations between sedimentary toxic metals and the physicochemical parameters. The correlation coefficient of the water sample parameters used in this investigation is shown in Fig. 16.4. There are colors that distinguish the significance's direction and intensity. These findings show that Na^+:K^+ ($r = 0.72$), Mg^{2+}:Na^+ ($r = 0.54$), and Ca^{2+}:Na^+ ($r = 0.46$) have positive associations. These demonstrate that they originate from the same sources, including waste discharges, industries, and agricultural sources. For the connections between Mg^{2+}:Na^+ and Ca^{2+}:Na^+. Ca^{2+} and Mg^{2+} may have had an impact on the Na^+ concentration through mineral solubility or reactions between water and rock. For both Ca^{2+} and K^+ ($r = 0.44$), geochemical sources may have had an impact on Ca^{2+}. Ca^{2+}, K^+, and Na^+ connections were also reported by Egbueri et al., (2019). Because Ca^{2+} and Mg^{2+} contribute to water hardness, the results of TH:Ca^{2+} ($r = 0.36$) and TH:Mg^{2+} ($r = 0.30$) are expected. The negative correlations between TDS:Mg^{2+} ($r = -0.26$) and TDS:K^+ ($r = -0.18$) suggest that an increase in ions did not play a substantial role in the solids' dissolutions. The fluctuation of TDS might have been more influenced by other causes. Water hardness did not significantly affect total solids, as seen by the modest negative connection between TH and TDS ($r = -0.12$). Similar to our research, Sahu et al. (2024) found a relationship among TH and TDS. Since water hardness has detrimental impacts on plumbing and most industrial uses, it is critical to continuously check Ca^{2+} and Mg^{2+} in water samples. Measures

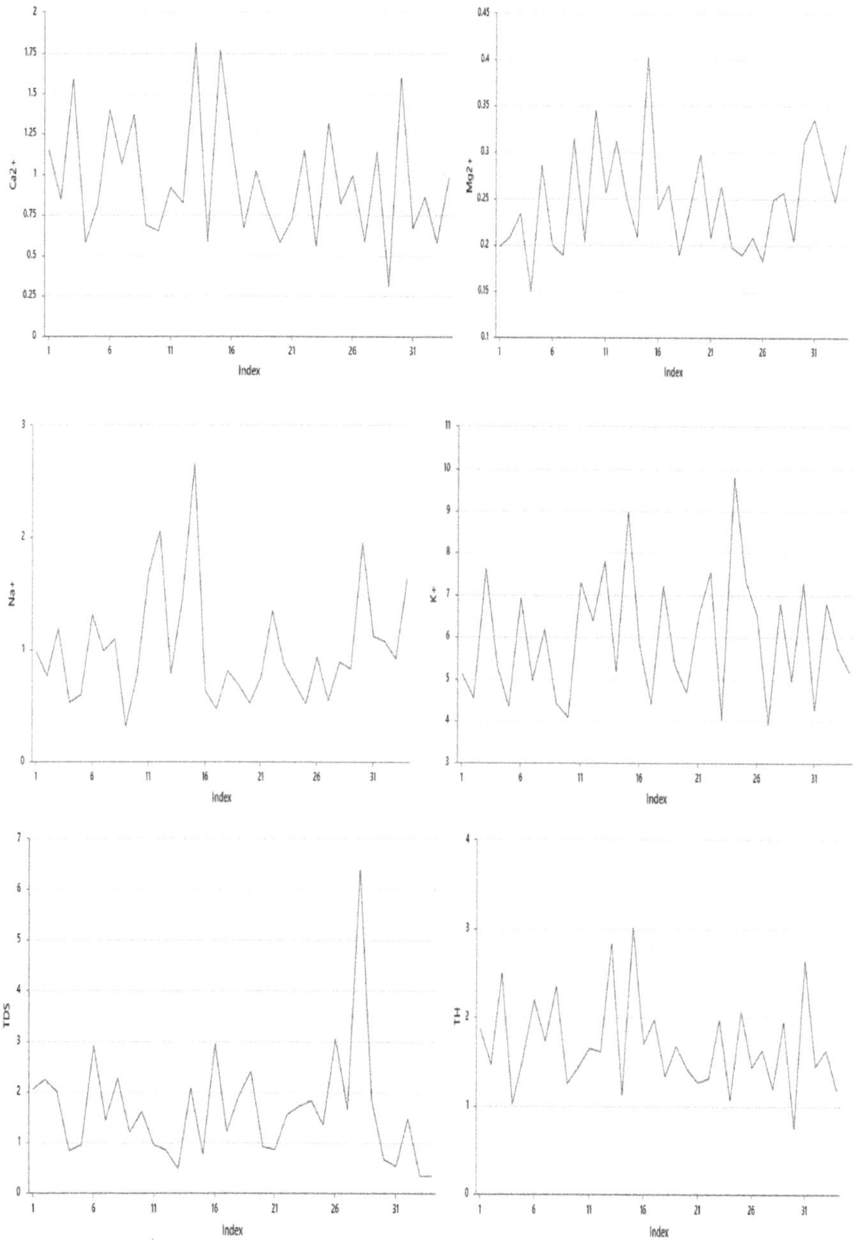

Fig. 16.3 Time series of the different parameters of the water samples

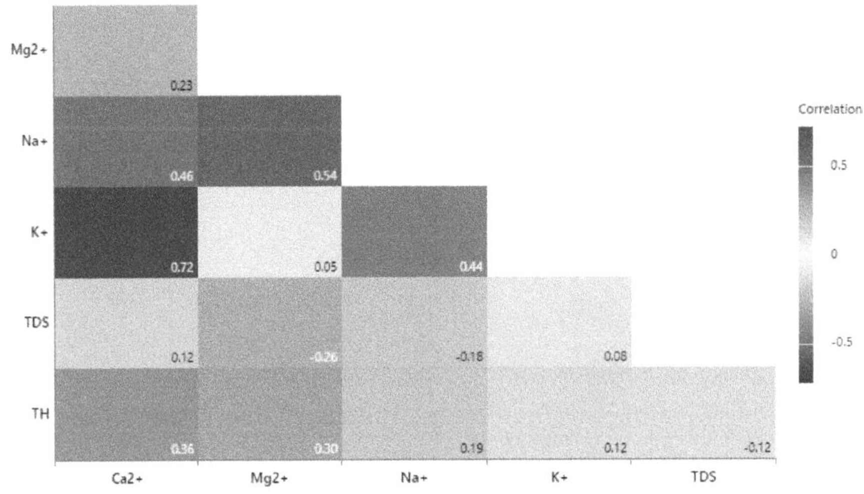

Fig. 16.4 Correlation coefficient of the results obtained in the study

should be taken to control sources from industrial waste and fertilizer use in order to mitigate the effects of Na^+ and K^+.

The factor loadings, communalities, and explained variances of the water samples are displayed in Table 16.3. Ca^{2+}, Na^+, K^+, Mg^{2+}, and TH are as follows in Factor 1 (variance: 40.5%): 0.823, 0.789, 0.711, 0.588, and 0.508, respectively. Since significant ions are present in groundwater and contribute to water hardness, this factor is known as the total mineralization or geogenic effect. High loadings of TDS, K^+, and Ca^{2+} with respective values of 0.737, 0.526, and 0.407 are found in Factor 2 (variance: 23.5%). The presence of TDS and K^+, which may have anthropogenic and agricultural inputs, makes this factor known as the dissolved load. The TH (− 0.808) loading of Factor 3 (variance: 14.6%) is strong. Factors 4–6 (variance: 12.1%–3.6%) have low and moderate variations, such as TDS and Mg^{2+}, which may have been caused by human activity. Communality is represented by 1. Perfect reconstructions by the extracted factors are shown in the figure; this could be the result of rotation or a standardized solution. The variance was 78.6% due to the first three components. Since Ca^{2+}, Na^+, and K^+ predominate, factor 1 represents natural mineral contents. Since TDS and K^+ are present, factor 2 indicates pollution or human influence. As a contrasting component, the third factor, which is mainly TH, is an inverse factor of scaling potential or a softness-hardness gradient.

Figure 16.5 displays the boxplot of the sample parameters. The distribution of Ca^{2+}, Mg^{2+}, Na^+, K^+, TDS, and TH concentrations as well as the sample quality indicators are shown in the figure. The charts show the six parameters' medians (central values), outliers, and IQR (spread). K^+ values stand out the most because it has the broadest IQR. Between six and ten is the median whisker. This suggests that there is a considerable amount of K^+ fluctuation in the water samples. Various pollution sources have an impact on this. The outcomes for Mg^{2+} are the opposite of

Table 16.3 The factor loadings, communalities, and variances of the water samples

Variable	Factor 1	Factor 2	Factor 3	Factor 4	Factor 5	Factor 6	Communality
Ca^{2+}	0.823	0.407	− 0.098	− 0.086	0.196	− 0.320	1.000
Mg^{2+}	0.588	− 0.588	0.056	0.444	0.312	0.103	1.000
Na^{+}	0.789	− 0.181	0.367	0.192	− 0.411	− 0.059	1.000
K^{+}	0.711	0.526	0.197	− 0.287	0.072	0.303	1.000
TDS	− 0.145	0.737	− 0.196	0.626	− 0.059	0.052	1.000
TH	0.508	− 0.222	− 0.808	− 0.093	− 0.157	0.080	1.000
Variance	24.300	14.118	8.776	7.249	3.381	2.176	6.0000
% Var	40.50	23.50	14.60	12.10	5.60	3.60	1.000

those for K^{+}, i.e., the lowest values, indicating limited sample variability (minimum Mg^{2+} present). Ca^{2+}, Na^{+}, TDS, and TH boxplots fall within the moderate range in terms of concentration and variability. While Na^{+} has more than two outliers, Ca^{2+} and Na^{+} have close IQRs. The very identical distribution of TDS and TH indicates that the same environmental contamination activities are probably responsible for their elevations. Seasonal variations, industrial wastes, and agricultural runoff could be the cause of the TH, TDS, and K^{+} outliers that are over their whiskers. Because of the substantial ion variability in the water samples, K^{+} does not show any significant outliers.

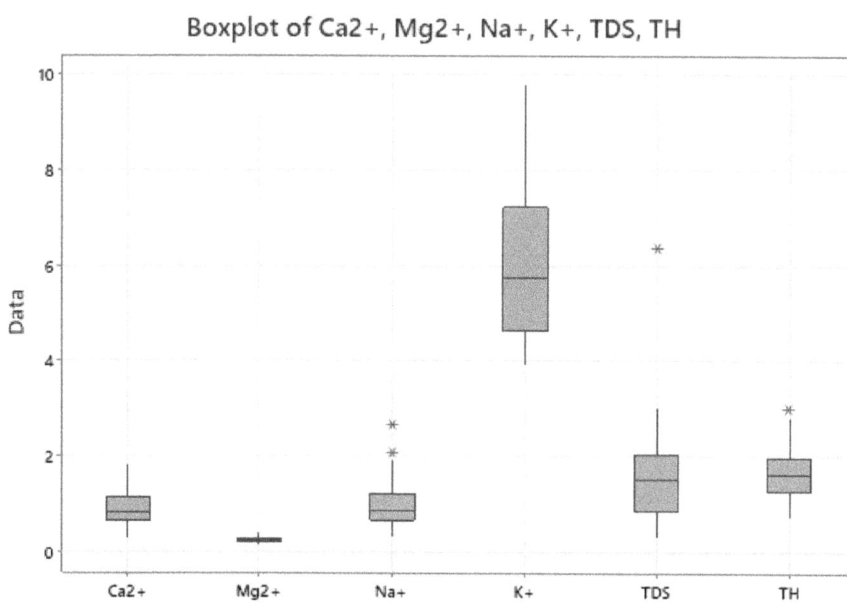

Fig. 16.5 The boxplots of the sample parameters

The loading plot and biplot of the water sample parameters are shown in Fig. 16.6. Ca^{2+} and K^+ are closely related and have a positive relationship with the first factor, while the positions of Na^+ and TH show a moderate correlation. Mg^{2+} is directly related to Ca^{2+} and K^+, indicating an inverse correlation; this may be because of the different origins of the two ions. TDS is perpendicular to Ca^{2+}, K^+, Na^+, and TH, and has a poor (weak correlation) with the first primary factor. The overall pattern of ion concentration in the water samples is shown by the First Factor (x-axis). The primary variables involved in this study are Ca^{2+}, K^+, and Na^+, which have substantial loadings in the plot. The result's interpretation indicates that the high values on this x-axis represent the high ion levels. Ionic variability resulting from various sources is depicted in the second factor. TDS has a significant y-axis load, indicating that other ions (cations) influence the variability. In contrast to Ca^{2+} and K^+, Mg^{2+} exhibits a negative correlation at the y-axis, suggesting a distinct impact. The first (x-axis) and second (y-axis) factors are displayed in the biplot. The study's individual findings are represented by the blue dots, each of which is based on the subjects' scores on the first and second PCs. Each variable's strength and direction are indicated by the arrows. The near proximity of Na^+, TH, and Ca^{2+} indicates their positive associations. The TDS behaves differently from other variables, which is known as deviance. The first factor seems to be tightly aligned with Mg^{2+} and Na^+.

The Dendrogram in Fig. 16.7 shows the hierarchical clusters of the parameters in the water samples. The variables (Ca^{2+}, K^+, TDS, Mg^{2+}, TH, and Na+) are based on correlation coefficient distance and full linkage. In Cluster 1 Ca^{2+} and K^+ there is a high correlation, which is corroborated by the correlation coefficient statistics in Fig. 16.4. The datasets are almost identical to one another. This illustrates how the parameters originated from related sources, such as geogenic processes or mineral dissolution. The pairings in Cluster 2 (Mg^{2+} and Na^+) are rather comparable, indicating that they vary together in the data set. This combination may be related to human activities such as the use of fertilizer and the release of garbage. Once more, TH joins (Mg^{2+} and Na^+) in cluster 2. This is because Mg^{2+} directly contributes

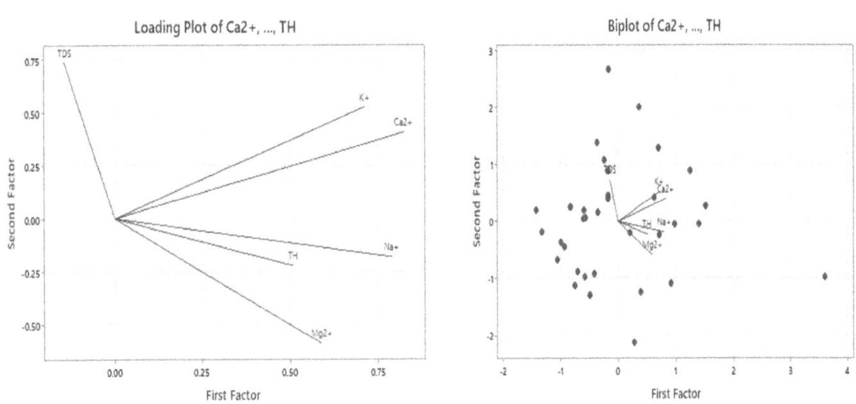

Fig. 16.6 The loading plot and biplot of water samples parameters

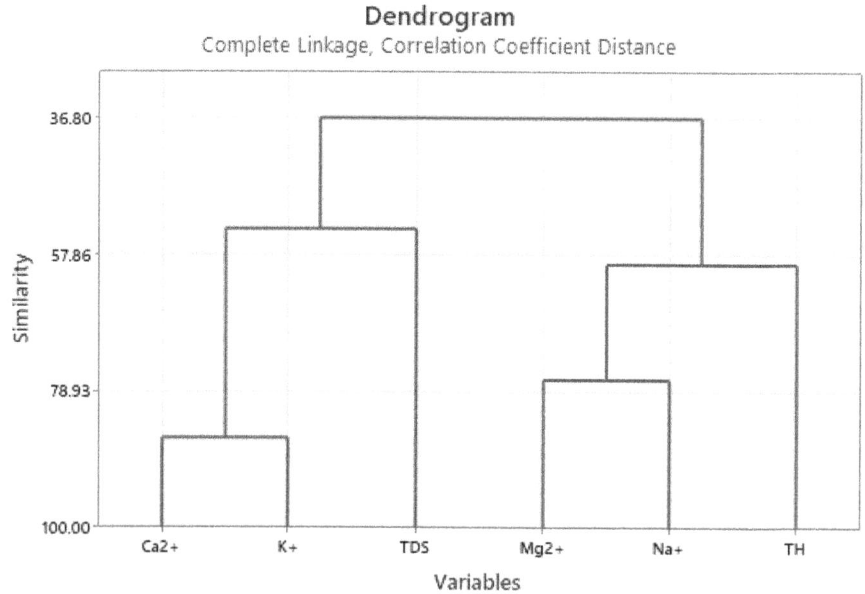

Fig. 16.7 The Dendrogram diagram showing the hierarchical clusters of parameters in water samples

to hardness, whereas Na^+ had a moderate correlation with TH due to the strong correlation between ions and water hardness. Although there is a weak correlation between TDS and (Ca^{2+} and K^+) in Cluster 1, the overall pattern of variation across the samples is consistent with Ca^{2+} and K^+, even though the individual correlations are low. The implication is that TDS could be influenced by a broader suite of ions including these two. The correlation matrix comprises the combinations of Ca^{2+}, K^+, and TDS and Mg^{2+}, Na^+, and TH this reveal distinct sources and pathway. Both anthropogenic activity (Cluster 2) and rock weathering (Cluster 1) may be the cause.

16.3 Conclusion

In order to understand the hydrochemical characteristics of water sources, this study evaluates the quality of the water. In this study, 34 samples are used to assess the relationships between Ca^{2+}, Mg^{2+}, Na^+, K^+, TDS, and TH using PCA and FA. Significant ion fluctuation is shown by the descriptive statistical data, which most likely point to a human-caused impact. While K^+ and TDS dominate a separate pollution/anthropogenic component, the FA supports it with Ca^{2+}, Na^+, and Mg^{2+} contributing to a mineralization factor. However, TH showed moderate variance, loading negatively on a distinct factor, indicating a complicated interplay amongst hardness contributors. The presence of both naturally occurring geogenic and man-made processes affecting water quality is generally supported by the multivariate analysis. Even though the ions, TDS, and TH levels were below the WHO-recommended thresholds, it is crucial to regularly check the water from several sites throughout the year.

References

Abulude, F. O., Adeoya, E. A., Arifalo, K. M., Adeyemi, O. O., & Jiddah-Kazeem, B. S. (2023a). Physiochemical quality assessment of surface and groundwater sources: A case study of Southwest Nigeria, *1*(1), 157–173.

Abulude, F. O., Akinnusotu, A., Adeoya, E. A., Mabayoje, S. O., Oluwagbayide, S. D., Arifalo, K. M., & Adamu, A. (2023b). Quality of surface and ground water in three states of Nigeria: Assessment of physicochemical characteristics and selected contamination patterns. *Environmental Sciences Proceedings, 25*(1), 48. https://doi.org/10.3390/ECWS-7-14258

Adesina, O. A., Onatunji, A. P., Akinlade, Y. O., & Ogunjobi, E. O. (2020). Multivariate modelling of water quality parameters in Nigeria. *International Journal of Research in Environmental Science (IJRES), 6*(3), 49–54.

Alshahrani, A., Ahmad, M., Laiq, M., et al. (2025). Geostatistical analysis and multivariate assessment of groundwater quality. *Scientific Reports, 15*, 7435. https://doi.org/10.1038/s41598-025-91055-3

Alzahrani, H., Basaloom, A., & Mosaad, S. (2025). Geochemical-based appraisal of karst groundwater quality, West Nile Valley, Central Egypt, for drinking and irrigation. *Journal of Hydrology: Regional Studies, 57*, 102152. https://doi.org/10.1016/j.ejrh.2024.102152

Anh, N. T., Can, L. D., Nhan, N. T., Schmalz, B., & Luu, T. L. (2023). Influences of key factors on river water quality in urban and rural areas: A review. *Case Studies in Chemical and Environmental Engineering, 8*, 100424. https://doi.org/10.1016/j.cscee.2023.100424

Awomeso, J. A., Ahmad, S. M., & Taiwo, A. M. (2020). Multivariate assessment of groundwater quality in the basement rocks of Osun State, Southwest, Nigeria. *Environment and Earth Science, 79*, 108. https://doi.org/10.1007/s12665-020-8858-z

Balogun, O., Adeaga, I. S., Akoteyon, R. O. O., & Adegun, B. (2012). Multivariate analysis of groundwater quality in parts of Lagos-Nigeria. *Global Journal of Pure and Applied Sciences, 18*(1&2), 43–52.

Behmel, S., Damour, M., Ludwig, R., & Rodriguez, M. J. (2016). Water quality monitoring strategies—A review and future perspectives. *Science of the Total Environment, 571*, 1312–1329.

Bogardi, J. J., Gupta, J., Nandalal, K. D., Salamé, L., Van Nooijen, R. R. P., Kumar, N., et al. (2021). *Handbook of water resources management: Discourses concepts and examples*. Springer. https://doi.org/10.1007/978-3-030-60147-8

Connor, R., & Coates, D. (2021). The state of water resources. In *The United Nations world water development report: Valuing water* (pp. 11–16). UNESCO.

Cosgrove, W. J., & Loucks, D. P. (2015). Water management: Current and future challenges and research directions. *Water Resources Research, 51*, 4823–4839.

Dallas, H. F., & Day, J. A. (2004). *The effect of water quality variables on aquatic ecosystems: A review* (p. 224). Pretoria, South Africa.

Egbueri, J. C., & Agbasi, J. C. (2022). Performances of MLR, RBF-NN, and MLP-NN in the evaluation and prediction of water resources quality for irrigation purposes under two modeling scenarios. *Geocarto International, 37*(26), 14399–14431. https://doi.org/10.1080/10106049.2022.2087758

Egbueri, J. C., Mgbenu, C. N., & Chukwu, C. N. (2019). Investigating the hydrogeo-chemical processes and quality of water resources in Ojoto and environs using integrated classical methods. *Modeling Earth Systems and Environment, 5*, 1443–1461. https://doi.org/10.1007/s40808-019-00613-y

Eyankware, M. O., Akakuru, O. C., Ulakpa, R. O. E., et al. (2021). Sustainable management and characterization of groundwater resource in coastal aquifer of Niger delta basin Nigeria. *Sustainable Water Resources Management, 7*, 58. https://doi.org/10.1007/s40899-021-00537-5

Gautam, V. K., Pande, C. B., Moharir, K. N., Varade, A. M., Rane, N. L., Egbueri, J. C., & Alshehri, F. (2023). Prediction of sodium hazard of irrigation purpose using artifcial neural network modelling. *Sustainability, 15*(9), 7593. https://doi.org/10.3390/su15097593

Gebre, T., & Gebremedhin, B. (2019). The mutual benefits of promoting rural-urban interdependence through linked ecosystem services. *Global Ecology and Conservation, 20*, e00707. https://doi.org/10.1016/j.gecco.2019.e00707

Gharbi, A., Ali, Z. I., & Zairi, M. (2019). Groundwater suitability for drinking and agriculture purposes using irrigation water quality index and multivariate analysis: Case of Sidi Bouzid aquifer, Central Tunisia. *Environment and Earth Science, 78*, 692. https://doi.org/10.1007/s12665-019-8733-y

Haener, P. (2018). *The handbook on water information systems administration, processing and exploitation of water-related data* (p. 116). UNESCO.

Ingrao, C., Strippoli, R., Lagiola, G., & Huisingh, D. (2023). Water scarcity in agriculture: An overview of causes, impacts and approaches for reducing the risks. *Heliyon, 9*(8), e18507. https://doi.org/10.1016/j.heliyon.2023.e18507

Irenosen, O., Festus, A., & Coolborn, A. (2012). Water quality assessment of the Owena multi-purpose dam, Ondo state, southwestern Nigeria. *Journal of Environmental Protection, 3*, 14–25. https://doi.org/10.4236/jep.2012.31003

Ishaku, J. M., Ankidawa, B. A., & Pwalas, A. J. D. (2016). Evaluation of groundwater quality using multivariate statistical techniques, in Dashen area, north eastern Nigeria. *British Journal of Applied Science & Technology, 14*(6), 1–17. https://doi.org/10.9734/BJAST/2016/24069

Kallon, H. D. S. (2024). Hydrochemical characteristics, controlling factors and water quality evaluation of groundwater quality in Kono, Sierra Leone. *Advances in Earth and Environmental Science, 5*(2), 1–17. https://doi.org/10.47485/2766-2624.1044

Khoshnam, Z., Sarikhani, R., Ghassemi Dehnavi, A., & Ahmadnejad, Z. (2017). Evaluation of water quality using heavy metal index and multivariate statistical analysis in Lorestan Province, Iran. *Journal of Advances in Environmental Health Research, 5*(1), 29–37.

Levallois, P., & Villanueva, C. M. (2019). Drinking water quality and human health: An editorial. *International Journal of Environmental Research and Public Health, 16*, 631.

Limgis. (2001). *Methodology: Physico-chemical and biological analyses of water*. Linnology.

Mehmood, H. (2019). Bibliometrics of water research: A global snapshot. In *UNU-INWEH report series, issue 06* (p. 24). United Nations University Institute for Water, Environment and Health.

Meybeck, M., Kimstach, V., & Helmer, R. (1996). Strategies for water quality assessment. In D. Chapman (Ed.), *Water quality assessments—A guide to use of biota, sediments and water in environmental monitoring* (2nd ed., p. 609). CRC Press.

Millogo, C., Sagnon, B., Kafando, S., & Nakolendoussé, S. (2024). Hydrochemical assessment of groundwater quality in the City of manga and the surrounding area, Burkina Faso, West Africa. *American Journal of Water Resources, 12*(4), 139–148. https://doi.org/10.12691/ajwr-12-4-4

Mohammed, A. I. (2017). Irrigation water resources and suitability for crops in Egypt. *Merit Research Journal of Agricultural Science and Soil Science, 5*(2), 040–053.

Moriarty, P., Batchelor, C., Laban, P., & Fahmy, H. (2007). *INWRDAM—The EMPOWERS approach to water governance: Background and key concepts*. Inter-Islamic Network on Water Resources Development and Management (INWRDAM).

Muniz, D. H. F., & Oliveira-Filho, E. C. (2023). Multivariate statistical analysis for water quality assessment: A review of research published between 2001 and 2020. *Hydrology, 10*(10), 196. https://doi.org/10.3390/hydrology10100196

Odebunmi, E. O., Olutona, G. O., Akintunde, E. A., & Oluwaniyi, O. O. (2014). Characteristics and quality assessment of drinking water in some major towns of Osun state, South-Western Nigeria. *Nigerian Society for Experimental Biology (NISEB) Journal, 14*(1), 51–55.

Oginni, F. A., & Ojoawo, S. O. (2014). Assessment of sanitation levels of sources of water in Osun state capital, Nigeria Int. *Journal of Engineering Research and Applications, 4*(12) (Part 1), 94–104.

Ojekunle, Z. O., Adeyemi, A. A., Taiwo, A. M., Ganiyu, S. A., & Balogun, M. A. (2020). Assessment of physicochemical characteristics of groundwater within selected industrial areas in Ogun state, Nigeria. *Environmental Pollutants and Bioavailability, 32*(1), 100–113. https://doi.org/10.1080/26395940.2020.1780157

Olajire, A. A., & Imeokparia, F. E. (2001). Water quality assessment of Osun River: Studies on inorganic nutrients. *Environmental Monitoring and Assessment, 69*(1), 17–28. https://doi.org/10.1023/a:1010796410829

Olayiwola, O. A., Onianwa, P. C., & Olayiwola, W. B. (2019). Spatial and water quality classification of tributaries of Osun River network in Nigeria. *International Journal Irrigation and Water Management, 6*(2), 1–25.

Olowe, B. M., Oluyege, J. O., & Famurewa, O. (2016). An assessment of drinking water quality using water quality index in ado-Ekiti and environs, Nigeria. *American Chemical Science Journal, 12*(2), 1–7.

Oluwafemi, O. K., Oluwadare, O., & Temidire, D. (2015). Water quality assessment of river elemi and Ureje in ado Ekiti, Nigeria. *Journal of Civil and Environmental Engineering, 5*, 199. https://doi.org/10.4172/2165-784X.1000199

Oluwagbayide, S. D., & Abulude, F. O. (2022). Assessment of the quality of dug well water samples in Nigeria and their suitability for drinking and irrigation purposes. *Sustainable Water Resources Management, 8*, 149. https://doi.org/10.1007/s40899-022-00736-8

Oluwagbayide, S. D., Akinnusotu, A., Arifalo, K. M., Adamu, A., Abulude, F. O., Mabayoje, S. O., & Kenni, A. M. (2025). Assessment of water quality for irrigation purpose: A case study of three states. *Environmental Earth Sciences, 84*, 212. https://doi.org/10.1007/s12665-025-12221-4

Popoola, L. T., Yusuf, A. S., & Aderibigbe, T. A. (2019). Assessment of natural groundwater physico-chemical properties in major industrial and residential locations of Lagos metropolis. *Applied Water Science, 9*, 191. https://doi.org/10.1007/s13201-019-1073-y

Saana, S. B., Fosu, S. A., Sebiawu, G. E., Jackson, N., & Karikari, T. (2016). Assessment of the quality of groundwater for drinking purposes in the upper west and northern regions of Ghana. *Springerplus, 5*(1), 2001. https://doi.org/10.1186/s40064-016-3676-1

Sahu, M., Shrivastava, A., Jhariya, D. C., Diwan, S., & Subuhadarsini, J. (2024). Evaluation of correlation of physicochemical parameters and major ions present in groundwater of Raipur using discretization. *Measurement: Sensors, 34*. https://doi.org/10.1016/j.measen.2024.101278

Singh, G., Singh, J., Wani, O. A., Egbueri, J. C., & Agbasi, J. C. (2024). Assessment of groundwater suitability for sustainable irrigation: A comprehensive study using indexical, statistical, and machine learning approaches. *Groundwater for Sustainable Development*. https://doi.org/10.1016/j.gsd.2023.101059

SIWI (Stockholm International Water Institute. (2020). Principles and practice of integrated water resources management, the first manual in the workplace-based professional training on integrated water resources management. IWRM_Manual1_final.pdf. Accessed 17 May 2025.

Soetan, O., Nie, J., & Feng, H. (2022). Preliminary environmental assessment of metal-contaminated sediment dredging in an Urban River, New Jersey, USA. *Marine Pollution Bulletin, 184*, 114212. https://doi.org/10.1016/j.marpolbul.2022.114212

Sulaiman, F. R., Hafz, M. F., Shamshudin, C. M., Rahim, M. R. A., & Sulaiman, N. (2022). Assessment of metal concentrations from recreational rivers in a tropical region (Jengka, Malaysia). *Applied Water Science, 12*, 127. https://doi.org/10.1007/s13201-022-01643-1

Unigwe, C. O., Egbueri, J. C., & Omeka, M. E. (2022). Geospatial and statistical approaches to nitrate health risk and groundwater quality assessment of an alluvial aquifer in SE Nigeria for

drinking and irrigation purposes. *Journal of the Indian Chemical Society*. https://doi.org/10.1016/j.jics.2022.100479
UN-Water. 2016 Water and sanitation interlinkages across the 2030 agenda for sustainable development, Geneva. Available online: http://www.unwater.org/publications/water-sanitation-interlinkages-across-2030-agenda-sustainable-development.pdf/. Accessed on 12 May 2021
Wahab, A. B., Omojola, S. O., & Fasogbon, S. (2012). Assessment of quality and supply of Ede water supply network, Osun State, Nigeria. *Journal of Applied sciences, 7*(8–12), 490–493.
Zigde, M., & Tsegaye, E. (2019). Multivariate analysis of water quality and identification of potential pollution sources of Lake Hawasa, Ethiopia. *Elixir Pollution, 126*, 52424–52429.

Chapter 17
Conclusion

Mariam Iyabo Adeoba

17.1 Introduction

Nigeria's water resources present a paradox of abundance amid scarcity. Despite vast rivers, lakes, and aquifers, millions of Nigerians still lack access to safe drinking water and adequate sanitation. This book's analysis reveals that the challenge is not resource availability but systemic issues such as fragmented governance, aging infrastructure, climate variability, and socio-economic inequities that undermine water security and public health.

Key Insights:

- Governance and Policy Gaps: Overlapping mandates and weak enforcement hinder integrated water management. Legal reforms such as the National Water Resources Bill aim to harmonize frameworks but require stronger stakeholder engagement and protection of customary rights.
- Climate Change as a Risk Multiplier: Intensifying floods, droughts, and erratic rainfall disrupts water supply, damage infrastructure, and amplify contamination risks, demanding climate-resilient planning.
- Water–Energy Nexus: Hydropower and thermal plants depend on reliable water, while water systems rely on energy for pumping and treatment. Failures in one sector cascade into the other, emphasizing the need for coordinated planning.
- Public Health and Equity: Unsafe water and poor sanitation perpetuate waterborne diseases, disproportionately affecting rural communities, women, and children, and reinforcing cycles of poverty.

M. I. Adeoba (✉)
UNISA Biomedical Research Group, Department of Mechanical, Bioresources and Biomedical Engineering, College of Science, Engineering and Technology (CSET), University of South Africa, Johannesburg, Florida, South Africa
e-mail: mariamsalami@yahoo.co.uk

M. I. Adeoba et al. (eds.), *Water Resources Management in Nigeria*, World Water Resources 10, https://doi.org/10.1007/978-3-032-22813-0_17

- Data and Monitoring: Multivariate analyses of hydrochemical parameters (Ca^{2+}, Mg^{2+}, Na^{+}, K^{+}, TDS, TH) highlight the need for continuous monitoring and adaptive management rather than static assessments.

Pathways Forward:

Nigeria must adopt integrated, climate-smart strategies that combine technological innovation (solar-powered pumps, smart irrigation), nature-based solutions (wetland restoration, watershed management), and robust governance reforms. Financing models should blend public and private investments with affordability safeguards, while community participation and gender inclusion remain central to sustainability. Continuous monitoring, transparent data systems, and adaptive operations will ensure resilience against climate shocks and improve service reliability.

Achieving Sustainable Development Goals 6 (clean water), 7 (clean energy), and 13 (climate action) requires a holistic approach that aligns infrastructure, policy, and people. By leveraging its natural endowments and embracing integrated water-energy planning, Nigeria can transform water from a source of vulnerability into a driver of health, equity, and economic growth. The future of water security lies in collaboration, innovation, and resilience ensuring that every Nigerian has access to safe water and sanitation for generations to come.

GPSR Compliance
The European Union's (EU) General Product Safety Regulation (GPSR) is a set of rules that requires consumer products to be safe and our obligations to ensure this.

If you have any concerns about our products, you can contact us on

ProductSafety@springernature.com

In case Publisher is established outside the EU, the EU authorized representative is:

Springer Nature Customer Service Center GmbH
Europaplatz 3
69115 Heidelberg, Germany

www.ingramcontent.com/pod-product-compliance
Ingram Content Group UK Ltd.
Pitfield, Milton Keynes, MK11 3LW, UK
UKHW020935180926
475692UK00003B/51
9783032228123